Nanoparticles' Promises and Risks

Mihai Lungu • Adrian Neculae
Madalin Bunoiu • Claudiu Biris
Editors

Nanoparticles' Promises and Risks

Characterization, Manipulation, and Potential Hazards to Humanity and the Environment

Springer

Editors
Mihai Lungu
Faculty of Physics
West University of Timisoara
Timisoara, Romania

Adrian Neculae
Faculty of Physics
West University of Timisoara
Timisoara, Romania

Madalin Bunoiu
Faculty of Physics
West University of Timisoara
Timisoara, Romania

Claudiu Biris
High Performance Computing Centre
West University of Timisoara
Timisoara, Romania

ISBN 978-3-319-11727-0 ISBN 978-3-319-11728-7 (eBook)
DOI 10.1007/978-3-319-11728-7
Springer Cham Heidelberg New York Dordrecht London

Library of Congress Control Number: 2014951026

Printed on acid-free paper

Springer is part of Springer Science+Business Media (www.springer.com)

This book is dedicated to all who care about our wonderful Planet Earth.

Foreword

This book is long overdue.

Much of the air pollution that we face today is presumed to originate from traffic infrastructure. While much certainly does come from traffic, particularly diesel traffic, much of it does not. I know because I researched it in the AEM (Analytical Electron Microscope) over 30 years ago. I was then "only" looking for asbestiform bodies and I found very few. What I did find was vast amounts of unregulated siliceous dusts and soot. The siliceous dusts came from coal combustion and the soot from coal and diesel traffic. Today in the AEM we can analyze (nondestructively) to atom dimensions. This was done with a single silicon atom sitting on graphene in 2012 by Lovejoy et al. This research employed the NW Superstem and engaged with EDS X-ray analysis and with PEELS, for corroboration.

I repeat that today, with out-of-date, inadequate, and arbitrary air pollution law, this book is long overdue.

I copy a line of text which I wrote as part of the conclusion of a research proposal into air pollution, public health, ill health, and dementia:

"We now know that combustion products cause cancer and heart attacks. We are increasing anthropogenic nanoparticle emissions, with no adequate control for any particle size smaller than PM2.5. Thirty-five years ago, in 1977, it was realised by Manchester University asbestos 'expert' Professor Jack Zussman, that '*any material to which people are exposed on a large scale needs to be tested for its physiological effects*'. The material to which he referred was processed crushed mineral material—that is small particles of material exposed to the public on a massive scale!

The potential for causing harm to human health, from unregulated nanoparticles, many of which are anthropogenic, is immense, and insufficient research is being applied to this problem. Most of the reasons I have been given, privately, are financial. I simply want to know who will be paying the bill in the future for the failure to act now.

We are supposed to learn from history, but by the time something is done, we may well be just too late for the many innocent victims of a failure to act today?"

These observations, although increasingly true today, met with no response—no response at all from regulators! Except to go away and do more research for which I cannot get the funding.

Air particle regulations have not yet engineered appropriate controls for the remarkable physical properties of the very small. Inhaled particles smaller than PM2.5 are without adequate regulation. This book addresses these issues, providing academic evidence of the causes of air pollution, particularly with nanotechnology.

We have here the sources of nanotechnology described, as well as the effects on human health from exposure and the characterization and detection of those particles. We should not make assumption about relative safety—we should actively do the research to learn the truth!

My alma mater, the University of Manchester, knows the problems and I am assured that nanotechnologies employed there will ensure public health safety. I want to see that increasingly to be the case after this book is published. Those in charge of this issue cannot deny that they have been told—the evidence they need is contained in this book.

I hope that readers of this book will give air pollution much greater attention than it gets today!

School of Earth, Atmospheric and Environmental Sciences Graham Cliff
University of Manchester
Manchester, UK

Preface

The advent of nanotechnology and nano-process engineering has brought about a new era of research in materials science, optics, chemistry, biology, and medicine among others. While the benefits, applications, and future possibilities offered by this new nano-science have been made center stage from their very inception, little has been said about the dangerous side effects both created by these new nanoscale processes and revealed by our deeper understanding of the nano world.

This collection of articles proposes to do just that, present the cutting-edge state of research in the harmful effects of human process-driven nanotechnology and propose realistic implementations of solutions to these issues. Specifically, we focus in this volume on nanoparticles and nanoparticle aggregates.

Nanoparticles as a by-product of human society are well known, ranging from, the worryingly common, nanoparticle aggregates present in combustion gases to the less encountered side effects of industrial processes such as metallic nano shavings or chemical nano residues. To gain a rigid scientific understanding of what these industrial side products entail, an exhaustive classification of the types of nanoparticles and nanoparticle aggregates that are a result of these processes is needed. The first part of this volume proposes to do just that and provides a detailed analysis of the properties, sources, and distribution of harmful nanoscale products in human society.

Second, it is also important to look at the effects of nanoparticles on the human body from a medical perspective. While medical science itself is no stranger to nanotechnology, much of the published body of work in this field focuses on the benefits that nanoscale processes can bring to diagnosis and treatment. While not attempting in any way to detract from the crucial importance of such breakthroughs, this volume will attempt to look at the harmful side effects that nanotechnology can have on human health. Consequently, the second part of the work focuses on this issue.

Thirdly, while characterization and analysis of nanoparticles and aggregates in controlled conditions is a challenge in itself, designing techniques that can work in a "real-life" environment requires even greater effort. In the third part of this volume, we look at innovative techniques for nanoparticle characterization ranging from wide-spectrum atmospheric spectroscopy to localized sensing and characterization of nanoparticles in low concentration. By presenting these techniques, we hope to provide a blueprint for future technologies that will allow for the implementation of cheap, portable, and easy-to-use nanoscale detection and classification systems.

Finally, the issue of how these problems can be addressed and redressed is raised. In a positive twist of fate, it seems that nanotechnology itself provides the solution to the very dangers it can create. The fourth and final part of the volume presents several new techniques for nanoparticle sorting, manipulation, and separation that can limit the hazardous effects of these products on the environment and human health. Ranging through different methods, we attempt to provide a broad base as possible for those interested in the state of the art in this field.

We envision this book as a helpful primer for this crucial new era of nanoscience and invite all those interested to use it as a trusted source of information, be it in research, study or decision making. Covering such a large and diverse area of science and technology in an exhaustive manner, it is the authors' and editors' hope that this volume can become a crucial stepping stone for those interested in both entering this fascinating and important topic as well as those looking to further develop the scientific methods and related technical aspects.

Timisoara, Romania
August, 2014

Mihai Lungu
Adrian Neculae
Madalin Bunoiu
Claudiu Biris

Acknowledgements

Primarily we would like to acknowledge the support of Ania Levinson, Editor, Springer Science+Business Media, and for inviting us to write this book.

We give our special thanks to the reviewers:

- Prof. Dr. Graham Cliff, Honorary Senior Research Fellow, University of Manchester, England,
- Prof. Dr. Costel Darie, Professor & Lead, Biochemistry & Proteomics Group Department of Chemistry & Biomolecular Science, Clarkson University 8 Clarkson Avenue, Potsdam, NY,
- Dr. Roxana Maria Ghiulai, Researcher "Victor Babes" University of Medicine and Pharmacy Timisoara, Department of Pharmacy; Pharmaceutical Chemistry Laboratory; Timisoara, Romania,
- Dr. Periasamy Anbu, Department of Biological Engineering, Inha University, Incheon 402-751, Korea,
- Prof. Dr, Thanganadar Ashokkumar, Indira Gandhi Institute of Engineering and Technology for Women, India,
- Professor Emeritus Dr. Nicolae Avram, Department of Physics West University of Timisoara, Romania,
- Dr. Ioan Grozescu and Dr. Stefan Novaconi, Senior researchers, Department of Applied Physics, National Institute on Research and Development for Electrochemistry and Condensed Matter, Timisoara, Romania,
- Dr. Mintu Pal, School of Medicine, University of Colorado, USA,
- Dr. Tapan Giri, NSHM College of Pharmaceutical Technology, Kolkata University, India,
- Prof. Dr. Ashok Vaseashta, Director of Research Institute for Advanced Sciences Convergence & Int'l Clean Water Institute NUARI, Herndon, USA,
- Prof. Dr. B. Ben-Nissan, Faculty of Science, University of Technology, Sydney, Australia.

for taking the time to read the manuscript and give valuable reviews. We would also like to thank the authors for their contributions to the book. Thanks must also go to Professor Ronald Pethig, School of Engineering, University of Edinburgh, and Graham Cliff, Honorary Senior Research Fellow, University of Manchester, England, for stimulating and encouraging discussions.

Mihai Lungu especially would like to acknowledge that this work was supported by a grant of the Romanian National Authority for Scientific Research, CNCS—UEFISCDI, project number PN-II-ID-PCE-2011-3-0762.

Finally, if there are any errors or omission, then of course the faults are entirely ours, but please inform us.

Bocsa
August, 2014

Project manager
Mihai Lungu

Contents

Part I
Nano-particle Classification and Sources in Human Society

Chapter 1
Nanoparticles: Definition, Classification and General Physical Properties

Nicolae Strambeanu, Laurentiu Demetrovici, Dan Dragos, and Mihai Lungu

Abstract This introductory chapter gives relevant historical information about nanoparticles and the attempts to define the nanoparticle size (1–1,000 nm), as well as their most important general properties, which vary with size: surface area, optical properties, uniformity, fictionalisation, quantum confinement and other fractal-type characteristics.

1.1 General Overview

Nanoparticles are considered a discovery of the twentieth century, but a brief overview of the field reveals that artisans in Mesopotamia used finely divided materials of this type as early as the ninth century BC, to obtain a glittering effect on the surface of ceramic vessels.

In the Middle Ages and the Renaissance, the production of glittering metallic films led to methods of covering glassy surfaces developed in various Far Eastern or European centres which became famous thanks to these methods that are largely employed even today.

During the Renaissance and in later times, the development of visual arts (painting in particular) and printing and engraving methods contributed to the technique of producing fine inorganic and organic dust particles, close to nanoparticles in size, and dispersions that were stable in different solvents and used as dyes or ink [1].

N. Strambeanu (✉)
Pro Air Clean Ecologic, 37, Vadul Crisului Street, Timisoara 300613, Romania
e-mail: ns_arana@yahoo.com

L. Demetrovici
Pro Air Clean Ecologic, Timisoara, Romania

D. Dragos
Department of Pharmacy, "Victor Babes" University of Medicine and Pharmacy, Timisoara, Romania

M. Lungu
Department of Physics, West University of Timisoara, Blv. V. Parvan No. 4, 300223 Timisoara, Romania

M. Lungu et al. (eds.), *Nanoparticles' Promises and Risks*,
DOI 10.1007/978-3-319-11728-7_1

Michael Faraday was the first to provide a scientific description of the optical properties of nanometric metal particles, in a paper he published in 1857 [2]. Turner further revealed that leaving very thin gold or silver films on glassy surfaces heated at 500 °C changed both the properties of the deposited metals and those of the glass, so that white light crossed the metallic film; this caused a marked reduction of reflection, while electrical resistivity increased significantly [3]. Ten years later (1867), James Clark Maxwell suggested a series of concepts of differentiation in nanotechnology, but without using the word "nanotechnology" to define thin, monomolecular layers [4].

The first accurate observations and measurements were made much later, by Richard Adolf Zsigmondy, who used dark field ultramicroscopy which allows the visualisation of particles smaller than monochromatic light wavelength (1914). With this method, Zsigmondy was able to see 1/1,000,000-mm particles and he was the first to apply the term "nanoparticles" explicitly to such particles [5].

Irving Langmuir and Katharine B. Blodgett, dealing with nanoparticle characterisation and related phenomena that define interface in colloid science (1920), introduced the concept of monolayer, a layer of material one molecule thick. Langmuir received the Nobel Prize for his theoretical contributions in chemistry in 1932 [6, 7].

The shortest definition of nanoparticles [8–10], which is probably the most intuitive one, takes into consideration only their size, which is limited conventionally to about 100 nm in any direction. This definition cannot be exhaustive, as it does not give net values. But without a classification, no matter how general, it is difficult to differentiate between the molecular and atomic field on the one hand and the nanoparticle field on the other.

Nanoparticles can be classified based on the following criteria [11, 12]:

- Origin:
  - Natural
  - Anthropogenic
- Size (as already shown):
  - 1–10 nm
  - 10–100 nm
  - Over 100 nm
- Chemical composition:
  - Inorganic substances
  - Organic substances
  - Elements of the living kingdom.

These criteria are presented synthetically in Fig. 1.1 [13].

The attempt to give a comprehensive definition of nanoparticles is generally the object of physical chemistry, as nanoparticle behaviour depends both on directly measurable quantities (mass, volume, electric charge) and their property

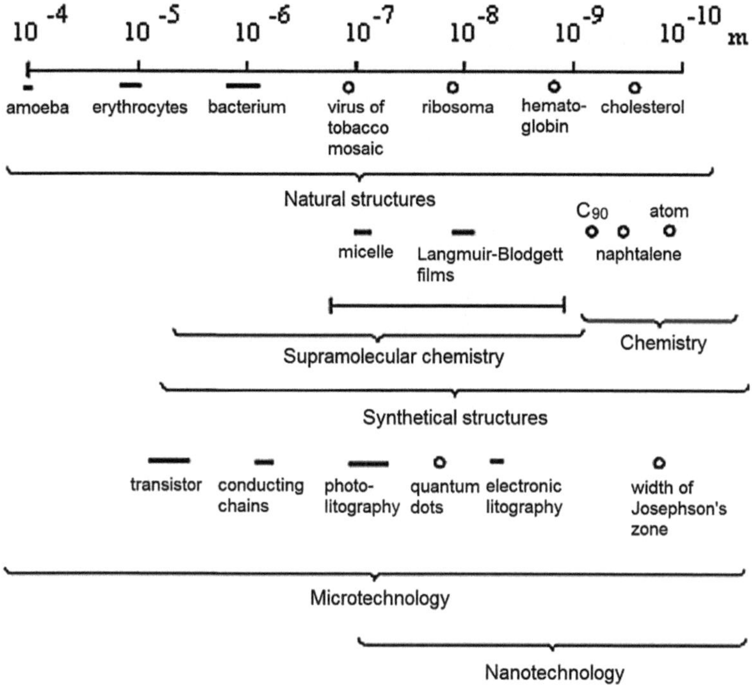

Fig. 1.1 Intuitive representation of micro and nanoparticle fields, with examples

to self-organise based on their high reactivity. These properties characterise fractal systems known as clusters, aggregates or filaments [13, 14].

Nanoparticles are of great scientific interest, as they are a bridge between bulk substances and their molecular or atomic structure. A bulk substance has constant physical and chemical properties irrespective of its size; nevertheless, at the nanoscale, these properties depend on more or less discreet molecular or atomic phenomena. Regardless of the nature of nanoparticles, their most important physical properties are the following:

1. Surface area. It has been found that properties vary with particle size. In sub-micrometre particles, forces that govern the atomic or molecular universe dominate to the detriment of statistical aspects, which are revealed at the macroscale.
2. Optical properties. Nanoparticles often have particular optical properties, as they are small enough to limit the thickness of the common electron layer of metals; this phenomenon generates quantum effects. Although it is common knowledge that gold is yellow and silicon is grey, gold and silicon nanoparticles are bright red to black. Moreover, gold nanoparticles melt at a much lower temperature (~300 °C, 2.5 nm size) than gold slabs that melt at 1,064 °C. Solar energy absorption in photovoltaic cells made of silicon-based nanomaterials is much

higher than in thin films of the same materials. The smaller the particles, the higher the absorption efficiency.

3. Uniformity. Clusters, aggregates or filaments, in other words, the molecular or atomic assemblies that form nanoparticles, are defined by the interaction of forces among the molecules or atoms of a particle and the interaction forces among particles.
4. Functionalisation. Nanoparticles of any type can be linked to microbiological entities randomly, through natural processes occurring in atmosphere, water or at the surface of the Earth. Nanoparticles are then directed to living organisms, organelles within the cells and individual protein or RNA molecules. This property is related both to the harmful effects of nanoparticles on the living kingdom and the pharmaceutical or biochemical studies conducted voluntarily, to the level of peptide molecules.
5. Quantum confinement. Changes in size-dependent properties also include quantum confinement, a phenomenon which causes spontaneous properties of semiconductivity, conductivity or electric insulation for neighbouring particles less than 10 nm in diameter [14].

In most cases, the dispersion medium of nanoparticles that come in close contact with man or other components of the living kingdom is air or water. According to their size, nanoparticles can be classified as ultradispersion systems (1–50 nm), high dispersion systems (50–100 nm) or sedimentable particles (100–1,000 nm). Sedimentable particles consist of either large particles or fractal-type particles [14].

Very fine colloidal suspensions present in water can be destabilised and turned to sedimentable materials easily.

Nanoparticles have always circulated in the atmosphere of the planet. They derive both from volcanic eruptions or fine desert dust lifted by air currents, and interstellar dust as such or resulting from meteorite disintegration.

Nanoparticles are also the result of various human activities, from the primary exploitation of quarry materials to exhaust gas, thermal power or cement plant stacks, engine jets or fireworks.

Chemically speaking, nanoparticles are of infinite diversity, as their diversity borrows both the nature of dust or fluids existing on the planet or generated naturally and all materials produced by human activities. Therefore, nanoparticles can be metal or non-metal particles in an elementary state, as well as inorganic and organic substances resulting from their combination.

Fine aerosols consisting of metal particles are usually spherical. They can be crystals, but in ionic state they form sub-colloidal or colloidal systems. These fields include combinations of about 50–60 metal atoms (for $1 < d < 10$-nm nanoparticles), but they can reach even 500–2,000 atoms.

The intuitive classification of metal particles is given in Fig. 1.2 [13].

In a finely divided state, many metals react instantaneously with the main atmospheric components (oxygen, nitrogen, carbon dioxide), even with those existing in very small concentrations, namely in a similar dispersion state. In this way, the chemical nature of nanoparticles of a certain type present in the

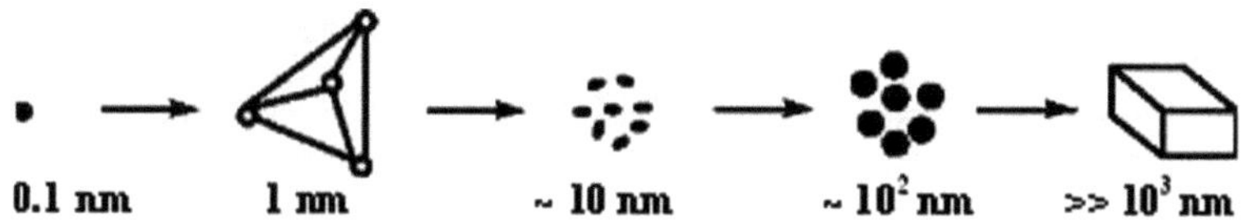

Fig. 1.2 The main stages of the transformation of metal atoms into a bulk metal

atmosphere changes in successive stages, under the influence of random environmental factors, until they agglomerate, become sedimentable and no longer react with these factors.

This process can take years, because many metal ions, having changed their chemical composition, become colloidal, through coordination with water molecules, and acquire an electric charge. As a result, coalescence with particles of the same type is possible only after the random destabilisation of the system in an electric field or through colloidal ageing.

At least partially, the theory exposed briefly above is also valid for very large molecules or organic macromolecules, because they can float in the atmosphere due to Brownian forces or the forces that move the air and react with its components, among themselves or with inorganic particles, changing their nature and state of matter.

The nanoparticle-size elements of the living kingdom are at the border between chemistry and biology (cholesterol, haemoglobin, hemocyanin, chlorophyll, etc.), as they are part of the life cycle, although most of them can be synthesised as well.

Similar in size to the largest nanoparticles, viruses are the simplest forms capable of reproduction. However, if one considers the primary composition (mainly carbon, nitrogen and oxygen) of these representatives of the living kingdom, the changes they undergo follow both the food chain and the sequence of perpetual change in the physical and chemical characteristics of matter.

Acknowledgements This work was supported by a grant of the Romanian National Authority for Scientific Research, CNCS—UEFISCDI, project number PN-II-ID-PCE-2011-3-0762.

References

1. Radu M, Repanovici A (2004) O istorie a tiparului şi a tipăriturilor. Editura Universităţii Transilvania, Brasov
2. Faraday M (1857) Experimental relations of gold (and other metals) to light. Phil Trans Roy Soc Lond 147:145–181
3. Turner T (1908) Transparent silver and other metallic films. Proc R Soc Lond A 81:301–310
4. Dattu RJ (2010) Engineering physics. Tata McGraw-Hill Education, New Delhi
5. Zsigmondy RA (1966) Properties of colloids, Nobel lectures, chemistry 1922–1941. Elsevier Publishing Co., Amsterdam
6. Langmuir I (1917) The constitution and fundamental properties of solids and liquids. J Am Chem Soc 39:1848–1906

7. Blodgett KB (1935) Films built by depositing successive monomolecular layers on a solid surface. J Am Chem Soc 57:1007–1022
8. Petrov Y (1986) Picosecond resonance Raman scattering study of methylviologen reduction on the surface of photoexcited colloidal cadmium sulfide crystallites, Nauka Ed, Moscow. http://pubs.acs.org/doi/abs/10.1021/j100276a014
9. Rossetti R, Brus L (1986) Picosecond resonance Raman scattering study of methylviologen reduction on the surface of photoexcited colloidal cadmium sulfide crystallites. J Phys Chem 90:558–560
10. Akimov IA, Denisyuk IY et al (1992) From molecule to molecular aggregation: clusters and crystals of clusters. Opt Spectrosc 72:1026. http://pubs.acs.org/doi/abs/10.1021/ar00038a003
11. Braga D, Grepioni F (1994) From molecule to molecular aggregation: clusters and crystals of clusters. Acc Chem Res 27:51
12. Klabunde KJ (1980) Chemistry of free atoms and particles. Academic, New York
13. Aitken RJ, Creely KS et al (2004) Nanoparticles: an occupational hygiene review, research report 274. Institute of Occupational Medicine for the Health and Safety Executive, Edinburgh
14. Nalwa HS (1999) Handbook of nanostructured materials and nanotechnology. Academic, New York

Chapter 2
Natural Sources of Nanoparticles

Nicolae Strambeanu, Laurentiu Demetrovici, and Dan Dragos

Abstract This chapter deals with the major natural sources of nanoparticles in the atmosphere: volcanic eruptions, desert surfaces, dust from cosmic sources located in the solar system or outside it. Details are given about the composition of very fine particles according to their type, the successive or parallel chemical transformations that can occur among them or when they meet the major components of the atmosphere (nitrogen, oxygen, carbon dioxide). The authors specify possible ways of evolution from the mineral kingdom to organic substances through nanoparticles originating from cosmic dust, based on suppositions that are provided in the literature of the field and accepted based on physical and chemical computational models.

2.1 Introduction

Earth, cosmic and weather-dependent phenomena on the planet produce particulate matter that is lifted in air through volcanic eruptions, air currents generated by storms or strong winds, the disintegration of meteorites entering the atmosphere or the accumulation of cosmic dust.

Hurricanes cause huge amounts of water to rise in the atmosphere from the surface of the planetary ocean. On evaporation, water releases both the saline and spore contents of algae and other unicellular organisms. Due to these complex phenomena which most of the time occur simultaneously, at different heights and distances, the atmosphere is constantly filled with nanoparticles that cause various reactions affecting the biosphere.

N. Strambeanu (✉)
Pro Air Clean Ecologic, 37, Vadul Crisului Street, 300613 Timisoara, Romania
e-mail: ns_arana@yahoo.com

L. Demetrovici
Pro Air Clean Ecologic, Timisoara, Romania

D. Dragos
Department of Pharmacy, "Victor Babes" University of Medicine and Pharmacy, Timisoara, Romania

M. Lungu et al. (eds.), *Nanoparticles' Promises and Risks*,
DOI 10.1007/978-3-319-11728-7_2

A very good example of nanoparticle sources is the eruption of Krakatoa on August 27th, 1883. The smoke column reached 80 km in height and the dust thrown into the ionosphere not only caused strange optical effects visible in North America and Europe, but also acted as a solar radiation filter, decreasing global temperature with about 1.5 °C in the next 2 years. In addition, the eruption threw a huge amount of gas into the atmosphere which later air currents carried around the planet, a phenomenon that led to an unprecedented elevation of acid concentration under the form of very fine drops in the cirrus clouds at high altitude. Acid rains were the obvious consequence everywhere in the world [1–7].

Another example is the Tunguska event [8–11], whose probable cause was the collision of a gigantic meteorite or comet fragment with Earth on June 30th, 1908. For several weeks after the event, luminescent yellowish green clouds that made reading possible even during moonless nights could be seen above Europe and North Africa, at the extremity of the troposphere [12–14].

Some authors [15, 16] hold that the northern lights are very bright optical phenomena seen on the night sky in the areas near the polar regions, caused by the interactions between the ionosphere nanoparticles and the solar wind particles, under the influence of Earth's magnetic field. In the northern hemisphere, the phenomenon is known as *aurora borealis*, as Galileo Galilei named it. The northern lights appear in September–October and March–April. In the southern hemisphere, *aurora australis* was seen for the first time by James Cook, in his failed attempt to reach the South Pole.

Auroras are not exclusive earthly events. They can also be seen on other planets of the solar system, such as Jupiter, Saturn, Mars and Venus. Although natural, they can be reproduced experimentally [16].

2.2 Volcanic Eruptions

Ash released during volcanic eruptions can reach temperatures over 1,400 °C and has a very complex structure consisting of solid and liquid particulate matter lifted by the hot gas current [17]. Following the paroxysmal phase of the volcano, as the ash spreads into the atmosphere, gas temperature lowers and gas composition changes; leading to the accumulation in deposit of particles either clusters through chemical reactions or based on electrostatic forces of attraction.

Volcanic gas emissions vary with thermodynamic and kinetic conditions (pressure, temperature, speed of reaction and diffusion, etc.) and the nature of magma (Fig. 2.1) [17]. Most volcanoes on Earth throw basaltic lava and erupt most frequently along the ocean ridges, at depth, having direct contact with the atmosphere only in several places of the world (Iceland and the Azores for the Atlantic Ridge).

Basaltic magma is rich in magnesium and iron and poor in silica. It generally has low viscosity and a reduced gas concentration consisting mainly of carbon dioxide and sulphur dioxide. Hydrogen sulphide (H_2S) and hydrochloric acid (HCl) prevail in gas emissions (Table 2.1) [18–21].

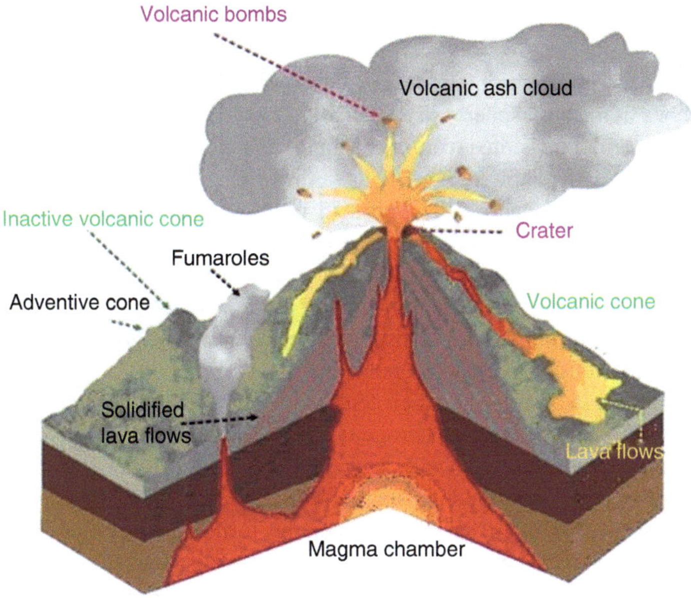

Fig. 2.1 Intuitive representation of active and inactive volcanic phenomena [17]

Table 2.1 Approximate composition of volcanic gas [18–21]

Component	H_2O	CO_2	SO_2	H_2S	COS	CS_2	HCl	HBr	HF
% vol	50–90	1–40	1–25	1–10	10^{-4}–10^{-2}	10^{-4}–10^{-2}	1–10	10^{-4}–10^{-2}	$<10^{-3}$

As hydrochloric acid is highly soluble in water, it is quickly removed from the atmosphere through acid rains that usually start during eruptions, when water vapour condenses on cooled ash particles, in higher atmospheric layers. Hydrosulphuric acid has the same contribution, after it is oxidised by the ozone in the troposphere:

$$H_2S + 2O_3 \rightarrow H_2SO_4 + O_2 \tag{2.1}$$

The hydrobromic acid (HBr) concentration is relatively low [22, 23]. However, its contribution, alongside hydrosulphuric acid oxidation, is significant for the direct decrease in both the ozone and the diatomic oxygen concentration, as it forms derivatives with the diatomic oxygen much easier than the hydrochloric acid:

$$HBr + O_3 \rightarrow HBrO_3 \tag{2.2}$$

$$2HBr + 3O_2 \rightarrow 2HBrO_3 \quad (2.3)$$

There is little information about the hydrofluoric acid concentration eliminated by volcanoes in the atmosphere. In some cases (Laki 1783 or Mount Hudson 1990), hydrofluoric acid emissions were very high and contaminated the vegetation, killing animals and people. As with the other halogen acids, hydrofluoric acid spreads around the volcanoes through acid rains, so that excessive fluoride ions are one of the major causes of animal mortality on large areas, even a long time after the eruption. Fluoride intoxication becomes chronic and eventually lethal, if its concentration in the freshly grown vegetal matter exceeds 250 ppm [24].

The chemical composition of forming ash is easily identified if one studies the permanent emissions of latent volcanoes.

Dry fumaroles lack water and their temperature exceeds the critical temperature of water (374 °C). They contain nitrogen (N_2), carbon oxide (CO), hydrogen (H_2), methane (CH_4), and sodium chloride vapour (NaCl), potassium chloride (KCl) and copper oxide (CuO). Through deposition on the edge of the crater, on the surface of the bombs thrown during eruption or under the form of ash, fumaroles generate particulate matter of halite (NaCl), sylvite (KCl), and tenorite (CuO) [17].

Acid fumaroles consist of hydrogen sulphide (H_2S), sulphur dioxide (SO_2), carbon dioxide (CO_2) and water vapour. Their temperature is close to the critical temperature of water. They also contain iron chlorides ($FeCl_2$, $FeCl_3$), copper chlorides (CuCl, $CuCl_2$) or iron oxides, mostly magnetite (Fe_2O_3), if acidity is not very high. In the presence of hydrogen sulphide (especially in submarine fumaroles), iron is found under the form of marcasite and pyrite (FeS, FeS_2) [17].

Alkaline fumaroles are made of ammonium chloride (NH_4Cl), ammonium hydroxide (NH_4OH), in the presence of ammonia (NH_3), in contact with water vapour. Ammonium chloride takes the form of micrometre or sub-micrometre particles. The temperature of these fumaroles is 100–400 °C.

Solfataras is a natural volcanic steam vent in which sulphur gases are the dominant constituent along with hot water vapours. The fumaroles releasing gases contain sulphur, hydrosulphuric acid (H_2S) and or sulphur dioxide (SO_2) and, in the presence of water vapour, sulphurous acid (H_2SO_3) and sulphuric acid (H_2SO_4), the latter being more stable. Realgar, orpiment (As_2S_3), ferrous, and copper pyrites, iron sulphate ($FeSO_4$) and native sulphur resulted from vapour condensation have been identified as solid state of matter [17].

It must be specified that most metal and non-metal ores in Earth's crust were formed throughout the geological eras, because of volcanic activity. Consequently, any chemical element existing as such or in one of its combinations may be found in the atmosphere under the form of nanoparticles during eruptions or immediately after them.

2.3 Desert Surfaces of the Planet

Earth's large deserts are a major permanent source of nanoparticles which air currents lift in the atmosphere. The first observations about dust storms carrying dust from the Gobi Desert to the Yellow Sea coast were made in the fifth year of Di Xin of the Shang Dynasty (about 1150 BC) [25].

Dust storms seem to be the most important source of nanoparticles in the atmosphere. The long-range migration of both mineral dust and anthropogenic pollutants on the continents has been the focus of recent investigations which have revealed that about 50 % of aerosols in the troposphere are minerals originating from deserts [26] (Figure 2.2).

The chemical composition of very fine desert sand dust varies with origin and the anthropogenic activities developed in the adjacent areas that the air currents cross.

Analytical determinations of the particulate matter composition during a dust storm, performed in China and South Korea, reveal high silicon concentrations, aluminium, calcium and iron traces (Table 2.2). Many toxic substances like heavy metals (Hg and Cd) and PAHs (Polynuclear Aromatic Hydrocarbons) usually produced by coal burning have also been identified, but without specifying their origin in the stratospheric particle clouds which were the subject of the experiments [31, 32].

Recent simultaneous determinations conducted in the area around Xian, a city located at 400-m altitude, and at the summit of Mount Hua (2,060 m) during a massive atmospheric dust transport from the Gobi desert, have revealed sulphate, nitrate and ammonium ions of anthropogenic origin, besides a high carbon and organic nitrogen concentration, probably resulted from the atmospheric microbiological activity. The size of the wind-carried particles varied between 80 and 1,000 nm.

In the past decades, the type and behaviour of aerosols originating in Sahara have been studied intensively both in the Sahel region, in several archipelagos in the Atlantic Ocean, and Southern Europe.

Saharan dust contribution to PM_{10} was determined by gravimetric evaluation of Al, Si, Fe, Ti, nns (non-sea salt) Ca, nssNa and nss K oxides [33]. In most cases, the microcrystal content accounts for over 75 % of the oxide content identified as PM through measurements taken in Italy during Saharan-type dust storms, for several years. According to the multiannual means, the microcrystal content in Southern Italy atmosphere, expressed as PM_{10}, is 5.42 μg/m^3, reaching a value as high as 67.9 μg/m^3 when air currents are strong. The amount of microcrystal aerosols and their contribution to PM_{10} indicate non-significant season dependence.

Optical measurements of particles, taken in parallel with composition measurements, show that within the 400–600-nm range, dust contains mostly soluble compounds of Fe, K and Co.

An estimation of the ionic balance of the atmosphere indicates that, on the one hand, the reactions involving anthropogenic acids and microcrystal particulates can play a major role in the sedimentation of nanoparticle systems; on the other hand,

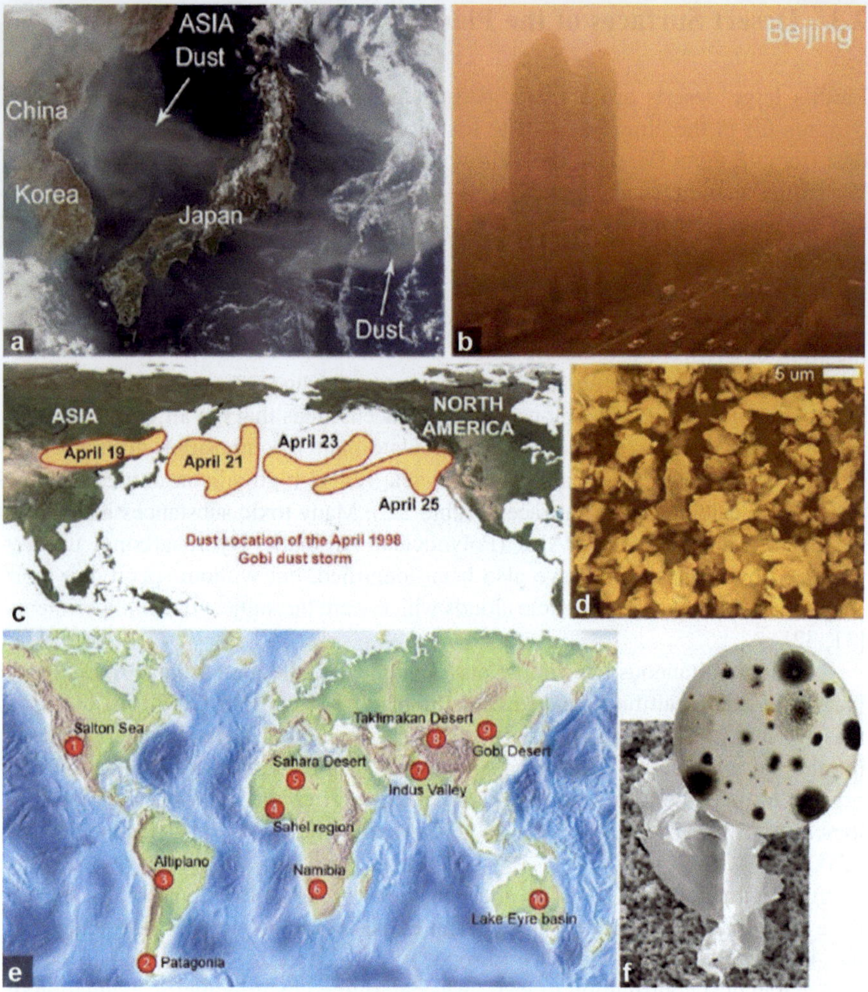

Fig. 2.2 Sand storms viewed at micro and macroscale. (**a**) Satellite images of the dust storm in mainland China over the Sea of Japan and the Pacific in April 2002. (**b**) Dust storm in Beijing [27]. (**c**) Approximate propagation of the dust cloud from a storm in the Gobi desert in April 1998 (models based on satellite photos) [28]. (**d**) Dust sample collected in South-Eastern Asia during a storm in the Gobi desert, on March 16th, 2001 [29]. (**e**) Ten major dust sources in the world [32]. (**f**) Bacteria collected from Sahara dust samples that reached North America [30]

Table 2.2 Approximate composition of particulate matter in the Gobi desert, in China and Korea [31, 32]

Component	Si	Al	Ca	Fe
Nanoparticle content (% mass)	24–32	5.9–7.4	6.2–12	<1

they can influence the introduction of the mentioned substances in the cellular structures of living organisms [33].

Other studies conducted in the Cape Verde Islands [33] and the Island of Sal near the Senegalese coast [30], during fine particle advection caused by the Harmattan, were based on collecting fine dust samples on board of aircraft, at altitudes between 500 and 11,000 m. The results showed that while moving, the main cloud consisted of different particle currents, sedimented according to their mass (0.5–1 to 4–4.5 km altitude), based on statistical distribution.

Beyond 4,500 m, air opacity and O_3 concentration decrease, which highlights the O_3 reaction with sub-micrometre particles. The supposition is based on specific Al_2O_3 concentrations containing up to 54 μg/m^3, measured during very intense dust transport.

The same studies [34, 36] identified nanoparticles of compounds resulting from reactions of their components with anthropogenic aerosols, especially ammonium sulphate (NH_4HSO_4). Samples collected at high altitudes were associated with air masses coming from Europe and travelling over North African desert areas. This hypothesis is much more plausible than that based on the great influence of local biomass combustion emissions, as the nitrate concentration did not increase. At high altitude, old pollution plumes resulted from fossil fuel combustion of possible North American origin were encountered.

The geochemical signature of dust particles resulted from the above-mentioned determinations was consistent with the previous results obtained in the area. According to these results, Si, Fe, and Ti concentrations were not changed, reflecting the soil composition faithfully, while the concentration of other elements like Ca and S increased from one decade to another, because of industrial activities developed in the African deserts [35].

2.4 Cosmic Dust

The total influx of cosmic dust on Earth is about 40,000 t/year, of which the major fraction originates in the interplanetary dust cloud surrounding the Sun, the planetoids between Mars and Jupiter or the comets [37]. The components of the minor fraction can be classified according to their astronomic origin from beyond our solar system: interstellar dust, dust from meteorite clouds existing between the stellar systems (Ephemeris, Perseids, and Leonids) that the Sun and its planets cross periodically or those in the Kuiper Belt (30–50 AU far from the Sun) and the very large Oort cloud. Figure 2.3 gives a brief presentation of cosmic dust sources on Earth [37].

Regardless of their size, cosmic objects cross the space at high speed, sometimes over 150,000 km/h. As they approach the Earth, they slow down because of the friction with the atmosphere, which causes meteorites to burn up and give their characteristic glow. Smaller particles produce friction that reduces their speed through the atmosphere, but they do not burn up and finally reach the surface of the Earth.

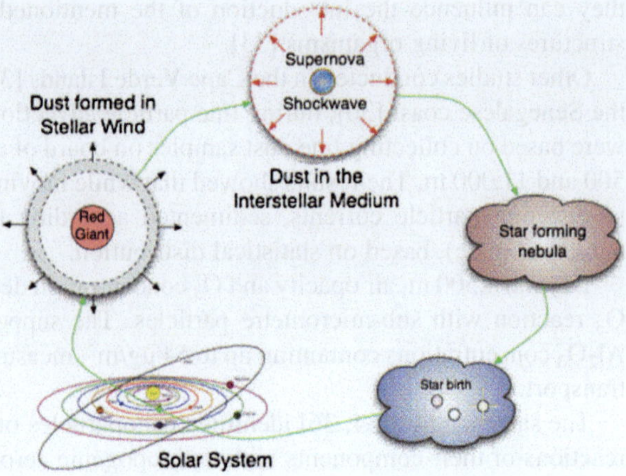

Fig. 2.3 Sources of cosmic dust on Earth [38]

Interplanetary dust composition and size were measured based on IR detection, during space missions like Pioneer 10 and 11, Giotto and Galileo. Similar measurements were taken under the EURECA programme, with satellites in circumterrestrial orbits situated at different distances. The studies conducted in the field were completed with data collected in the exosphere, at an altitude of 20–25 km.

This older or more recent research work reveals that cosmic dust consists of microparticles, nanoparticles and their aggregates. These are irregular-shaped and their porosity varies from spongy to compact. Their composition, size and physicochemical properties are origin-dependent. For instance, the grains in dense matter interstellar clouds are covered in ice; consequently, they are much larger than dust particles in diffuse interstellar medium. Interplanetary dust particles (IDP) are usually the largest.

The largest fraction of the extra-terrestrial dust influx falling on Earth comes from meteorites in the 50–500 μm size range, 2.0 g/cm^3 average density and 40 % porosity. The density of most IDP captured in the atmosphere ranges between 1 and 3 g/cm^3, with an average of 2.0 g/cm^3 [38].

Interstellar dust clouds contain carbon monoxide, silicon carbide, amorphous calcium silicate, water ice and polynuclear aromatic hydrocarbons (PAH) or other simpler organic structures, while in diffuse interstellar medium carbon and silicon particles have been identified.

Comet dust differs in composition from dust resulted from asteroid disintegration. The former is closer to interstellar clouds, while the latter contain large amounts of silicon and iron [39–41].

A recent study [42] has identified the proportions of different element forms—hydrogen, oxygen, carbon, nitrogen—in comet dust samples collected during space missions. The relations between these elements provide information about the distance between the Sun and the place where comets were formed and what type of existence comets had (a high temperature is an indication that they passed close to the Sun).

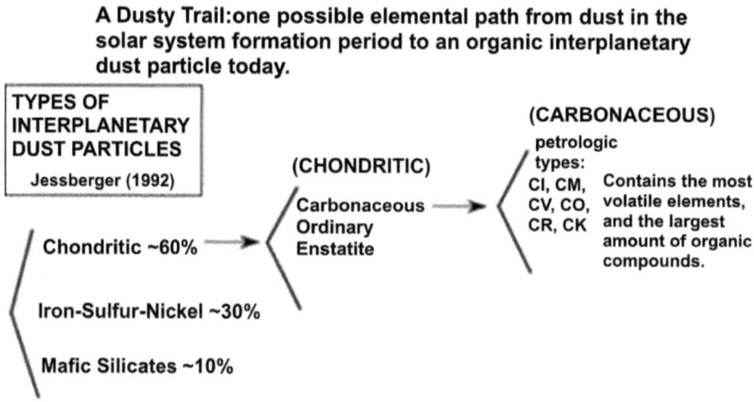

Fig. 2.4 A possible evolution path from the mineral kingdom to organic substances through cosmic dust nanoparticles [42]

It has also been revealed that cosmic dust contains organic substance complexes under the form of aromatic–aliphatic mixtures, which may be created spontaneously in the Universe, as shown in Fig. 2.4 [42].

Computational chemistry studies about complex organic molecules that forms the bases of life have led to the theory that such molecules might have been formed in the protoplanetary disc, in the nanoparticles that revolved around the Sun, before the formation of the Earth [43]. According to computer models, the process could be the same around any star that acquires planets through a mechanism similar to their formation around the Sun [44].

Other studies [45, 46] have revealed that under conditions similar to those in the interstellar medium, polynuclear aromatic hydrocarbons (PAH) change to amino acids and nucleotides through hydrogenation, hydroxylation or partial oxidation. The database for tracking PAH in the universe has been updated and enriched. According to very recent estimations, over 20 % of the carbon compounds can be tied up in PAH. It has even been asserted that PAH may have been formed only a short time after the Big Bang, are spread throughout the universe and can be associated with the formation of new stars and exoplanets [47].

Studies concerned with nanoparticle properties provide valuable information about planetary systems formation and the origin and role of water and organic substances in producing the first self-reproductive systems that have generated life.

References

1. Aber JS (2006) Climatic controls of glaciations. ES 331/767 Ice age environments. Emporia State University, Kansas, USA. Course web site: http://academic.emporia.edu/aberjame/ice/icehome.html
2. Mandia A. Scott (2005) Possible causes for climate change, Smithtown Sciences Bldg

3. Houghton JT, Ding Y (2002) Climate change 2001: the scientific basis. IPCC. Int J Climatol 22(9):1144. http://onlinelibrary.wiley.com/doi/10.1002/joc.v22:9/issuetoc
4. Jones PD, Mann ME (2004) Climate over past millennia. Rev Geophys 45(2). http://onlinelibrary.wiley.com/doi/10.1002/rog.v42.2/issuetoc
5. Garrett C (1997) Volcanoes and climate change. http://tiger.chm.bris.ac.uk/
6. Robock A (1979) The little ice age: northern hemisphere average observations and model calculations. Science 206:1402–1404
7. Weber G (2005) Toba volcano, The Andaman Association
8. Gasperini L, Bonatti E, Longo G (2007) 100 years later, Tunguska remains mysterious. Terra Nova 19(4):245–251
9. Morgan J, Ranero CR et al (2004) Contemporaneous mass extinctions, continental flood basalts, and 'impact signals': are mantle plume-induced lithospheric gas explosions the causal link? Earth Planet Sci Lett 217:263–284
10. Yu A (2003) Ol'khovatov. Earth, moon and planets 93:163–173
11. Rubtsov V (2009) The Tunguska mystery. Springer, New York, NY
12. Verma S (2005) The Tunguska fireball: solving one of the great mysteries of the 20th century. Icon Books, Cambridge
13. Verma S (2006) The mystery of the Tunguska fireball. Icon Books, Cambridge
14. Las Heras A (2008) Enigma Tunguska. Editura Litera Internaţional, Bucureşti
15. Eather RH (1980) Majestic lights: the aurora in science, history, and the arts. American Geophysical Union, Washington, DC
16. Savage C (2001) Aurora: the mysterious northern lights. Sierra Club Books/Firefly Books, San Francisco, CA
17. Juravle D (2012) Geologia României, vol 1. Editura Universitatăţii, Iaşi
18. Symonds RB, Rose WI et al (1994) Volcanic gas studies: methods, results and applications. Rev Mineral 30:1–66
19. Symonds RB, Rose WI et al (1988) Contribution of Cl- and F-bearing gases to the atmosphere by volcanoes. Nature 334:415–418
20. Cadle RD (1980) A comparison of volcanic with other fluxes of atmospheric trace gas constituents. Rev Geophys Space Phys 18:746–752
21. Chin M, Davis DD (1993) Global sources and sinks of OCS and CS2 and their distributions. Global Biogeochem Cycles 7:321–337
22. McElroy MB, Salawitch RJ et al (1992) The changing stratosphere. Planet Space Sci 40:373–401
23. Bureau H, Keppler H et al (2000) Volcanic degassing of bromine and iodine: experimental fluid/melt partitioning data and applications to stratospheric chemistry. Earth Planet Sci Lett 183:51–60
24. Textor C, Graf HF, Timmreck C, Robock A (2003) Emissions from volcanoes (chapter 7). In: Granier C, Reeves C, Artaxo P (eds) Emissions of chemical compounds and aerosols in the atmosphere. Kluwer, Dordrecht
25. Tungsheng L, Xiongfei G et al (1981) Desert dust: origin, characteristics, and effect on man. Geological Soc Am 186:149–157
26. Shi Z, Shao L, Jones T, Lu S (2005) Microscopy and mineralogy of airborne particles collected during severe dust storm episodes in Beijing, China. J Geophys Res. p 301
27. USA today, 18361–18370, Sept 19 2005
28. Husar RB (2001) The Asian dust events of April 1998. J Geophys Res 2001:233–243
29. Buseck PR, Pósfai M (1999) Airborne minerals and related aerosol particles: effects on climate and the environment. Proc Nat Acad Sci 96:3372–3379
30. Taylor DA (2002) Dust in the wind. Environ Health Perspect 110(2):A80–A87, http://www.ncbi.nlm.nih.gov/pubmed/?term=Taylor%20DA%5Bauth%5D
31. Du Xiaodan (2007) Northern dust brings dirty skies in Shanghai. J Geophys Res
32. Chun Y, Cho H et al (2008) Historical records of Asian dust events (Hwangsa) in Korea. Bull Am Meteorol Soc 89:823–827

33. Wang GH, Zhou CB et al (2012) Atmos Chem Phys Discuss 12:21355–21397
34. Marconi M, Sferlazzo DM et al (2013) Saharan dust aerosol over the central mediterranean sea: optical columnar measurements vs. aerosol load, chemical composition and marker solubility at ground level. Atmos Chem Phys Discuss 13:21259–21299
35. Formenti P, Elbert WJ (2003) J Geophys Res Atmos 108
36. Goudie AS, Middleton NJ (2001) Earth Sci Rev 56:179–204
37. Zook HA (2001) Accretion of extraterrestrial matter throughout Earth's history. pp 75–92
38. Outreach site of Herschel Space Observatory. http://sci.esa.int/herschel/
39. Love SG, Joswiak DJ et al (1992) Densities of stratospheric micrometeorites. Icarus 111: 227–236
40. Humphreys W, Roberta M et al (1972) Spectroscopic and photometric observations of M Supergiants in Carina. Astrophys J 172:75
41. Donald D, Clayton W et al (1999) Condensation of carbon in radioactive supernova gas. Science 283:1290–1292
42. Donald D, Clayton W (2011) A new astronomy with radioactivity: radiogenic carbon chemistry. New Astronomy Rev 55:155–165
43. Starkey N (2013) Insight into the silicate and organic reservoirs of the comet forming region. Gnocchi et Cosmochim Acta 105:73–91
44. Kwok S, Zhang Y (2011) Mixed aromatic: aliphatic organic nanoparticles as carriers of unidentified infrared emission features. Nature 479:80–83
45. Moskowitz C (2012) Life's building blocks may have formed in dust around young sun. Space. com
46. Gudipati MS, Yang R (2012) In-situ probing of radiation-induced processing of organics in astrophysical ice analogs: novel laser desorption laser ionization time-of-flight mass spectroscopic studies. Astrophys J Lett 756(1). http://iopscience.iop.org/2041-8205/756
47. Hoover R (2014) Need to track organic nano-particles across the universe? NASA's got an app for that. http://www.nasa.gov/ames/need-to-track-organic-nano-particles-across-the-universe-nasas-got-an-app-for-that/#.VC2hdmeSzSs

Chapter 3
Anthropogenic Sources of Nanoparticles

Nicolae Strambeanu, Laurentiu Demetrovici, and Dan Dragos

Abstract This chapter deals with anthropogenic sources of nanoparticles in atmosphere. The stationary and diffuse sources are described first: mining and extraction of fossil fuels, ferrous and non-ferrous ores and other mineral resources. Brief references are then made to secondary stationary sources of particulate matter which originate in thermal energy and power generation, ferrous and non-ferrous metallurgy, chemical industry and construction materials (cement, lime) production technologies. A separate section is dedicated to particulate matter resulted from municipal, toxic and hazardous waste incineration and its effect on the environment and human health. The authors also speak about emissions from point sources of nanoparticles, such as the residential heating sector that makes a major contribution to the total amount of nanoparticles produced by anthropogenic activities. Mobile nanoparticle sources include engines that are most frequently used in the present: plane engines, rocket engines, engines for road and water transport. Particulate matter emissions can be reduced by using alternative fuels or other forms of propulsion energy.

3.1 Introduction

Most human activities that take place in open air or produce emissions in the environment generate, directly or indirectly, fine powders lifted by air currents.

As it has already been shown [1], the major fraction of particulate matter (PM) floating in the atmosphere comes from space and it does not have a definite origin that astrophysicists and astrogeologists can explain [2]. Nevertheless, nanoparticles of anthropogenic origin are definitely among the most harmful particulates, not only to human and animal health, but also to the environment.

N. Strambeanu (✉) • L. Demetrovici
Pro Air Clean Ecologic, 37, Vadul Crisului Street, 300613 Timisoara, Romania
e-mail: ns_arana@yahoo.com

D. Dragos
Department of Pharmacy, "Victor Babes" University of Medicine and Pharmacy, Timisoara, Romania

M. Lungu et al. (eds.), *Nanoparticles' Promises and Risks*,
DOI 10.1007/978-3-319-11728-7_3

And if the impact of natural particulates has determined the specific adaptation of living organisms along the time, most anthropogenic nanoparticles were generated in the nineteenth and twentieth centuries, when large amounts of countless harmful and change-resistant particulates were released in the atmosphere. Among these are cancer-causing, mutagenic and teratogenic chlorinated pesticides [3] that were largely used in agriculture until recently, heavy metal oxide PM of heavy metals, PM from the synthesis or chemical purification of various chemical compounds, fuel combustion or deliberate burning of vegetation for deforestation purposes in underdeveloped countries.

None of the above-mentioned categories has been the subject of long-term studies about their impact on health and the environment. The effects of nanoparticles on the living kingdom have been analysed only recently.

One cannot ignore the unprecedented development of scientific research into nanoapplications through the synthesis, characterisation and industrial production of such materials [4–9] which in turn generate various waste types in a similar state of matter.

Anthropogenic sources of nanoparticles are classified as primary, represented by ore exploitation, and secondary—PM resulted from industrial activities and energy production and transport activities.

Secondary sources can be stationary or mobile. The former are thermal power, chemical, milling, metallurgic or other industrial plants that are continuous and diffuse and, unlike the latter, produce and release large amounts of PM. The latter are mostly terrestrial, marine and air vehicle engines and rockets launched to the extra-atmospheric space. These are considered discontinuous point sources.

As far as global anthropogenic emissions are concerned, the amounts of nanoparticles released by stationary sources are equal or less than those released by engines, since the latter total the daily equivalents which account for 40 % of global pollution [10].

Table 3.1 is a synthetic presentation of several Australian anthropogenic sources of nanoparticles [11].

3.2 Primary Nanoparticle Sources

Depending on the materials extracted from quarries or by underground mining, PM concentration released during technological processes can have characteristics similar to those of natural nanoparticle emissions, as it results from the exploitation of mineral reserves accumulated in time, through orogenesis [12].

Consequently, the origin of nanoparticles is the direct exploitation of rocks by surface mining, evacuation through mine shafts or waste heaps produced either directly, by the separation of the useful fraction, or indirectly, through decantation, sedimentation or flotation, following the concentration of the useful fraction in ores. In either case, waste heaps comprise millions of PM which are more or less stable when rising in the atmosphere.

Table 3.1 A synthesis of sources of nanoparticles resulted from anthropogenic activities in several representative Australian areas [11]

		Particulate matter <10 μm		
Source Group	Source Type	Annual emissions (tonnes/year)	Proportion of annual anthropogenic emissions (%)	Cumulative (%)
GMR				
Domestic-Commercial	Solid fuel combustion	5,986	19.6	19.6
Off-Road Mobile	Industrial off-road vehicles and equipment	5,191	17.0	36.6
Industrial	Coal mining	4,154	13.6	50.3
Industrial	Generation of electrical power from coal	1,708	5,6	55.9
Industrial	Primary iron and steel production	1,444	4.7	60.6
On-Road Mobile	Exhaust emissions light duty—diesel	1,073	3.5	64.1
On-Road Mobile	Exhaust emissions heavy duty commercial—diesel	1,002	3.3	67.4
On-Road Mobile	Exhaust emissions passengers cars petrol	972	3.2	70.6
Industrial	Crushing, grinding or separating works	918	3.0	73.6
Industrial	Other land-based extraction	799	2.6	76.2
All	Other	7,253	23.8	100
Sydney				
Domestic-Commercial	Solid fuel combustion	4,503	34.3	34.3
Off-Road Mobile	Industrial off-road vehicles	1,152	8.8	43.1
On-Road Mobile	Exhaust emissions light duty—diesel	840	6.4	49.5
Industrial	Crushing, grinding or separating works	807	6.2	55.6
On-Road Mobile	Exhaust emissions passengers cars petrol	797	6.1	61.7
On-Road Mobile	Exhaust emissions heavy commercial duty—diesel	681	5.2	66.9
Industrial	Ceramics production (excluding gas)	606	4.6	71.5
Industrial	Other land-based extraction	418	3.2	74.7
Commercial	Poultry farming (meat)	237	1.8	76.5

(continued)

Table 3.1 (continued)

Source Group	Source Type	Particulate matter <10 μm Annual emissions (tonnes/year)	Proportion of annual anthropogenic emissions (%)	Cumulative (%)
Industrial	Petroleum refining	237	1.8	78.3
All	Other	2,848	21.7	100
Newcastle				
Domestic-Commercial	Solid fuel combustion	381	22.6	22.6
Industrial	Production of ammonium nitrate	207	12.3	34.8
Off-Road Mobile	Industrial off-road vehicles and equipment	125	7.4	42.2
Industrial	Bitumen pre-mix or hotmix production	114	6.8	49.0
Industrial	Primary aluminium production	111	6.6	55.6
Off-Road Mobile	Commercial ships	80	4.7	60.3
Industrial	Production of phosphate fertiliser	78	4.6	64.9
Industrial	Coal mining	71	4.2	69.1
On-Road Mobile	Exhaust emissions light duty—diesel	60	3.6	72.6
On-Road Mobile	Exhaust emissions heavy duty commercial—diesel	52	3.1	75.7
All	Other	409	24.3	100
Wollongong				
Industrial	Primary iron and steel production	1,442	65.8	65.8
Domestic-Commercial	Solid fuel combustion	248	11.3	77.1
Off-Road Mobile	Industrial off-road vehicles and equipment	187	8.5	85.6
On-Road Mobile	Exhaust emissions heavy duty commercial—diesel	40	1.9	87.5
On-Road Mobile	Exhaust emissions light duty—diesel	36	1.7	89.1
Industrial	Coke production	33	1.5	90.6
On-Road Mobile	Exhaust emissions passenger cars petrol	32	1.4	92.1
Industrial	Coal loading	19	0.9	93.0

(continued)

Table 3.1 (continued)

Source Group	Source Type	Particulate matter <10 μm Annual emissions (tonnes/year)	Proportion of annual anthropogenic emissions (%)	Cumulative (%)
Off-Road Mobile	Commercial ships	14	0.7	93.6
Industrial	Coal mining	14	0.6	94.2
All	Other	126	5.8	100.0
Non-Urban				
Industrial	Coal mining	4,056	30.1	30.1
Off-Road Mobile	Industrial off-road vehicles and equipment	3,727	27.6	57.7
Industrial	Generation of electrical power from coal	1,708	12.7	70.3
Domestic-Commercial	Solid fuel combustion	855	6.3	76.7
Industrial	Other land-based extraction	368	2.7	79.4
Industrial	Hard-rock gravel quarrying	360	2.7	82.1
Commercial	Poultry farming (meat)	272	2.0	84.1
Off-Road Mobile	Commercial off-road vehicles and equipment	261	1.9	86.0
On-Road Mobile	Exhaust emissions heavy duty commercial—diesel	228	1.7	87.7
Industrial	Primary aluminium production	202	1.5	89.2
All	Other	1,456	10.8	100.0

On the long run, countries rich in mineral deposits (see Fig. 3.1) accumulate the largest number of waste heaps and consequently release the largest amounts of nanoparticles in the atmosphere. The most vulnerable are the underdeveloped countries where ore is extracted with obsolete technology and old tools [13].

3.2.1 *Fossil Fuel Exploitation*

Regardless of their nature, fossil fuels can be exploited on the surface, in quarries or through underground extraction. High carbon concentrations (anthracite, pit coal) in fossil deposits require exploitation at considerable depth [14, 15].

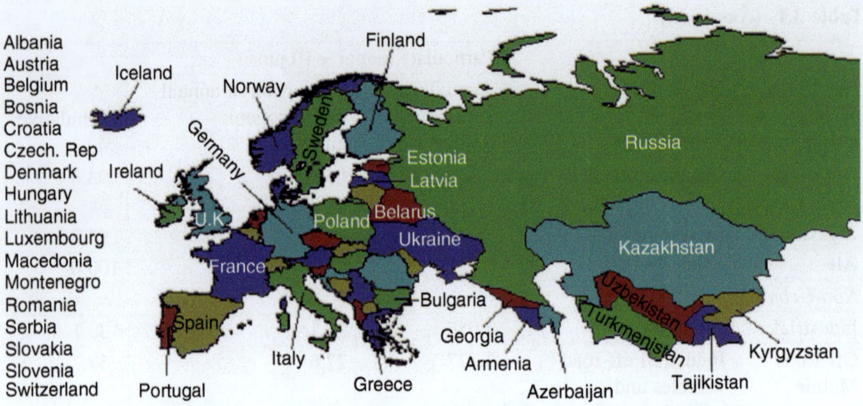

Fig. 3.1 Countries with the most important mineral resources in Europe and Central Eurasia [13]

Coal extraction from quarries, a method that is almost generalised at the moment, is presented intuitively in Fig. 3.2.

When the soil has been stripped and the deposit uncovered (1), coal is excavated and transported either with large vehicles or via conveyor belts. Steps (2) and (3) represent different stages of quarry exhaustion. This technique, though very productive, is also very polluting, especially because of slowly sedimenting PM.

Generally, studies about toxic agents and PM released during surface mining [16] deal with the PM issue superficially; the concentration of micropollutants under the form of nanoparticles is not specified directly, but with global indicators of the TSP type (total suspended particulates), which are irrelevant for the subject of this book [17–19].

Recent studies [19, 20] give more specific details about toxic agent determinations in coal plants during daytime, using more precise indicators such as PM_{10} or $PM_{2.5}$.

Figure 3.3 indicates that in the air samples collected within the perimeter of the two quarries, the highest concentrations are those of the slowly sedimenting particles represented by PM_{10}, or $PM_{2.5}$ which seem to follow the same specific law of concentration variation, while coarser particles sediment immediately, so their concentration is less relevant.

As far as the $PM_{2.5}$ variation is concerned, daily concentrations recorded in the studied period in several places around the coal exploitations vary between 4.5 and 10.2 μg/Nm^3, while the average value of the whole period was about 7 μg/Nm^3.

The toxic gaseous agents found in the deposits (methane, nitrogen oxides, hydrogen sulphide, etc.) must be mentioned among the sources of nanoparticles reformed in the atmosphere. When released, these agents cause degradation through oxidation-reduction, hydrolysis, and neutralisation; in other words, they generate nanoparticles with a different chemical nature than that of the initial pollutants [18–20] (Table 3.2).

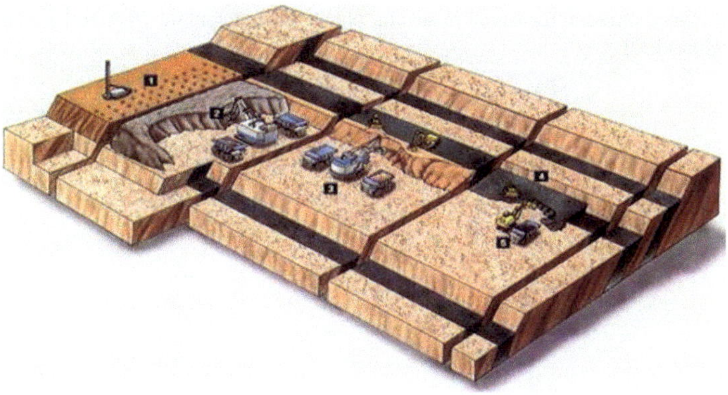

Fig. 3.2 Open-pit coal mining at Greenhills schematic [16]

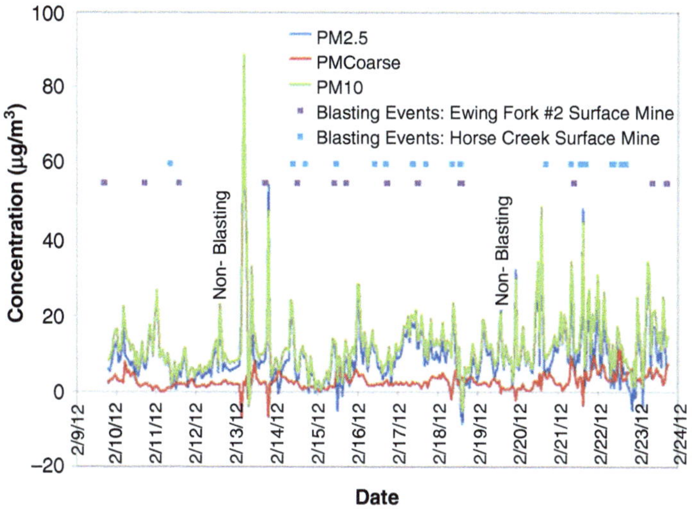

Fig. 3.3 PM_{10} and $PM_{2.5}$ variation near coal quarries in the Appalachians (West Virginia) [20]

Air pollutants are usually the same when coal is extracted through underground mining: PM and gas, including also methane (CH_4), sulphur dioxide (SO_2), nitrogen oxides (NO_x), hydrogen sulphide and carbon monoxide (CO) [21].

The mixture of these constituents is called *firedamp* and its presence in mine galleries can cause explosions and death by intoxication [22].

Accurate nanoparticle measurements expressed as $PM_{2.5}$ were taken under a thorough study conducted for several years in several mines of the Appalachian region. The research results are given in Fig. 3.4 [11].

The above results show that the most important fraction of fine particles expressed as $PM_{2.5}$ is not evacuated through airshafts; it originates in open-air

Table 3.2 $PM_{2.5}$ variation measured in several places around coal quarries in the Appalachians (West Virginia) [20]

End data	PM2.5 concentration (μg/m³)			
	Residence (integrated)	School (integrated)	Beckley (integrated)	TEOM (continuous)
2/10/12	NS	NS	13.1	8.3
2/11/12	NS	NS	5.5	8.9
2/12/12 (Non-blasting)	5.7	NS	3.9	6.1
2/13/12	NS	5.5	4.7	14.2
2/14/12	7.7	6.9	8.7	9.4
2/15/12	5.2	5.6	4.5	3.7
2/16/12	8.9	8.9	5.8	8.9
2/17/12	10.2	9.4	8.4	12.0
2/18/12	8.9	8.8	10.6	8.1
2/19/12 (Non-blasting)	NS	6.7	7.4	8.9
2/20/12	8.0	7.4	7.4	12.1
2/21/12	8.4	6.3	7.3	13.1
2/22/12	6.9	6.1	5.5	9.3
2/23/12	9.2	NS	4.9	10.0
Average	7.3	7.1	7.0	9.5
2010 annual average for beckley	10.3			

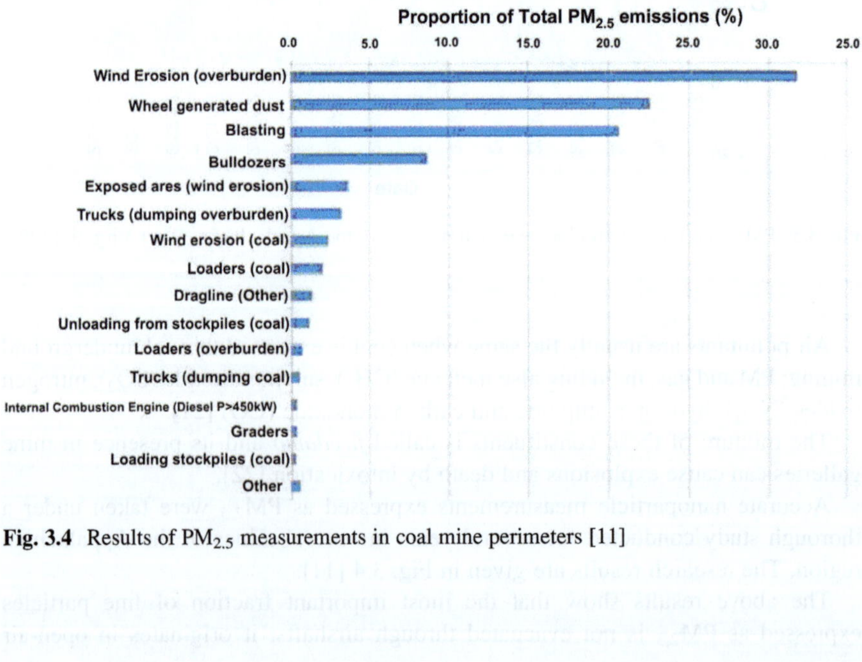

Fig. 3.4 Results of $PM_{2.5}$ measurements in coal mine perimeters [11]

Table 3.3 Mining and separation technologies employed at present [23]

Resource	Mining techniques	Extraction techniques
Iron	Surface mining (open pit mining)	Smelting and chemical reduction
Gold	Underground mining (shaft mining) and surface mining	Gold cyanidation is used. The ore is chemically treated to extract the gold
Copper	Surface mining (open pit mining)	Leaching is used to extract the copper using an acid
Platinum	Underground mining (shaft mining)	Chemical methods and as a byproduct of copper mining
Zinc	Underground mining (shaft mining)	Smelting and leaching
Chromium (Chrome)	Surface mining (open pit mining) and underground mining (shaft mining)	Smelting, redox reactions
Manganese	Surface mining (open pit mining) and underground mining (shaft mining)	Smelting and chemical processes
Diamonds	Surface mining (alluvial deposits) and underground mining (pipe mining)	Diamonds are extracted from rocks and in almost pure form
Phosphate	Surface mining (open pit mining)	Treatment with acid
Asbestos	Surface mining (open pit mining)	Extracted in fairly pure form

activities related to mining, performed especially in the coal storage area, where almost one-third of the whole measured quantity is recorded.

3.2.2 *Ferrous and Non-ferrous Ore Extraction*

Table 3.3 provides a synthetic presentation of the current mining techniques employed in the extraction of metals and minerals [23].

Iron ore is extracted both by surface and underground mining. The most important surface mining facilities are in South America (Bolivia and Brazil in particular), Western Australia, China, India, Ukraine and Canada. In the past years, these countries have replaced other countries with traditional iron ore mining on the market, such as France, Sweden or Germany. The most important iron ore deposit is at El Mutùn, in Bolivia, with an estimated amount of 40 billion tons of iron ore, of which over 50 % is made of magnetite (Fe_3O_4) and hematite (Fe_2O_3) [24]. Pyrite (FeS_2) is another mineral with significant industrial applications in iron and sulphuric acid production.

After extraction and crushing (and roasting, in case of pyrite) and before the metallurgic process starts, ore concentration occurs through sinterisation, producing a porous conglomerate which is essential for furnace processing, because it allows air circulation [25].

PM determination in an Indian open-pit mining facility [26] has provided the following $PM_{2.5}$ values (Fig. 3.5).

Logically, the highest concentrations of PM (about 20 μg/Nm^3) were measured in point A1, corresponding to the office building of the mining facility, and point

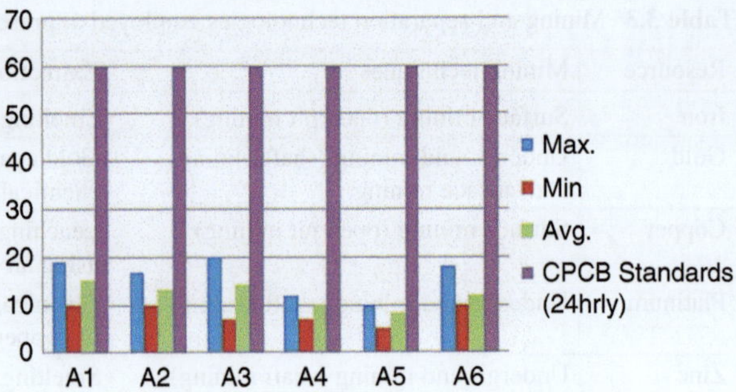

Fig. 3.5 $PM_{2.5}$ monthly averages in Baliparbat-Daitari, India

A3, near the OMC (the company that took the measurements) office building. However, similar concentrations were measured outside the mining facility, in various places of local social activity, which indicates that nanoparticles do no show sedimentation tendencies.

The amounts of metallic copper found in nature cannot cover the market demand. Therefore, it is exploited by mining and successive industrial processing. The most important copper ores are copper matte CuO(OH), cuprite Cu_2O, tenorite CuO, algodonite Cu_6As, paramelaconite $(Cu^{+}Cu^{2+})_2O_3$ and chalcocite or copper pyrite CuS.

A long-term study on PM emissions in copper mining facilities provides relevant results in the field (Table 3.4) [27]. For these emissions, it is easy to notice a general decreasing tendency in the concentrations of all indicators characterising PM, which may be the consequence of both technological improvement along the past two decades and the state-of-the-art monitoring equipment, besides the specific legal limitations. $PM_{2.5}$ variations along the years—an increase of the initial value followed by a constant decrease towards the end of the studied period—may be the result of the intensive/extensive copper ore extraction activity depending on the metal demand on the market.

Studies similar to those conducted for copper extraction in India were carried out for a chromite mine belonging to the same company. The PM_{10} values were between 65–85 μg/Nm^3 within the quarry perimeter and 45–76 μg/Nm^3 in Bangur, a locality near the quarry [28]. Although within the admitted range, most of these values are closer to the upper limit. The most important chromite ores, used almost exclusively for chromium production, are the chromite ($FeCr_2O_4$) and the crocoite $PbCrO_4$ [29].

A comparison of the values recorded for chromite ore excavation (Table 3.5) with those recorded within the iron ore quarry (Table 3.6) will show that the former are higher without exception. This means that either the excavation tools crush the rocks into finer particles or the chromite mining facility is at a much lower altitude than the iron mining facility, and since there are no air currents to spread the PM, they accumulate in the extraction caldera. Such tendencies are also present in countries that use advanced technologies in surface mining [30].

Table 3.4 Average $PM_{2.5}$ values measured in a long-term study conducted in an open-pit mining facility in a copper ore deposit in Rosemont (Virginia, USA) [27]

Pollutant	Period	Year 1	Year 5	Year 10	Year 15	Year 20
PM/TSP	lb/h	1,273.17	1,544.97	1,376.67	1,361.96	1,009.35
	tpd	9.66	12.93	10.91	10.73	6.50
	tpy	2,923.53	3,370.34	2,859.10	2,787.85	1,658.96
PM_{10}	lb/h	443.43	513.41	470.14	464.83	371.71
	tpd	2.79	3.63	3.11	3.05	1.93
	tpy	852.11	961.72	830.18	801.94	495.68
$PM_{2.5}$	lb/h	54.88	61.98	57.66	56.84	46.64
	tpd	0.45	0.54	0.49	0.48	0.36
	tpy	141.60	152.52	139.39	135.64	101.67
CO	lb/h	3,516.38	3,516.38	3,516.38	3,516.38	3,516.38
	tpd	2.13	2.13	2.13	2.13	2.13
	tpy	644.83	615.22	611.73	447.98	184.94

Table 3.5 PM_{10} values measured in an open-pit mining facility for chromite extraction in Bangur (Uttar Pradesh, India) [28]

Mine Office (A1)				
		Results in μg/m³		
Sample No.	Date	RPM	SO_2	NO_x
1	03.04.11	85	<4.0	11.2
2	07.04.11	58	4.0	14.3
3	11.04.11	57	<4.0	19.5
4	15.04.11	65	<4.0	22.3
5	18.04.11	65	4.0	20.8
6	21.04.11	75	7.6	22.6
7	25.04.11	73	4.3	27.8
8	30.04.11	78	5.1	21.0
Bangur village (A2)				
1	03.04.11	59	6.1	16.9
2	07.04.11	57	6.7	13.8
3	11.04.11	45	6.0	10.2
4	15.04.11	51	<4.0	14.3
5	18.04.11	61	<4.0	14.1
6	21.04.11	73	5.0	18.4
7	25.04.11	53	<4.0	10.4
8	30.04.11	76	7.1	12.2
National ambient air quality standards	Annual	60	50	40
	24 h	100	80	80

Asbestos is an iron and sodium hydroxy silicate with the following formula: $Na_2Fe^{2+}{}_3Fe^{3+}{}_2Si_8O_{22}(OH)_2$. Its fibre size is compared with other sub-micrometre particles in Fig. 3.7 [32].

Table 3.6 Size of asbestos fibres revealed by measurements taken near an asbestos quarry in Quebec [32]

	Outdoor air MDDEP (Bisson and Couture, 2007) (f/ml)	Indoor air AVAQ (Marier et al. 2007) (PCMe fibres/ml)
Counting criteria	L >5 μm, D <3 μm, and ratio L/D >3:1	L >5 μm, D ≥0.25 μm and <3 μm and ratio L/D >3:1
Average	0.0029	0.0020
Minimum	0.00038	0.000553
Maximum	0.028	0.010
Upper limit of 95 % Cl	0.0035	0.0031

Concentrations vary from 6–22 μg/Nm^3 in indoor air to 28–96 μg/Nm^3 [33]

Once the use of asbestos in large-scale applications was limited because of its proven cancer-causing characteristic, asbestos mining decreased considerably, but did not stop. Consequently, in order to limit the danger of asbestosis and lung cancer, numerous studies have been carried out to establish strict and coherent rules for asbestos exploitation [31–34].

As shown in Fig. 3.6, asbestos fibres can vary in size from 80–90 nm to 10 μm. As a result, they are considered potentially dangerous for any kind of living cells [33].

Table 3.6 shows values of asbestos fibres measured inside and outside buildings situated near an asbestos quarry in Quebec (Canada).

3.2.3 Exploitation of Natural Construction Materials

Stone quarries and gravel pits are major sources of nanoparticle pollution, especially if stone crushers and/or sorting machines are used for sorting sand or crushed material by grain size.

Most technological processes of rock removal (or extraction by water bed dredging), transport (conveyor belts or trucks), grain sorting and storage take place in open air, therefore wind carries away stone dust and fine sand particles.

In a stone quarry located near Harare (Zimbabwe), in a usual working month, the level of PM expressed as respirable and inhalable PM varied with the distance from the crusher and the stone sorting machine, as shown in Fig. 3.7 [35].

The PM amounts in the two categories represent maximum and minimum values corresponding to discontinuous operations of unloading stone from the crusher. The obvious general tendency is a concentration decrease with the distance from the quarry machines. The smaller the particles are, the smaller the decreasing tendency is. PM_4 and $PM_{2.5}$ emissions in a silica sand quarry in Arizona were measured before and during plant operation with vibrating sieves [36]. The values exceeded the limits established in US EPA, but this is a general issue of any plants operating with particulate matter.

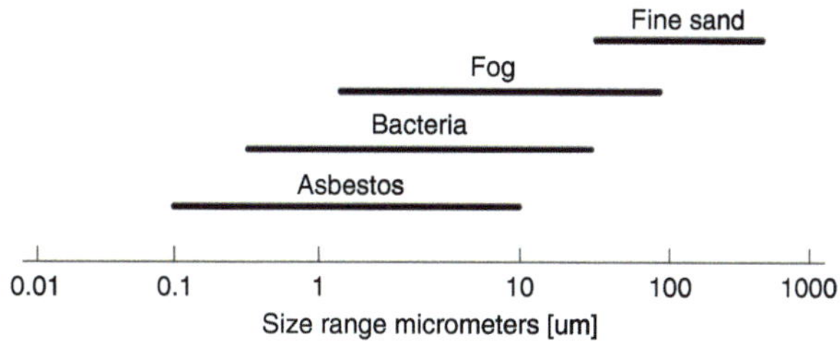

Fig. 3.6 Size of asbestos fibres compared with other materials

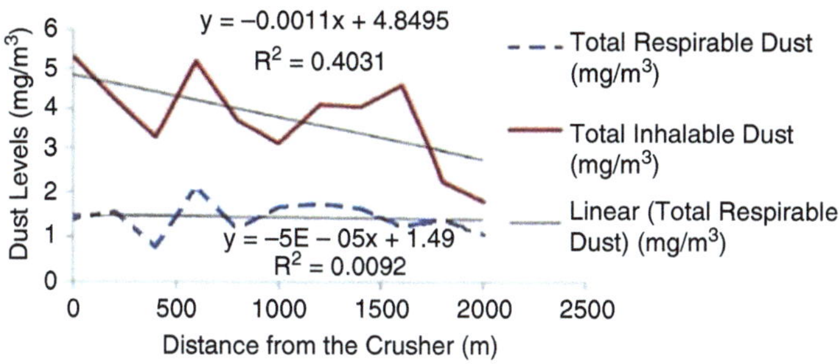

Fig. 3.7 Level of respirable and inhalable PM in the stone quarry at Pomona (Zimbabwe) in January 2011 [35]

3.3 Secondary Sources of Nanoparticles

Secondary sources of anthropogenic nanoparticles are among the most harmful sources, although their contribution is smaller by comparison with both natural PM and PM released during mining activities. If during the evolution of the biosphere its constituents adapted to the impact of natural PM, even to dust containing pathogens or toxic metals, through the parallel evolution of metabolism, PM resulting from industrial activities is more recent, therefore much more harmful to the living beings, both as a consequence of its diverse origins and its toxicity, since man has synthesised millions of substances so far.

Figure 3.8 [37] shows that the European regions with the highest concentration of atmospheric nanoparticles are grouped in the most developed Western European states: Leeds-Manchester industrial zone (Great Britain), Belgium, The Netherlands and Luxemburg (almost entirely); the Paris region and the lower course of the Seine (France), Ruhr Basin (Germany), the Turin-Milan-Venice-Florence area (Italy), the Bohemian Plateau (the Czech Republic) and the coal region of Silesia (Poland). In Eastern Europe, high concentrations of nanoparticles are found in the Donetsk Basin (Ukraine), where large iron and coal deposits are explored [38, 39].

Road transport is the major source of nanoparticle emissions released in the atmosphere (Fig. 3.9) [37].

Quite unexpectedly, the second major contribution to atmospheric pollution with PM comes from point residential sources, not from the energetic sector. Relevant

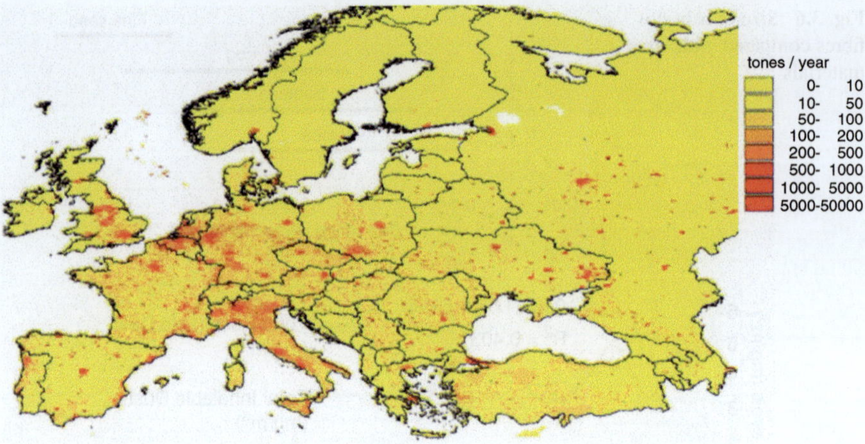

Fig. 3.8 Pan-European $PM_{2.5}$ emissions [37]

Fig. 3.9 Particle emissions in 15 European Union states, Iceland, Norway and Switzerland, in 2005 [37]

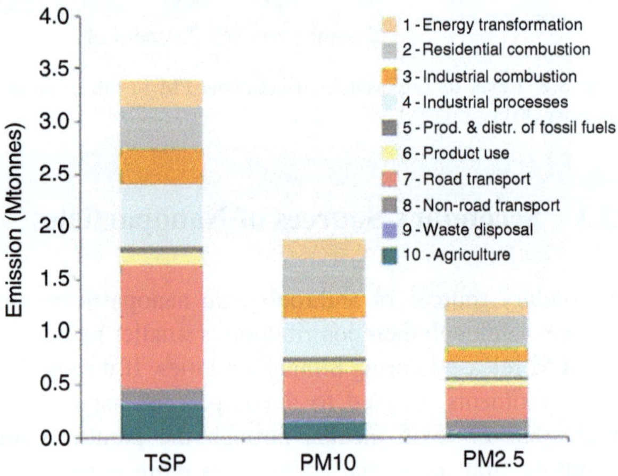

and almost equal contributions come from the industrial combustion sector (thermal and catalytic reformation of hydrocarbons, chemical conversion of coal, syngas production, etc.), other chemical processes, industrial processing of resources and the municipal and industrial waste incineration sector.

3.3.1 Stationary Sources of Nanoparticles

As mentioned at the beginning of this chapter, stationary sources of nanoparticles are industrial thermal power plants, chemical plants producing dust-containing products like fertilisers or solid pesticides, cement plants and municipal or special

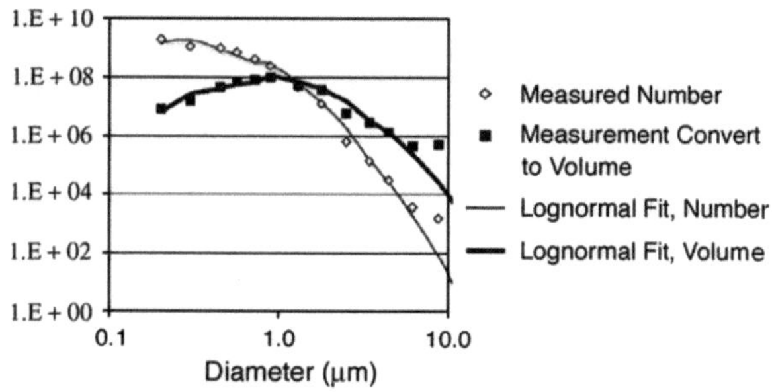

Fig. 3.10 Size distribution of $PM_{2.5}$ flying ash collected from the stack of a thermal power plant in California (USA) [47]

waste incineration plants. This category includes the point sources of the residential heating sector [40–44].

3.3.1.1 Thermal Energy and Power Generation Systems

Thermal power plants release PM in different ways: flying ash escaping from the electric filters mounted on the stack, bottom ash extraction, fine dust scattering and final ash storage, if ash is not dried and sprayed permanently [45, 46].

Studies on the $PM_{2.5}$ fraction of flying ash in a fluidised bed coal-fired thermal power plant have revealed the particle distribution shown in Fig. 3.10 [47].

The data presented so far indicate that about 50 % of the ash collected from the electrostatic filter bags is 100–1000 nm in size.

For the bottom ash resulted from fluidised bed combustion of crushed coal in several thermal power plants in Poland, the size distribution of nano- and microparticle-containing fractions is given in Fig. 3.11 [46].

The comparison of the size distribution of flying ash and bottom ash in modern thermal power plants in Poland and the USA indicates that $PM_{1-2.5}$ in the bottom ash are less than a quarter of the whole amount and half of the nanoparticle concentration in the flying ash.

The two types of ash differ very slightly in their chemical composition. The difference lies in the origin of the coal [46, 47]. Their typical chemical composition is given in Table 3.7 [47].

The composition was determined by high-resolution methods:

- Ion chromatography
- AAS spectrophotometry
- Total organic carbon
- X-ray spectroscopy

The current trend of the studies dealing with the recovery of unburnt carbon from coal ash, as well as the encouragement of renewable fuels in the energetic production require the use of vegetal pellet or dust (straw, wood) in combination with bottom ash [48].

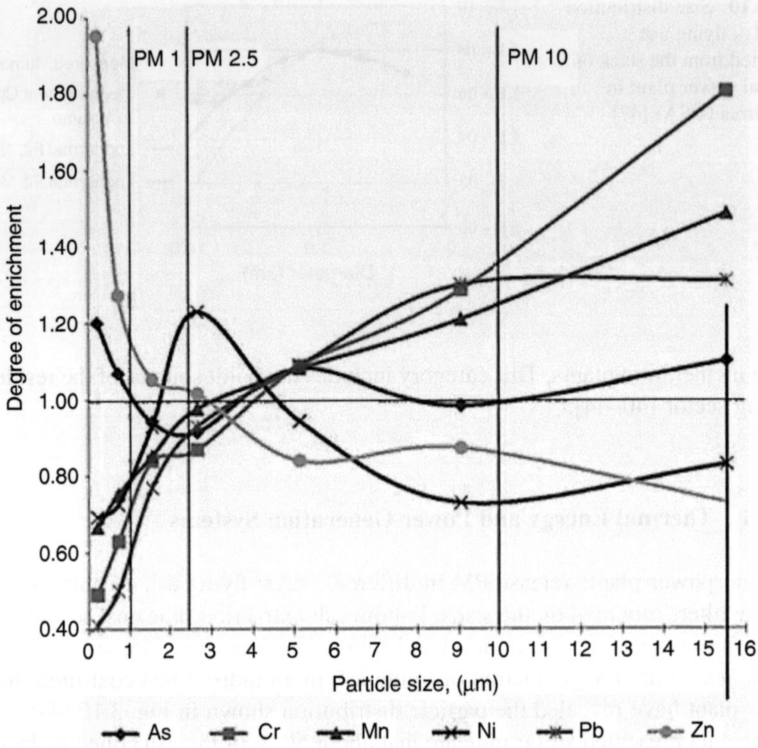

Fig. 3.11 Size distribution of bottom ash in a fluidised bed coal-fired thermal power plant

Under the circumstances, aerosol formation during combustion was studied in a fluidised bed 800 MW_th boiler, during combustion of finely divided wood, with and without coal or flying ash as alkaline capture additive.

Both the physico-chemical characteristics and the PM morphology of the aerosol samples were analysed. Sub-micrometre sizes were determined with the electron microscope (SEM/EDS and Met/EDS). Sub-micrometre particles (PM_1) with a concentration of 44–47 mg/Nm^3 were identified.

The Gaussian distribution of nanoparticles highlighted a characteristic maximum peak at about 200 nm. This peak consisted mainly of ultrafine aggregates (<100 nm) of particles rich in K, Cl and S formed through the homogeneous and heterogeneous vaporisation of solutions containing chloride, potassium, sodium and sulphate ions from the gaseous state.

In an experimental variant based on ash tree wood flour with the addition of alkaline flying ash, for a 1:2 or 1:4 mass ratio, the Gaussian distribution of sub-micrometre particles indicated sizes larger than those in the experiments without coal flying ash. The PM_1 concentration diminished from the above-mentioned value to 11–19 mg/Nm^3.

In this case, ultrafine particles (<60 nm) had a more regular, almost spherical shape, and the composition changed from the Na-K-Cl-S quaternary system to the

Table 3.7 Chemical composition of the same thermal power plant ash [47]

Soluble ions by atomic absorption or ion chromatography	
Chloride	0.003
Nitrate	ND[a]
Phosphate	ND
Sulphate	3.97
Ammonium	0.03
Sodium ion	0.37
Potassium ion	0.06
Calcium ion	3.67
Carbon fractions by thermal/optical reflectance	
Organic carbon 1	0.05
Organic carbon 2	0.29
Organic carbon 3	0.34
Organic carbon 4	0.46
Pyrolysed carbon	0.09
Elemental carbon 1	0.50
Elemental carbon 2	1.40
Elemental carbon 3	1.28
Carbonate carbon	0.35
Elements by X-ray fluorescence	
Sodium	1.00
Magnesium	0.14
Aluminium	4.92
Silicon	12.43
Phosphorus	0.22
Sulphur	1.26
Potassium	1.23
Calcium	5.64
Titanium	0.65
Vanadium	0.06
Chromium	0.02
Manganese	0.03
Iron	4.16
Cobalt	ND
Nickel	0.004
Copper	0.02
Zinc	0.04
Selenium	0.009

[a]Not detected

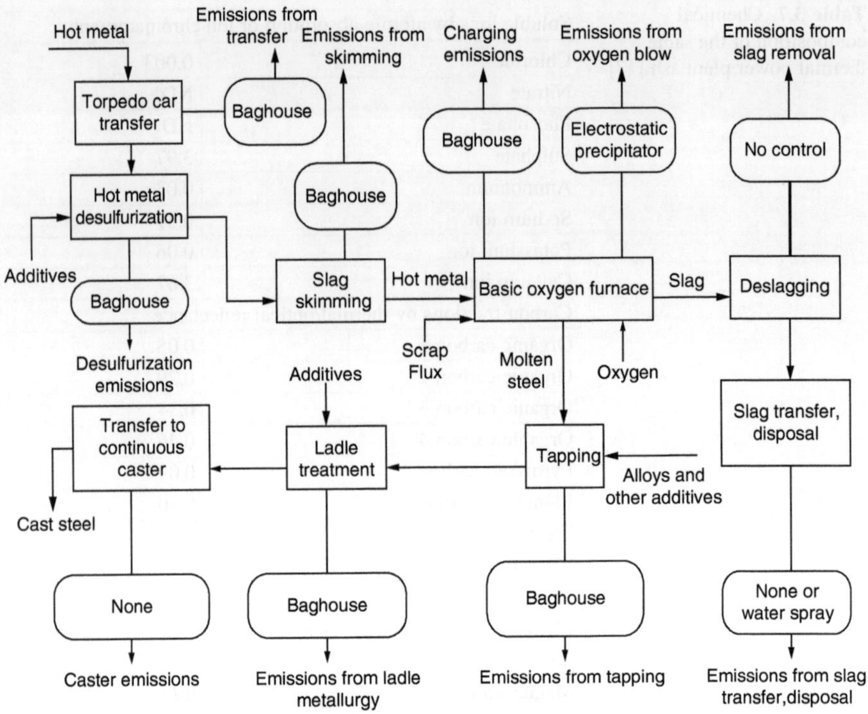

Fig. 3.12 Sources of nanoparticles in a steel plant in Michigan (USA) [51]

Ca-P-Si tertiary system, which demonstrated that the ions resulted mostly from the coal flying ash. This proved the optimisation/minimisation of the solid aerosol content of the coal flying ash added to the alternative dust fuel in fluidised bed coal-fired thermal power plants [48–50].

3.3.1.2 Ferrous and Non-ferrous Metallurgy

A recent thorough study on several steel plants with coke oven batteries in the Great Lakes area reveals both direct sources of $PM_{2.5}$ (Fig. 3.12) and sources that generate nanoparticles through *de novo* synthesis between PAH and the heavy metals released in the atmosphere (Fig. 3.13) [51].

The total amounts of $PM_{2.5}$ released from the steel plants are given in Table 3.8 [51].

The above-mentioned experimental data indicate that these amounts must not be ignored, especially because they include heavy metals under the form of both organic and organometallic compounds that are very dangerous for health and the environment.

As far as the orientative distribution of iron-containing nanoparticle concentration in large steel plant areas is concerned, the concentrations measured in various

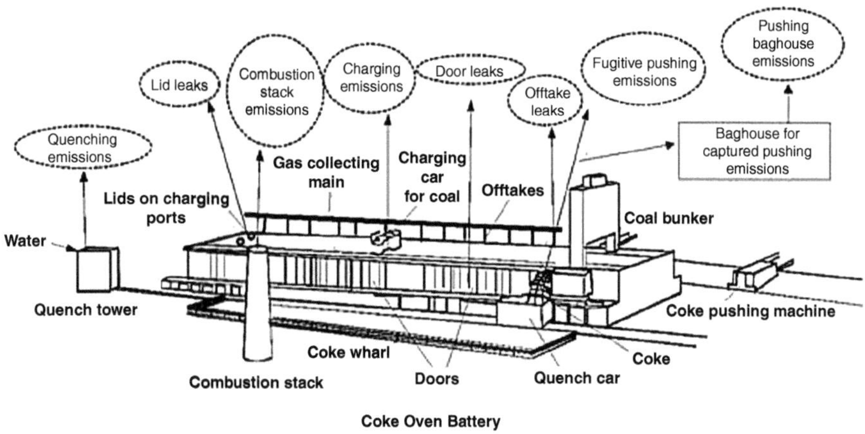

Fig. 3.13 Sources of nanoparticles in a coke oven battery in Michigan (USA) [51]

Table 3.8 Annual emissions of nanoparticles in a steel plant with a coke oven battery in Michigan (USA) [51]

Pollutant	*Emissions (tpy)*	*Percent*
SO_x	4,567	35
NO_x	5,616	43
PM-CON	1,876	14
$PM_{2.5}$-FIL	1,130	8
Total	13,189	100
HAP	*Total emissions (tpy)*	*HAP in PM2.5 (tpy)*
Manganese	13	7.2
Lead	1.9	0.7
Nickel	0.04	0.01
Chromium	0.2	0.1
Mercury	0.4	0.4
Total	15.5	8.4

adjacent points varied from 250 to 3,000 ng/Nm^3, depending on the environmental factors and the Saharan dust advection (Fig. 3.14) [52].

The days when the natural nanoparticle concentration was higher than that of the investigated emissions were days when Saharan dust was present in the atmosphere.

3.3.1.3 Chemical Industry

The sources of nanoparticles released by chemical product plants are so diverse that listing them here would go beyond the scope of the present monograph. This is because besides raw materials, half-finished and finished products under the form of PM with direct impact on air quality, any chemical plant releases, intentionally or unintentionally, gases that interact with other pollutants and generate PM.

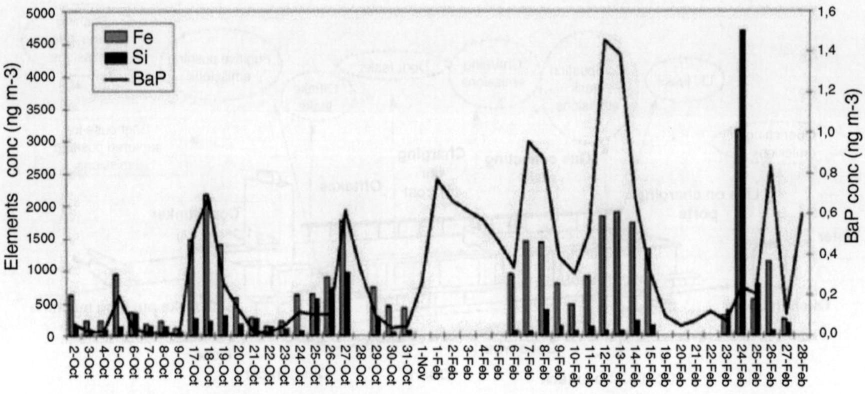

Fig. 3.14 Variation of iron and silicon concentration in PM in the Apulia (Italy) steel plant [52]

Consequently, fertilisers, pigments and pesticide production will be dealt with in short, as these are the most representative sub-branches that produce chemical dust particles.

A case of global investigation of pollutants generated on an industrial platform in South-Western Spain reveals the interaction between many metallic ions released from steel or other metallurgic plants with various constituents released in the atmosphere during phosphate rock processing under the form of NPK fertilisers or by road transport [54].

In a large NP (ammonium phosphate) fertiliser plant in India, air quality monitoring revealed $PM_{2.5}$ particles between 13 and 38 μg/Nm^3, values that are within the allowed local range (Table 3.9) [53].

Nanoparticles generated in pesticide production and large-scale uses in agriculture are harmful to man's health and the biosphere. Consequently, studies were conducted on their travel distance in air under the influence of environmental factors, their effect as nanoparticles or VOC in the atmosphere expressed as half-life [41].

The dyes and pigments industry releases fine inorganic or organic dust in the air [55–57]. Inorganic pigments are based on heavy metal salts whose ions have characteristic colours. Dyes that contain such pigments are mostly mixtures of pigments and siccative oils or other organic substances that can form polymerisable films. The most common ions found in inorganic pigments are chrome, nickel, iron, cobalt, lead, zinc and arsenic ions; as a result, the released nanometric dust is extremely dangerous [55].

A federal monograph of American industrial sources of nanoparticles highlights the annual PM concentrations, including inorganic pigments, and recommends measures to limit them these concentrations (see for example Table 3.10) [56].

As a rule, organic pigments are functional derivatives with chromofore groups that ensure the coloured perception of these substances by the human eye. Most of them (anilins, azoic derivatives, microdispersed pigments) are toxic and cause cancer. It has been proved that those pigments that are to be eliminated from current surface protection processes belong to the polychlorinated biphenyls class whose use was strictly limited long ago [57].

Table 3.9 Data about air quality monitoring at the fertiliser plant in Mangalore (Jaipur, India) [53]

Details		PM10 μg/ m^3	PM2.5 μg/ m^3	NH_3 μg/ m3	SO_2 μg/ m^3	NO_x μg/ m^3	CO mg/ m^3	Hydrocarbon μg/ m^3	F^- μg/ m^3
Jan. 2013	Max.	55	38	58	9	9	0.26	ND	ND
	Min.	22	15	<5	<5	<5	<0.1		
Febr 2013	Max.	58	36	57	9	9	0.23	ND	ND
	Min.	26	16	<5	<5	<5	<0.1		
March 2013	Max.	56	36	57	9	9	0.22	ND	ND
	Min.	25	13	<5	<5	<5	<0.1		

ND not detectable

3.3.1.4 Cement Industry

In the past 20 years, industrial systems of cement production have made great progress, so that they can set higher limits to environmental pollution with both gaseous constituents and PM. These very expensive technological changes were imposed by the conclusions founded on measurements taken in many cement factories around the world and mathematical models based on which the long-term effects of pollution under the influence of atmospheric factors were evaluated [58, 59].

The long-term measurements taken in a large cement factory in Harbin (China) during 1990 and 2010 indicate that, in the case of very fine dust, the $PM_{2.5}$ value dropped to half in less than 15 years (Fig. 3.15) [58].

Unfortunately, this obvious worldwide tendency of reducing atmospheric PM emissions does not characterise other pollutants released during cement production (CO, CO_2, NO_x, SO_2), as their value rose with production increase [59].

3.3.1.5 Waste Incineration

Waste incineration is one of the largely employed methods of waste removal both for reducing waste amounts and toxicity and energy recovering as a final alternative of thorough use. Depending on their destination, industrial incineration plants eliminate either municipal waste or flammable industrial, sanitary and toxic waste that must be destroyed with strictly controlled methods [60].

Municipal Waste Incineration

PM emissions in modern municipal waste incineration plants are limited under the provisions of EC Directive on waste incineration [61].

Table 3.10 Annual $PM_{2.5}$ concentrations in the St. Louis (USA) industrial area [56]

Facility Name	SCC	SCC L2	SCC L3	SCC L4	PM2.5-PRI emissions (tpy)	Control classification
American Commercial Terminals	30501011	Mineral products	Coal Mining, cleaning, and material handling (see 305319)	Coal transfer	1,052	Uncontrolled
Amerenue-Meramee Plant	101000226	Electric generation	Bituminius/sub-bituminous coal	Pulverised coal: dry bottom tangential (sub-bituminous coal)	754	Controlled
Pace Construction Co-Chesterfield	30500260	Mineral products	Asphalt concrete	Drum mix plant: rotary drum dryer/mixer, #2 oil-fired, counterflow	494	Uncontrolled
Dial Corp–Dial Corp	30113210	Chemical manufacturing	Organic acid manufacturing	Acetic acid via acetaldehyde	426	Uncontrolled
Dynegy Midwest Generation Inc	10100202	Electric generation	Bituminius/subbituminous coal	Pulverised coal: dry bottom (Bituminous coal)	350	Controlled
Dynegy Midwest Generation Inc	10100203	Electric generation	Bituminius/subbituminous coal	Cyclone furnace (Bituminous coal)	348	Controlled
Dynegy Midwest Generation Inc	10100203	Electric generation	Bituminius/Subbituminous coal	Pulverised coal: dry bottom	341	Controlled
Elementis Specialties Inc	30103553	Chemical manufacturing	Inorganic pigments	Pigment dryer	184	Uncontrolled
Elementis Specialties Inc	30103553	Chemical manufacturing	Inorganic pigments	Pigment dryer	184	Uncontrolled
Masterchem Industries Inc Imperial	30101401	Chemical manufacturing	Paint manufacturer	General mixing and handling	117	Uncontrolled
US Silica Company Pacific	30502511	Mineral products	Construction sand and gravel	Screening	113	Uncontrolled
Anheusr Busch Inc-St Louis	10200202	Industrial	Bituminius/Subbituminous coal	Pulverised coal: dry bottom	102	Controlled

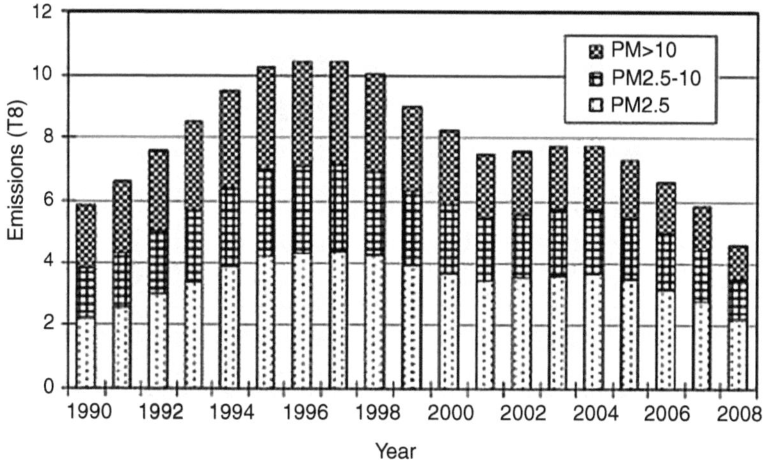

Fig. 3.15 Variation of PM values for dust emissions measured at the cement factory in Harbin (China) [58]

All municipal solid waste (MSW) incinerators are equipped with bag filters that retain particulate matter. In Great Britain, such incinerators contribute to atmospheric pollution with 0.042 % of the total $PM_{2.5}$ emissions [62].

Some studies show that, as far as municipal waste incineration is concerned, ultrafine particulate matter $PM_{0.1-1}$ can have significant effects on health and the environment. These studies hold that health risks are associated rather with the number of ultrafine particles than with their mass [63–65].

Measurements of nanoparticles released at the stack of a municipal waste incinerator with energy recovery located in Piacenza (Italy) [63] indicated that no particles larger than 2.5 μm in diameter were present in the emissions, but about 65 % of the measured $PM_{2.5}$ emissions were sub-micrometre particles (Fig. 3.16). Particles larger or smaller than 100 nm had a relatively equal distribution [63, 64].

The identification of the primary composition of stack emissions released by the same incinerator led to the conclusion that the improved particle retention systems were effective in removing 99.99 % of emissions, with similar efficiency in reducing particles of smaller diameter [65].

It has been found that the boiling points of the elements present in the incinerated waste are very important: higher boiling point elements like chromium were identified in larger particles, while lower boiling point elements like arsenic and cadmium were more present in ultrafine particles, making them more dangerous [65].

Nevertheless, the conclusions that several authors from North America and Japan and European researchers have drawn about the level of global pollution with ultrafine particles are unanimous: in the general context of anthropogenic

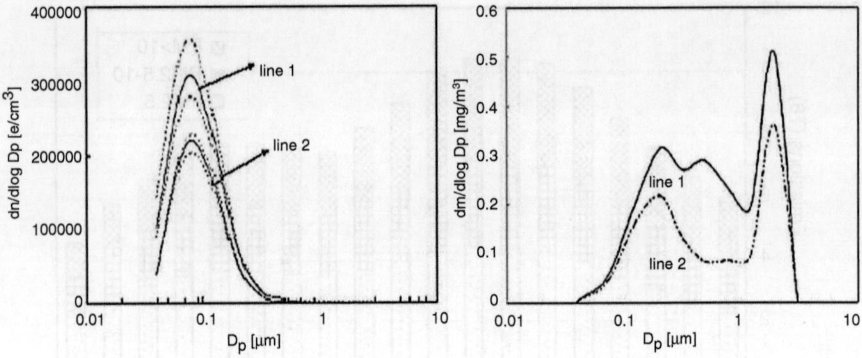

Fig. 3.16 Size distribution of stack nanoparticles in the municipal waste incineration plant in Piacenza (Italy)

emissions, the nanoparticle contribution resulted from MSW incineration plants is small [62].

Hazardous Waste Incineration

Hazardous waste is eliminated under specific regulations that are meant to guarantee the reduction to a minimum of direct emissions, through VOC or other forms of volatile or particulate waste release and stack emissions [66, 67].

Current norms regarding hazardous waste incineration require strict limits both for $PM_{2.5}$ emissions and indirect nanoparticle emissions under the form of heavy metals which are more or less volatile at high temperatures (Table 3.11) [62, 67].

Although incineration of any type of waste has a small contribution to anthropogenic nanoparticle emissions [62–65, 67], there are studies that attempt at finding a direct correspondence between thermal waste elimination and the statistical parameters of population health (Fig. 3.17) [68, 69].

According to these sources, waste incineration plants produce huge amounts of fine and ultrafine particles, because the current standard value of PM emissions (10 mg/m^3) specifies a filter retention rate of only 5–30 % for $PM_{2.5}$ and 0 (zero) for PM_1. In fact, most stack emissions are ultrafine particles of the most dangerous type. Fabric filters with Teflon membrane are not effective against particles under 200–300 nm which are very harmful to health [69]. In this case, what matters is the particle number and size, and not their weight [65, 68].

The distribution of particle size by weight will cause great errors, especially in hazardous waste incinerators with equipment for reducing nitrogen oxides, which may actually double the $PM_{2.5}$ emissions [69, 70].

Table 3.11 Emission limits recommended for incineration of hazardous waste and sewage sludge [67]

	Code	Name			
NFR source category	5.C.1.b.i, 5.C.1.b.ii, 5.C.1.b.iv	Industrial waste incineration including hazardous waste and sewage sludge			
Fuel	NA				
Not applicable	HCH				
Not estimated	Nh3, Cr, Cu, Zn, Se, Benzo(a)pyrene, Benxo(b)fluoranthene, Benzo(k) fluoranthene, Indeno(1,2,3-cd)pyrene, PCBs				
Pollutant	*Value*	*Unit*	*95 % confidence interval*		*Reference*
			Lower	*Upper*	
NO_x	0.87	kg/Mg waste	0.087	8.7	European Commission (2006)
CO	0.07	kg/Mg waste	0.007	0.7	European Commission (2006)
NMVOC	7.4	kg/Mg waste	0.74	74	Pasant (1993)
SO_2	0.047	kg/Mg waste	0.0047	0.47	European Commission (2006)
TSP	0.01	kg/Mg waste	0.001	2.3	European Commission (2006)
PM_{10}	0.007	kg/Mg waste	0.0007	0.15	US EPA (1996) applied on TSP
$PM_{2.5}$	0.004	kg/Mg waste	0.0004	0.1	US EPA (1996) applied on TSP
BC	3.5	% of $PM_{2.5}$	1.8	7	Olmez et al. (1998)
Pb	1.3	g/Mg waste	0.48	1.9	Theloke et al. (2008)
Cd	0.1	g/Mg waste	0.048	0.15	Theloke et al. (2008)
Hg	0.056	g/Mg waste	0.04	0.08	European Commission (2006)
As	0.016	g/Mg waste	0.01	0.019	Theloke et al. (2008)
Ni	0.14	g/Mg waste	0.048	0.19	Theloke et al. (2008)
PCDD/F	350	μg I-TEQ/ Mg waste	0.5	35,000	UNEP (2005)
Total 4 PAHs	0.02	g/Mg waste	0.007	0.06	Wild (1995)
HCB	0.002	g/Mg waste	0.0002	0.02	Berdowski et al. (1997)

Current modelling methods do not take into consideration the secondary particles reformed from atmospheric gas emissions, far from the incinerator stack, although a number of studies have shown that in 95 % of cases ultrafine heavy metal particles combine with polynuclear aromatic hydrocarbons (PAH) to produce PM_3 or smaller conglomerates. PAHs are toxic, mutagenic, teratogenic and cancer causing by nature and this kind of combinations increases lung cancer risk by almost eight times [69–72].

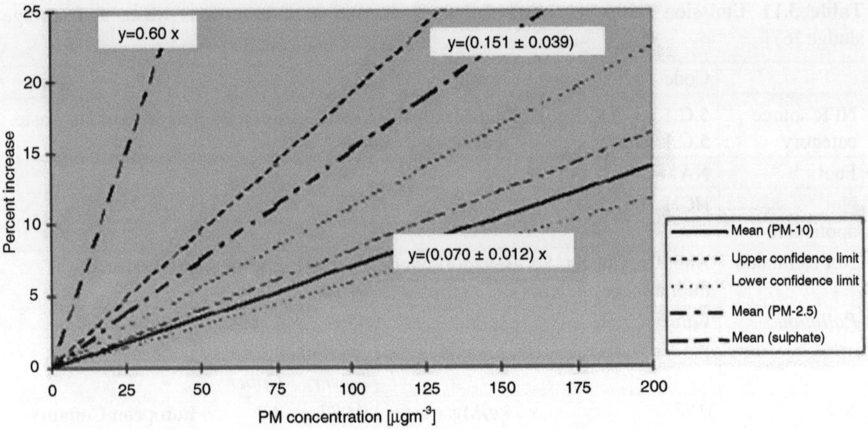

Fig. 3.17 Dependence of mortality on $PM_{2.5}$ and PM_{10} concentrations in Great Britain population [68]

3.3.1.6 Residential Heating

Irrespective of fuel nature, residential heaters have a major contribution to the anthropogenic accumulation of nanoparticles in the atmosphere (Fig. 3.18) [37].

Detailed studies on nanoparticles released by conventional residential and social heating sources (schools, offices, performance halls, etc.) in Fairbanks, Alaska indicate that of the total annual emission amount, the highest nanoparticle concentration comes from wood-fired heaters [73–75]. In order to limit these emissions in areas on the planet where residential heating is a long-term or constant necessity, programmes for the gradual reduction of wood combustion in favour of other low-emission fuels were suggested [75].

Given that fossil fuels obtained from exhaustible sources (coal, oil, natural gas) need to be replaced with renewable fuels (wood, straw, biodiesel, biobenzene), the policies of reducing nanoparticle emissions should be harmonised with the policies concerning the use of such fuel types.

3.3.2 *Mobile Sources of Nanoparticles*

Mobile sources of nanoparticles are engines. Regardless of fuel nature or mechanical drive, the modern classification of engines identifies two classes:

- External combustion engines;
- Internal combustion engines.

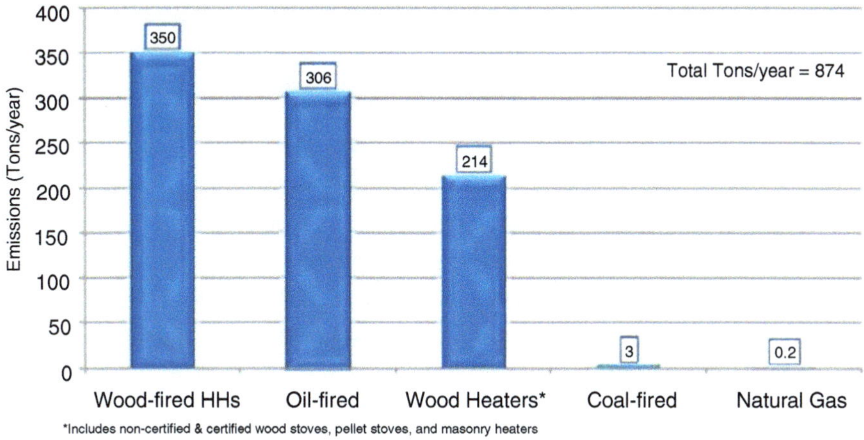

Fig. 3.18 Estimated $PM_{2.5}$ emissions caused by residential heating with various fuels in Fairbanks (Alaska) [75]

Functionally, external combustion engines produce vehicle propulsion through the adiabatic energy of gases that are burnt and evacuated at high pressure and speed—sometimes passing through a turbine. For this reason, many researchers do not include them in the class of proper engines. Additionally, rocket engines use other oxidisers, much stronger than oxygen, instead of air. This is because, on the one hand, rocket engines must produce as much energy as possible in a very short time, and on the other hand, air supply is no longer possible when leaving the atmosphere.

Most internal combustion engines are engines with spark (Otto) or compression (Diesel) ignition. If spark-ignition engines are almost exclusively petrol engines and are used for cars or smaller applications (boat engines, low-voltage generating sets, etc.), depending on their size, diesel engines use diesel oil, light or heavy fuel oil and even coal dust or carbon black (marine engines).

Historically speaking, one should also mention steam engines, either Watt-Carnot or Stirling, which were the basis for industrial and water and road transport development for almost 200 years. However, applications for historical machinery are anticipated for the future [76].

3.3.2.1 External Combustion Engines (Jet and Rocket Engines)

Almost without exception, medium or long-range wide body jet airliners use jet engines or turbojet engines. The distribution of nanoparticles released by jet engines in the northern hemisphere is given in Fig. 3.19. The analytical

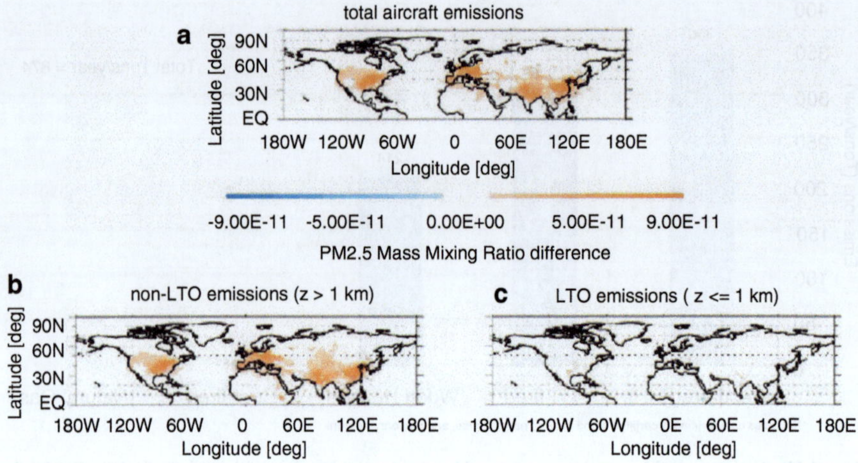

Fig. 3.19 $PM_{2.5}$ emissions from jet engines of planes travelling in the northern hemisphere [77]

determinations shown in these figures indicate that the most densely populated areas on the planet, with the largest number of airports and air routes, are also the most polluted with $PM_{2.5}$ resulting either directly, from kerosene combustion, or indirectly, from the interaction between exhaust gas constituents and the components in Earth's higher atmospheric layers, under the influence of solar radiation [77, 78].

A variation in emission concentration is also visible. Concentration varies with seasons and the altitude where the measurements were taken. It is known that the most frequent air routes for medium and long distances are in the troposphere, at 10,000–14,000 m (Fig. 3.20) [77].

Irrespective of their engine fuel, oxidiser or structure, rocket engines used in the present release, besides CO_2, significant amounts of carbon black in the stratosphere. The large carbon black volume is caused by the incomplete fuel combustion, given the low oxygen concentration in the upper atmospheric layers [79–81].

Recent models indicate that, given the current rate of 1,000 rocket launches per year, the effect of the warming potential caused by finely divided carbon particles—carbon black—is about 100,000 higher than the effect of CO_2 emissions. Although carbon nanoparticles do not always react with stratospheric ozone, through the solar radiation protection effect, they contribute to ozone depletion [79, 81]. Since carbon black absorbs radiation in the visible spectre and remains in the upper atmosphere for 5–10 years, its effects can lead, in the same period of time, to a temperature increase of 1 K at the two poles, causing changes in the pressure

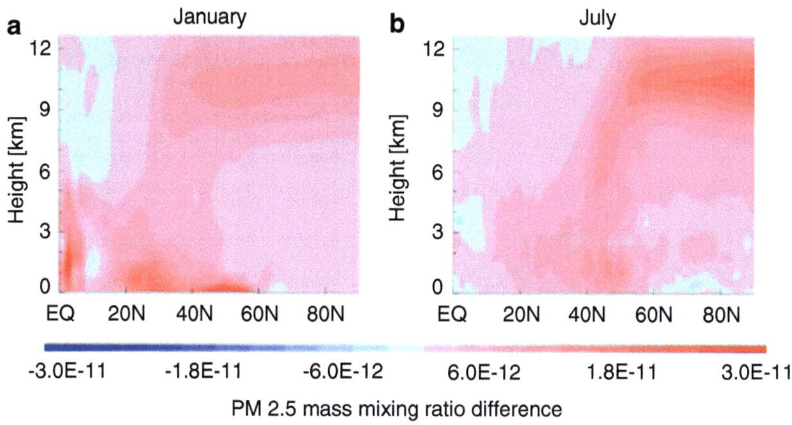

Fig. 3.20 Seasonal and altitude distribution of nanoparticles from aircraft in the northern hemisphere [77]

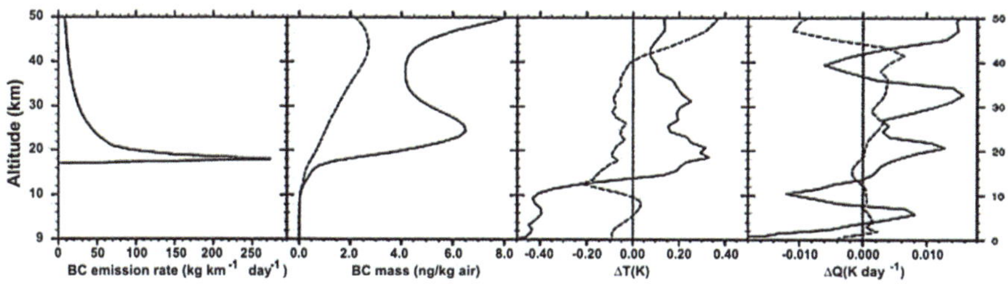

Fig. 3.21 Carbon black emissions in the atmosphere and temperature variations caused by rocket launches in space [79]

gradients [80]. Figure 3.21 shows both the variations in carbon emissions with altitude and the variation of the thermal gradient in relation to the same parameter [79].

3.3.2.2 Internal Combustion Engines

Despite the variety of internal combustion engines, most land and sea vehicles are equipped with petrol-fuelled stark-ignition (Otto) engines or diesel oil-fuelled compression engines. Large commercial ships are equipped with variants of diesel engines fuelled with heavy fuel oil or even coal dust.

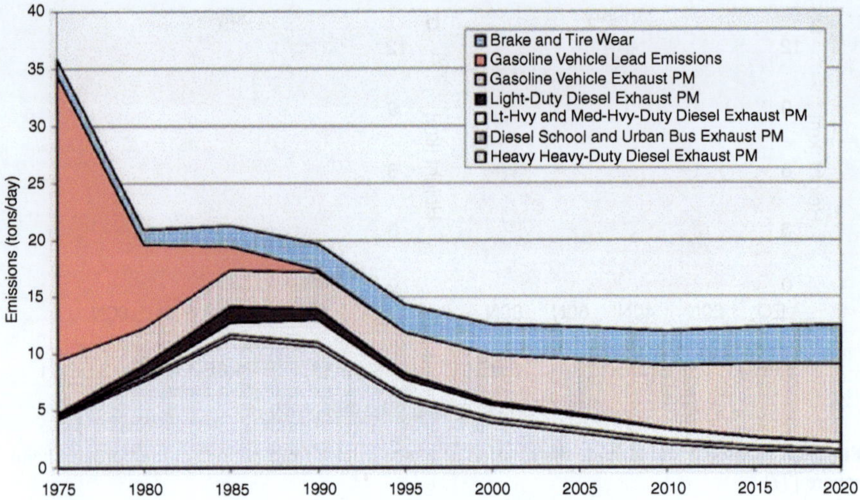

Fig. 3.22 Variation of $PM_{2.5}$ engine-released emissions along the time and in the near future, by fuel type [84]

Regardless of the fuel, these engines release nitrogen oxides, carbon monoxide (CO), particulate matter (PM), volatile organic compounds (VOC) and smaller amounts of other pollutants like sulphur dioxide (SO_2) and ammonia (NH_3).

Most non-electric trains and big trucks use diesel engines, unlike cars, which have petrol engines. Diesel engines release the same pollutants as petrol engines, but they produce much higher quantities of $PM_{2.5}$ and smaller amounts of VOC and CO (see for example the data presented in Fig. 3.22). Particulate matter released by diesel engines are considered very harmful, as the particles are extremely small and can be inhaled very easily [82–84].

Long distance, high capacity ships, as well as large-scale fish farming activities, naval military actions or large cruise liners are major producers of PM emissions. Accordingly, international agreements limit marine emissions to road traffic emission levels [84].

Nuclear-powered ships, used almost exclusively for military purposes, are vessels that do not release nanoparticles in the atmosphere.

A past and future perspective of $PM_{2.5}$ emissions released by both diesel and stark-ignition engines are presented in Fig. 3.23 [84]. The statistical data show that in the middle of the tenth decade of the past century, lead emissions disappeared completely, because the method of controlling autoignition with tetraethyl lead-based antiknock agents was abandoned.

The obvious decrease in nanoparticle emissions is the consequence of engine manufacturers' efforts to improve technical performances and the attempts to

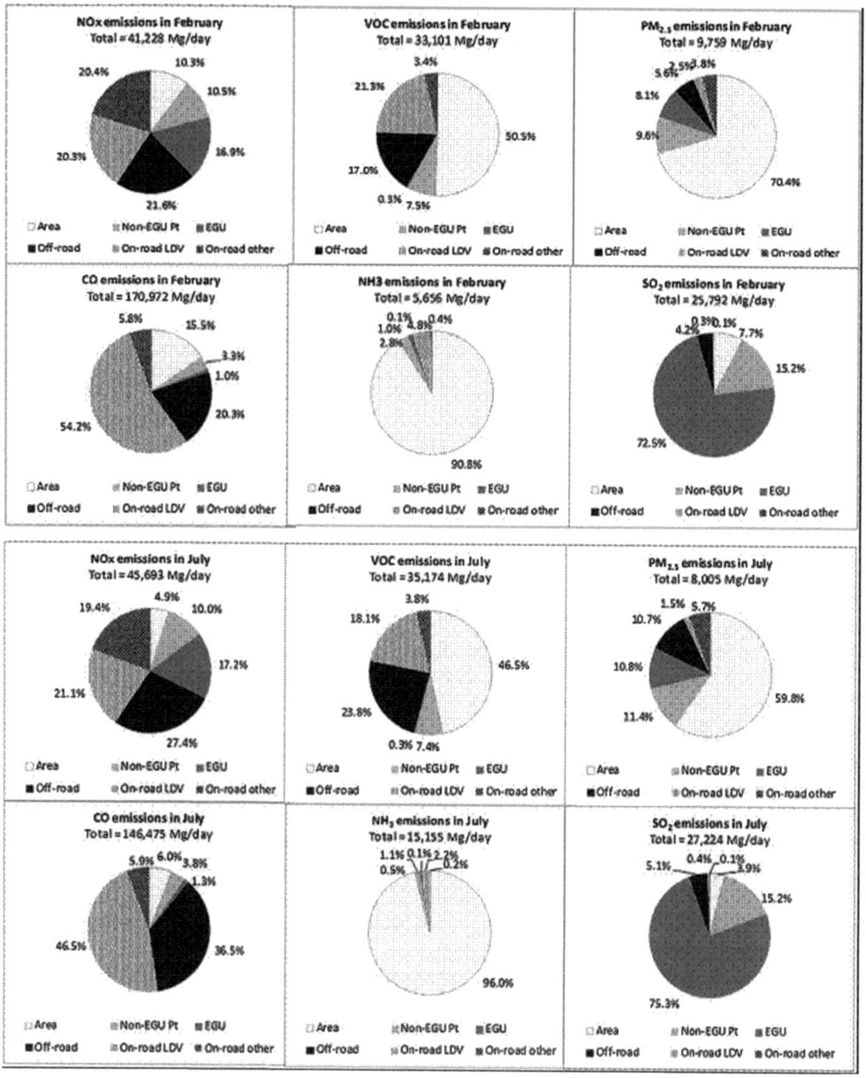

Fig. 3.23 Estimated polluting emissions; road traffic pollutants in central USA in February and July 2008 [83]

control global pollution through fuel cells, electrical engines or non-polluting fuel (water, hydrogen) engines, given the worrying perspective of fossil fuel depletion.

References

1. Starkey N (2013) Insight into the silicate and organic reservoirs of the comet forming region. Geochim Cosmochim Acta 105:73–91
2. Burchell M, Price M et al (2013) Sample return missions to minor bodies. Astron Geophys 54:3.28–3.32

3. Kinter WB, Merkens LS et al (1972) Studies on the mechanism of toxicity of DDT and polychlorinated biphenyls: disruption of osmoregulation in marine fish. Environ Health Perspect 1:169–173
4. Shubayev V, Pisanic T et al (2009) Magnetic nanoparticles for the agnostic. Adv Drug Deliv Rev 61:467–477
5. van der Linger E, Cortie MB et al (2001) Catalyst and method of producing catalyst. SA Patent 2001/5816
6. Niemeyer CM (2001) Nanoparticles, proteins, and nucleic acids: biotechnology meets materials science. Angew Chem Int Ed Engl 40:4128–4159
7. Cheng C (2008) A Multiquantum: dot-doped, fiber amplified with characteristics of broadband, flat gain and low noise. J Lightwave Technol 26(11):1404–1410
8. Qin D, He X, Wang K, Zhao XJ, Tan W, Chen J (2007) Fluorescent nanoparticle based indirect immunofluorescence microscopy for detection of *Mycobacterium tuberculosis*. J Biomed Biotechnol 2007
9. Sattler KD (2008) Handbook of nanophisics. CRC, Boca Raton, Fl, 20.01–27.01
10. Mazzi EA, Dowlatabadi H (2007) Air quality impacts of climate mitigation: UK policy and passenger vehicle choice. Environ Sci Technol 41:387
11. Sarah-Jane Donnelly, Andrew Balch, Andrew Wiebe, Natalie Shaw, Simon Welchman, Alex Schloss, Ella Castillo, Kim Henville, Andrew Vernon, John Planner. NSW coal mining benchmarking study: international best practice measures to prevent and/or minimise emissions of particulate matter from coal mining. Document reference: KE1006953_NSW Coal Mining Benchmarking Study_Volume I.doc
12. Almánzar F, Baker MS, et al. Mineral facilities of Europe: U.S. Geological Survey open-file report 2010–1257. 1 pl., 47-p. table
13. Safirova E, Brininstool M, Anderson ST, Perez AA, Newman HR, Wallace GJ, Wilburn DR (2012) U.S. Department of the Interior, U.S. Geological Survey, Europe and Central Eurasia, 2011 Minerals yearbook
14. Maraca I (1987) Tehnologii noi în extracţia cărbunelui energetic. Editura Tehnică, Bucureşti
15. In clean coal we trust – or do we? ParisTech Rev. http://www.paristechreview.com/2013/10/15/clean-coal-trust/
16. Greenhill's Canada (2014). http://www.mining-technology.com/projects/greenhills/
17. Popa RG, Drăguţ Gh (2011) Studiu privind efectele pulberilor rezultate din activitatea depozitului de cărbune Roşiuţa asupra populaţiei din zonă, Analale Universităţii "Constantin Brâncuşi", seria inginerie, 2, 142–149, Târgu Jiu, România
18. World Coal Institute (2009) Coal mining. www.worldcoal.org/.../coal_resource_overview_of_coal_report
19. (2012) Final report on West Virginia air quality assessment near a surface coal mine blasting operation
20. Final Report on West Virginia Air Quality Assessment near a Surface Coal Mine Blasting Operation, 2012
21. Das Sharma P (2009) Coal mining and pollution. Knoll Website, http://saferenvironment.wordpress.com/2009/09/01/environmental-pollution-problems-and-control-measures/
22. Blum Deborah (2010) Firedamp, speakeasy science. http://scienceblogs.com/speakeasyscience/tag/deborah-blum/.
23. (2014) Syvalula everything science: mining and mineral processing. http://cnx.org/contents/b2ceb903–8abe-4848–9ff6–6e59c9137d09@2.1:28/Siyavula_textbooks:_Grade_11_P
24. (2005) USGS – Mineral industry surveys, January 2005. http://minerals.usgs.gov/minerals/pubs/mcs/2005/mcs2005.pdf
25. Mitelea I, Lugscheider E et al (1999) Ştiinţa materialelor în construcţia de maşini. Editura Sudura, Timişoara
26. OMC (2013) Daitari iron ore mining project, environmental monitoring report, Odisha, India
27. JBR (2011) Rosemont Copper Co. emission inventory information, vol 1, 2
28. OMC (2011) Banghur chromite mining. Regular post clearance environmental monitoring report of April 2011, Ranchi, India
29. Neniţescu CD (1972) Chimie generală. Didactică şi Pedagogică Bucureşti, pp 983–984

30. Escarré J, Lefebvre C et al (2011) Heavy metal concentration survey in soils and plants of the Les Malines mining district (Southern France): implications for soil restoration. Pollution 216:485–504
31. Bourgault MH, Belleville D (2010) Presence of asbestos fibres in indoor and outdoor air in the city of Thetford mines: estimation of lung cancer and mesothelioma risks. INSP of Quebec, Montreal
32. (1978) Asbestos waste management guidance special report. U.S. EPA 450/2–78–014
33. (2009) International Agency for Research on Cancer (IARC). Lancet Oncol
34. Van Gosen B, Heather A, et al. (2010) A USGS study of talc deposits and associated amphibole asbestos within mined deposits of the Southern Death Valley region, California
35. Madungwe E, Mukonzvi T (2012) Assessment of distribution and composition quarry mine dust: case of pomona stone quarries, Harare. Atm Clim Sci 2:52
36. Pace TG (2012) Examination of the multiplier used to estimate PM2.5 fugitive dust emissions from PM10, US EPA report, 2012
37. Visskedijk A, Pacyna J et al. (2004) Coordinated European Particulate Matter Emission Inventory Program (CEPMEIP) EUR 21302 EN, JRC 2004. Proceedings of the PM emission inventories scientific workshop, Lago Maggiore, Italy. 18 Oct 2004, pp 163–174
38. Kurakov IV, Samofalov F et al (2010) Coal mining in the Russian Donetsk basin. Coke Chem 53:121–123
39. Grumau S (2002) Coal mining in Ukraine. Economic Rev 44
40. (2011) North American power plant air emission. Com Env Cooperation Rep
41. Kodiak R, Bürkle L, et al. (2008) Pesticides in air: consideration for exposure. Eur Com Health Consumer Prot. SANCO/ 10553 /2006, Rev 2
42. Oguntoke O, Awanu A et al. (2012) Impact of cement factory operations on air quality and human health in Ewekoro local government area, South-Western Nigeria. Int J Environ Stud 69. http://www.tandfonline.com/action/doSearch?Contrib=Awanu%2C+A+E
43. Moriske HJ, Drews M et al (1996) Indoor air pollution by different heating systems: coal burning, open fireplace, and central heating. Toxicol Lett 88:349–354
44. Schmidt RH, Rodrick GE (2003) Food safety handbook (USA-CANADA)
45. Friedlander SK (2000) Smoke, dust, and haze. Oxford University Press, New York, NY, pp 130–139
46. Juda-Resler K, Kowalczyk D (2012) Size distribution and trace elements contents of coal fly ash from pulverized boilers. Pol J Environ Stud 22:25–40
47. Smith KR, Veranth JM et al (2006) Acute pulmonary and systemic effects of inhaled coal fly ash in rats: comparison to ambient environmental particles. Toxicological Sci 93:390–399
48. Juul Damoe A, Wu H (2014) Impact of coal fly ash addition on combustion aerosols (PM 2.5) from full-scale suspension firing of pulverized wood. Am Chem Soc J
49. Verandah JM, Smith KR (2000) Coal fly ash and mineral dust for toxicology and particle characterization studies: equipment and methods for PM 2.5 and PM 1 enriched samples. Aerosol Sci Tech 32: 127–141
50. Cheremishinoff NP (2012) Clean electricity through advanced coal technologies. Elsevier, Oxford
51. (2006) Evaluation of PM 2.5 emissions and controls at two Michigan steel mills and a Coke Oven Battery. Final US EPA report, February, 7, 2006
52. Amodio M, Andriani E et al (2013) How a steel plant affects air quality of a nearby urban area: a study on metals and PAH concentrations. Aerosol Air Quality Res 13:497–508
53. Mangalore Chemicals and Fertilizers Ltd. (2013) Health, safety & environment rep
54. Fernandez-Camacho R, Rodriguez S, et al. (2012) Ultrafine particle and fine trace metal (As, Cd, Cu, Pb and Zn) pollution episodes induced by industrial emissions in Huelva, SW Spain. AEA 11554:1–11
55. US EPA (2012) US EPA report. 17 Oct 2012
56. (2005) Evaluation of potential PM2.5 reduction by improving performance of control devices: conclusions and recommendations. Pechan draft report, 30 Sept 2005
57. Winneke G, Ranft U (2014) Behavioural sexual dimorphism in school-age children and early developmental exposure to dioxins and PCBs: a follow-up study of the Duisburg Cohort. EHP Rev

58. Lei Y, Zhang Q, et al. (2010) An inventory of primary air pollutants and CO2 emissions from cement production in China, 1990–2020, Elsevier Atm Env
59. STAPPA (2006) Controlling fine particulate matter under the clean air act: a menu of options. State and Territorial Air Pollution Program Administrators (STAPPA)
60. Lungu M, Strâmbeanu N, et al. (2013) Separation of nanoparticles from combustion gases wastes of incinerators. Intern Symp Env Ind., Bucharest, 28–30 Oct 2013
61. EC directive 2000/78
62. AEA (2012) Rep, rev, res, into health effects of energy from waste facilities. AEA ED 57607001
63. Buonanno G, Ficco G et al (2009) Size distribution and number concentration of particles at the stack of a municipal waste incinerator. Waste Manag 29:749–755
64. Buonanno G, Stabile L et al (2010) Dimensional and chemical characterization of particles at a downwind receptor site of a waste-to-energy plant. Waste Manag 30:1325–1333
65. Buonanno G, Stabile L et al (2011) Chemical, dimensional and morphological ultrafine particle characterization from a waste-to-energy plant. Waste Manag 31:2253–2262
66. Strâmbeanu N, Demetrovici L, et al. (2009) Comparative calculation of incineration costs at 3,500 tons/year and 22,000 tons/years capacity. Intern. Symp Env. Ind, Bucharest, III, 66. 18–20 Oct 2009
67. Commisional decision CE 438, 28 Feb 2006
68. (2005) The health effects waste incineration. 4th Rep. British society for ecological medicine. 8–10 Dec 2005
69. Howard CV (2000) Health impacts of waste management policies. Hippocrates foundation, Kos, Greece. 12–14 Nov 1998. Acad. Publ
70. Espinosa AJ, Rodriquez MT et al (2001) Size distribution of metals in urban aerosols in seville (Spain). Atmos Env 35:2595–2601
71. Venkataraman C, Friedlander SK (1994) Source resolution of fine particulate polycyclic aromatic hydrocarbons: using a receptor model modified for reactivity. J Air Waste Manage 44:1103–1108
72. Zmirou D, Masclet P et al (2000) Personal exposure to atmospheric polycyclichydrocarbons in a general adult population and lung cancer assessment. J Occup Environ Med 42:121–126
73. Davis J, Misiuk D, et al. (2009) Reducing PM 2.5 emissions from residential heating sources in the Fairbanks North Star Borough: emissions estimates, policy option, and recommendations
74. Hartig L (2008) Rev. August 18th letter regarding EPA's PM2.5 designations for Alaska. 20 Oct 2008
75. Holdmann G, Murphy J (2008) Fairbanks North Star Borough baseline greenhouse gas emission inventory: base year 2007. Alaska Center for Energy and Power, University of Alaska Fairbanks
76. Homutescu CA, Savitescu G (2003) Introducere în motoare Stirling, 13–27. Editura Cerni, Iaşi
77. Lee H, Olsen SC et al (2013) Impacts of aircraft emissions on the air quality near the ground. Atmos Chem Phys 13:5505–5522
78. Tilmes S, Lamarque JF et al (2012) Technical note: Ozone sonde climatology between 1995 and 2011: description, evaluation and applications. Atmos Chem Phys 12:7475–7497
79. Ross, M, Mills, M, et al. (2010) Potential climate impact of black carbon emitted by rockets. Geophys Res Lett 37L24310
80. Ross M, Sheaffer P (2014) Radiative forcing caused by rocket engine emissions. Earths Future. doi:10.1002/2013EF000160
81. Ross M (2009) Potential climate and ozone impacts from hybrid rocket engine emissions. Am Geophys Union Fall Meeting, 2009
82. Brook JR, Poirot RL et al (2007) Assessing sources of PM 2.5 in cities influenced by regional transport. J Toxicol Env Health 70:191–199
83. Vijayaraghavan K, Lindhjem C (2012) Effects of light duty gasoline vehicle emission standards in the united states on Ozone and particulate matter. Atm Env 60:109–120
84. Schaeffer A (2010) Diesel technology and black carbon. Diesel technology forum. https://www.acs.org/content/.../blackcarbon/schaeffer-presentation.pdf

Chapter 4
Mechanisms of Nanoparticle Formation and Reformation in the Atmosphere

Nicolae Strambeanu, Laurentiu Demetrovici, and Dan Dragos

Abstract Several relevant models of nanoparticle formation and reformation in the atmosphere are underlined: the mechanism of forming fine ash particles in coal-fired thermal power plants; carbon particulates formed either directly or through partial oxidation intermediates; acid rain, bases, ammonia, saline or oxide particulates formation mechanism. For solid or liquid particulate matter of organic nature present in the atmosphere, the phases of the partial oxidation of organic compounds and the dioxin and furan generation in various combustion systems are presented. The mechanism of reforming these compounds from flue gas through de novo syntheses initiated through the conversion of carbon dioxide into formose is also described.

4.1 Introduction

Nanoparticles are entities that move freely until they combine with particles of similar nature or react with particles of different chemical nature. Such processes increase their specific mass up to a value that allows their sedimentation.

Figure 4.1 shows a physico-chemical method of forming fine ash particles in fluidised bed coal-fired thermal power plants [1].

It must be specified that in all combustion chambers, irrespective of the fuel used, carbon particulates can form directly, through the incomplete combustion of the organic substrate:

N. Strambeanu (✉)
Pro Air Clean Ecologic, 37, Vadul Crisului Street, 300613 Timisoara, Romania
e-mail: ns_arana@yahoo.com

L. Demetrovici
Pro Air Clean Ecologic, Timisoara, Romania

D. Dragos
Department of Pharmacy, "Victor Babes" University of Medicine and Pharmacy, Timisoara, Romania

M. Lungu et al. (eds.), *Nanoparticles' Promises and Risks*,
DOI 10.1007/978-3-319-11728-7_4

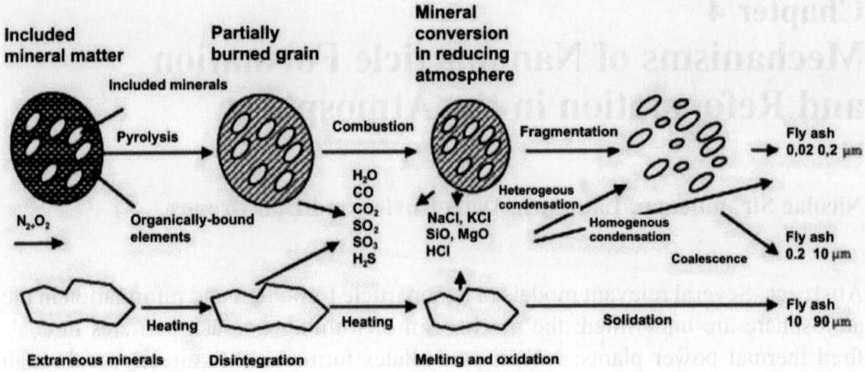

Fig. 4.1 Mechanism of fine ash particle formation in fluidised bed coal-fired thermal power plants [1]

$$C_nH_{2n+2} + (n+1)/2O_2 \rightarrow nC + (n+1)H_2O \tag{4.1}$$

$$C_nH_{2n} + n/2O_2 \rightarrow nC + nH_2O \tag{4.2}$$

$$C_nH_n + n/4O_2 \rightarrow nC + n/2H_2O \tag{4.3}$$

$$C_nH_{2n+2} + (n + 5/2)O_2 \rightarrow (n-3)C + 2CO + CO_2 + (n+1)H_2O \tag{4.4}$$

For any other organic compound used as fuel oil or fuel, the incomplete carbon oxidation reaction is identical, the functional groups on the main, aliphatic or aromatic chain generating various oxidation compounds of the C, CO or CO_2 type or organic intermediates with oxygen [2]. Several characteristics of particulate matter produced through incomplete combustion are given in Table 4.1.

4.2 Inorganic Acids and Bases

Atmospheric oxidation is mediated by free radicals. Unlike oxidation in flame regime, the energy source for atmospheric oxidation is solar radiation. In lower atmospheric layers, the photolysis of ozone molecules produces peroxy or hydroxyl radicals in the presence of water vapour [2, 3]:

$$O_3 + h\upsilon \rightarrow O\bullet + O_2 \tag{4.5}$$

$$O\bullet + H_2O \rightarrow H_2O_2 \tag{4.6}$$

$$O\bullet + H_2O \rightarrow 2HO^- \tag{4.7}$$

$$2O\bullet \rightarrow O_2 \tag{4.8}$$

Ozone molecules, like atomic oxygen and hydroxyl ions, have high reactivity to any reductive atmospheric compound.

Table 4.1 Physical properties of particulate matter depending on the amount of oxygen in the combustion system

Average particle diameter measured with the electron microscope (nm)	Specific surface area by the BET method ($m^2\ g^{-1}$)	Relation to carbon, in reaction (% mass)		Structure properties	
		Hydrogen	Oxygen	DBP index ($cm^3\ g^{-1}$)	Apparent specific volume for 50 kg $cm^3\ g^{-1}$
175–275	95–135	0.20–0.32	0.75–1.45	1.25–1.45	1.45–1.75
140–270	120–145	0.28–0.34	1.00–2.00	1.30—1.70	1.55–1.75
290–350	95–125	0.60–0.75	2.90–3.50	1.04—1.20	1.49–1.64
240–300	100–135	0.50–0.60	2.70–3.60	1.04–1.25	1.49–1.60
230–260	130–160	0.40–0.52	2.70–3.60	1.03–1.20	1.50–1.65
175–275	175–420	0.33–0.45	2.60–4.00	1.15–1.65	1.50–1.75

In general, free radicals have a short life in the atmosphere, but three types of radicals are abundant especially in industrial areas or heavy traffic areas around major urban agglomerations: $HO_{(x)}$, $NO_{(x)}$ and $XO_{(x)}$, (X = Cl, Br) [4–6]. Section 2.1 describes in detail the formation mechanism of $XO_{(x)}$-type of oxygenated acids of halogens in the atmosphere.

Consequently, the reaction of ozone or its above-mentioned radicals with the gaseous compounds produced by pollution leads to gas oxidation: SO_2 to SO_3, NOx to NO_2. Through the action of the fine water drops contained in clouds, these anhydrides become H_2SO_4 and HNO_3 respectively (Fig. 4.2) [7].

The two acids are most commonly seen both in low pH (below 6.5) precipitations, and reactions with metal compounds present under the form of nanoparticles in flying ash or with salts lifted by hurricanes from the planetary ocean, converting them into nanoparticles of different nature [8].

HX halogenated acids (HCl in particular) appear in the atmosphere following the incineration of chloride-containing waste or the reactions that chlorides contained in wind-carried aerosols from the planetary ocean produce in contact with H_2SO4 and HNO_3 [9]:

$$CH_3 - CH_2 - Cl + 3O_2 \rightarrow 2CO_2 + HCl + 2H_2O \tag{4.9}$$

$$2NaCl + H_2SO_4 \rightarrow Na_2SO_4 + 2HCl \tag{4.10}$$

$$NaCl + HNO_3 \rightarrow NaNO_3 + HCl \tag{4.11}$$

The stratospheric reactions between ozone or any other forms of active oxygen with compounds of the chlorofluorocarbons class (also known as freon) can generate both organic radicals and significant amounts of hydrochloric or hydrofluoric acid [10].

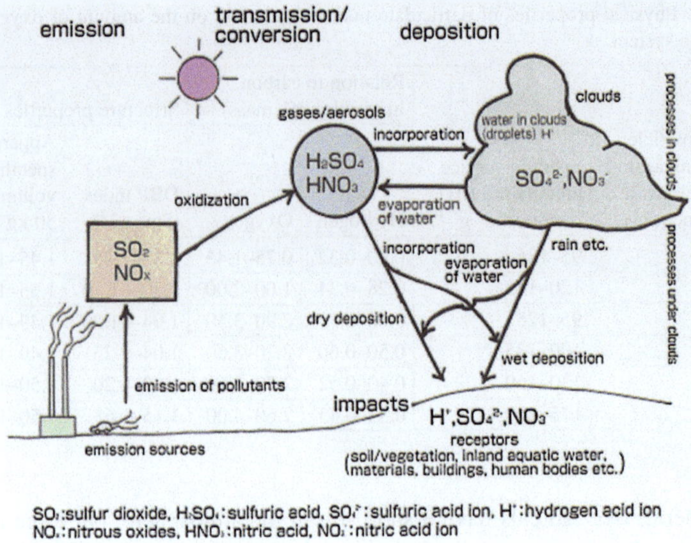

Fig. 4.2 Acid rain formation mechanism [7]

The regional increase of atmospheric pH is generally caused by terrestrial ammonia sources, among which the most important are agriculture and animal breeding [11].

The presence of ammonia can have a beneficial influence as a result of the neutralisation of free acidity in the air; at the same time, in higher atmospheric layers, with the contribution of solar energy, through oxidation reactions catalyzed by heavy metal compounds such as TiO_2 under the form of nanoparticles, ammonia can take different forms of NO_x [12]:

$$2NH_3 + (2x + 3)O\bullet \rightarrow 2NO_x + 3H_2O \tag{4.12}$$

The reaction of ammonia oxidation in the atmosphere is slow and plays only a minor role in reducing its concentration. The experimental data obtained for ammonia from the series of oxidation reactions under normal atmospheric conditions are based on kinetic models with elements of intermediary accumulation, as the oxidation reactions occur simultaneously with ammonia generation in the atmosphere, through the mass transfer processes that control the kinetics of parallel chemical and biochemical reactions [13].

The conversion and reconversion of nitrogen compounds to different oxidation states is a much more complicated process, because the reactions also involve atmospheric nitrogen, electrical phenomena, the soil and the biosphere with all its life forms capable of producing the nitrification–denitrification reactions that characterise the circulation of nitrogen in nature [13–15].

4.3 Oxides and Salts

The first source of oxide nanoparticles in the atmosphere is particulate resulted from waste heaps of oxide ore flotation plants or slag left over after ferrous and non-ferrous metallurgical processing. Secondary sources are the industrial coal combustion plants, mainly the flying ash, and the waste incineration plants that produce oxides through the flame oxidation of fusible metals at temperatures below 1,000 °C [16–18]:

$$\mathrm{Sn} + \mathrm{O_2} \rightarrow \mathrm{SnO_2} \tag{4.13}$$

$$\mathrm{2Hg} + \mathrm{O_2} \rightarrow \mathrm{2HgO} \tag{4.14}$$

$$\mathrm{2Zn} + \mathrm{O_2} \rightarrow \mathrm{2ZnO} \tag{4.15}$$

$$\mathrm{Pb} + \mathrm{O_2} \rightarrow \mathrm{PbO_2} \tag{4.16}$$

At nanoparticle level, ammonium salts are easily produced and represent the main constituents of smog (urban aerosols) which affects atmospheric opacity and increases the Earth's greenhouse effect.

$$\mathrm{NH_3} + \mathrm{H_2SO_4} \rightarrow \mathrm{NH_4HSO_4} \tag{4.17}$$

$$\mathrm{NH_4HSO_4} + \mathrm{NH_3} \rightarrow (\mathrm{NH_4})_2\mathrm{SO_4} \tag{4.18}$$

$$\mathrm{NH_3} + \mathrm{HNO_3} \rightarrow \mathrm{NH_4NO_3} \tag{4.19}$$

$$\mathrm{NH_3} + \mathrm{HCl} \rightarrow \mathrm{NH_4Cl} \tag{4.20}$$

Combined with acids present in the atmosphere through the series of reactions described above, very finely divided salts form metallic cations that occur in nanoparticles resulting from waste heaps or slag. Depending on salt density, wind speed and humidity, one can find, at various altitudes, almost all nitrates, sulphates and chlorides formed with the cations (in different oxidation states) of elements with stable isotopes or very long half-lives included in the periodic table [19]:

$$\mathrm{Me}^{n+} + n\mathrm{HNO_3} \rightarrow \mathrm{Me}(\mathrm{NO_3})_n + n/2\mathrm{H_2} \tag{4.21}$$

$$\mathrm{Me}^{n+} + n\mathrm{H_2SO_4} \rightarrow \mathrm{Me_2}(\mathrm{SO_4})_n + n\mathrm{H_2} \tag{4.22}$$

$$\mathrm{Me}^{n+} + n\mathrm{HCl} \rightarrow \mathrm{MeCl}_n + n/2\mathrm{H_2} \tag{4.23}$$

The data provided in this subsection show that nanoparticles tend to form constantly, as new amounts of materials derived from anthropogenic or natural activities (volcanic eruptions, cosmic dust, desert dust) rise in the atmosphere. These materials interact with one another or with the gaseous constituents in the air, but they also contribute to nanoparticle reformation through opposed tendencies of sedimentation/de-sedimentation, changes in their chemical nature and/or their previous state of matter.

4.4 Organic Oxidation Intermediates

Irrespective of fuel oil, fossil fuel or carbon-based fuel, oxidation intermediates occur in any combustion reaction, because a combustion reaction cannot be considered complete even when oxygen is in excess or the flame temperature is high [20–22].

As fuel combustion can be a longer or shorter process, one can assume that, depending on the nature of fuel, it consists of the following phases that always end in carbon dioxide formation [20].

4.4.1 Intermediate Oxidation Phases in Methane Combustion (Zero Carbon)

$$\overset{-4}{C}H_4 \xrightarrow{[O]} H_3\overset{-2}{C}-OH \xrightarrow{[O]} H_2\overset{0}{C}=O \xrightarrow{[O]} H-\overset{+2}{C}OOH \xrightarrow{[O]} O=\overset{+4}{C}=O \quad (4.24)$$

4.4.2 Intermediate Oxidation Phases in the Combustion of Primary Carbon Compounds

$$R-\overset{-3}{C}H_3 \xrightarrow{[O]} R-\overset{-1}{C}H_2-OH \xrightarrow{[O]} R-\overset{+1}{C}H=O \xrightarrow{[O]} R-\overset{+3}{C}OOH \quad (4.25)$$

4.4.3 Intermediate Oxidation Phases in the Combustion of Secondary Carbon Compounds

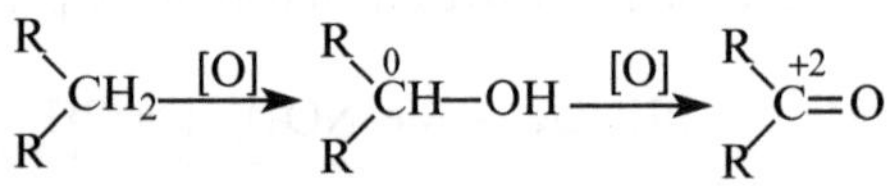

(4.26)

4.4.4 Intermediate Oxidation Phases in the Combustion of Tertiary Carbon Compounds

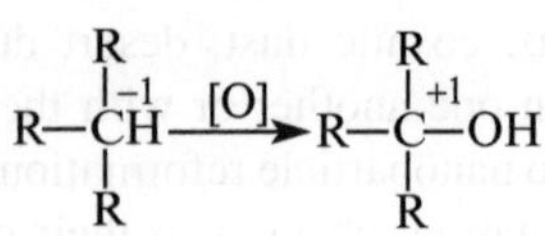

(4.27)

4.4.5 Intermediate Oxidation Phases in the Combustion of Quaternary Carbon Compounds

$$\begin{array}{c} \mathrm{R} \\ |^{0} \\ \mathrm{R{-}C{-}R} \\ | \\ \mathrm{R}\end{array} \xrightarrow{[O]} \mathrm{O{=}\overset{+4}{C}{=}O} \tag{4.28}$$

In any fuel combustion reaction, in the final oxidation phase, carbon changes to dioxide (maximum oxidation state of the carbon atom) regardless of its chain position, and the hydrogen contained in the organic molecule changes to water [21–23].

Reactions (4.24–4.28) show that the intermediate oxidation phases for the primary and/or secondary carbon compounds are the following:

Organic compound → alcohol → aldehyde (ketone) → carboxylic acid → carbon dioxide + water

Hydrocarbons containing secondary, tertiary or quaternary carbon atoms do not form all of the intermediate compounds; they skip some phases and change, through combustion, directly from ketone, alcohol and hydrocarbon to carbon dioxide and water.

Consequently, smoke produced by fuel combustion consists of a heterogeneous hot gas suspension containing traces of the initial (or reformed) hydrocarbon under the form of VOC (volatile organic compound), partial oxidation intermediates like alcohols, aldehydes and ketones, carboxylic acids, carbon dioxide and water vapour. As shown above in this chapter, smoke also contains other forms of incomplete combustion: carbon monoxide (CO) and carbon particulates. Depending on chain length, these intermediates exist in solid, nanoparticle, liquid, very fine aerosols or gaseous state [22].

4.5 Dioxins and Furans

In the past decades, owing to the toxicity and long effect of these compounds in the environment, extensive research has been carried out into the most important precursors of dioxins and furans (polyhalogenated cyclic derivatives, chiefly biphenyls and terphenyls, are mentioned frequently), and the mechanisms of formation, decomposition and reformation of dioxins and furans in combustion processes. Studies have identified 16 compounds whose emissions in the environment should be limited on the whole planet (Table 4.2) [24]:

TCDD is considered the most toxic of these chlorinated compounds. For this reason it was assigned a value of 1.000. The toxicity of all the other compounds in the two mentioned classes is expressed as an I-TEQ fraction to TCDD.

Table 4.2 Emissions of dioxins and furans under control [24]

No	IUPAC name	Common abbreviation	I-TEQ
1.	2,3,7,8 Tetrachlorodibenzodioxin	TCDD	1.000
2.	1,2,3,7,8 Pentachlorodibenzodioxin	PeCDD	0.500
3.	1,2,3,4,7,8 Hexachlordibenzodioxin	HxCDD	0.100
4.	1,2,3,7,8,9 Hexachlordibenzodioxin	HxCDD	0.100
5.	1,2,3,4,6,7,8 Heptachlorodibenzodioxin	HpCDD	0.010
6.	Octochlordibenzodioxin	OCDD	0.001
7.	2,3,7,8 Tetrachlorodibenzofuran	TCDF	0.100
8.	2,3,4,7,8, Pentachlorodibenzofuran	PeCDF	0.500
9.	1,2,3,7,8 Pentachlorodibenzofuran	PeCDF	0.050
10.	1,2,3,4,7,8 Hexachlorodibenzofuran	HxCDF	0.100
11.	1,2,3,6,7,8 Hexachlorodibenzofuran	HxCDF	0.100
12.	1,2,3,7,8,9 Hexachlorodibenzofuran	HxCDF	0.100
13.	2,3,4,6,7,8 Hexachlorodibenzofuran	HxCDF	0.100
14.	1,2,3,4,6,7,8 Heptachlorodibenzofuran	HpCDF	0.010
15.	1,2,3,4,7,8,9 Heptachlorodibenzofuran	HpCDF	0.010
16.	Octochlorodibenzofuran	OCDF	0.001

The largest amounts of polychlorinated dioxins and furans are produced mainly through the combustion of wood fuel, coal and many other organic substances with different, generally halogenated functional groups. For temperatures between 250 and 800 °C, the process is called "cold combustion" and releases dioxin compounds through intermediates resulted from the combustion of lignin or tannin—for wood, or intermediaries produced by cyclic and aromatic compounds—for coal [22, 23] (Figs. 4.3, 4.4, and 4.5).

Conversion through combustion of organic compounds with aromatic cycles to dioxins and furans is possible in the presence of any Cl^- ion donor, the reaction being catalyzed by some heavy metal ions, of which Cu^{2+} is mentioned most frequently [22, 24, 26].

If fuel and fuel oil combustion generally produces small specific amounts of dioxins, chloride-containing organic compounds generate dioxins and furans directly in the flame, for which reason they are known as *polychlorinated organic compounds* (POPs) and their use is strictly controlled or forbidden according to international conventions concerning pollution control [25].

This category includes biphenyls or polychlorinated terphenyls (PCB or PCT). Owing to their dielectric and thermo-mechanical properties, in the past they were largely used in high-voltage transformers and condensers. Currently, PCB or PCT-containing oils are forbidden in any field of activity, because they are dioxin precursors and very toxic. Besides these polynuclear compounds, the POPs list

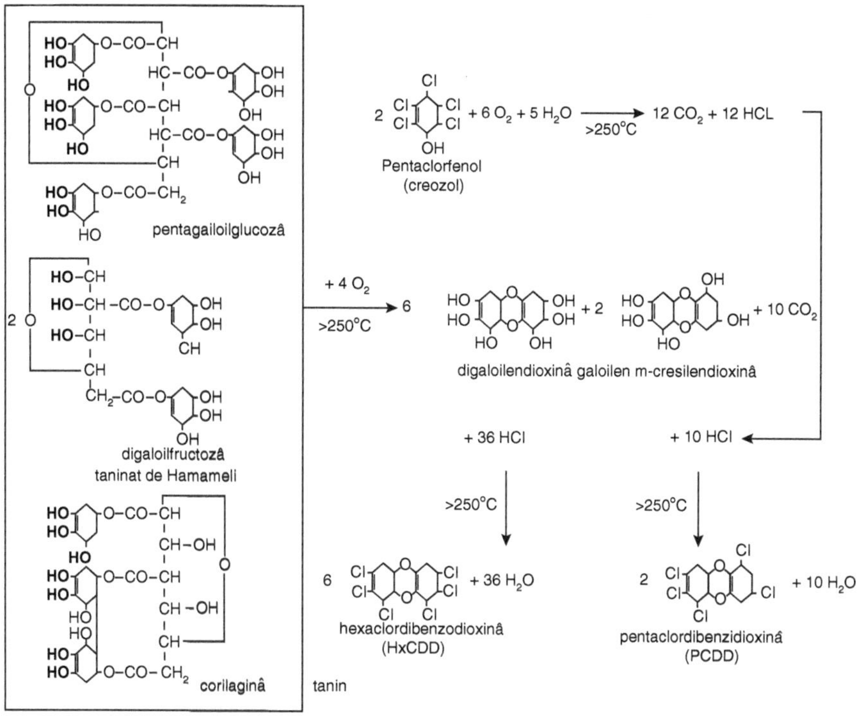

Fig. 4.3 Conversion of tannins to dioxins in the combustion of railroad ties made of creosote-treated wood [22, 23]

comprises a large number of halogenated solvents and pesticides of the HCH and DDT type [25, 26]. More recent research studies indicate other involuntary ways of generating POPs-type compounds [27] (Fig. 4.4).

Conversion through combustion of polychlorinated biphenyls (PCB and PCT) to dioxins and furans is shown in Fig. 4.5 [21, 22].

Combustion of highly chlorinated polymers such as polyvinyl chloride (PVC) may also produce dioxins and furans (Fig. 4.6) [22, 24, 27].

Dioxins and furans persist as particulate matter in the atmosphere, water or soil not only because they are chemically stable and easy to form during any combustion process, but also because they tend to reform from smoke gas, through the slow cooling of smoke (Figs. 4.4 and 4.7) [21, 22, 28].

Most certainly, the environment contains many other nanoparticle species of organic nature such as organometallic compounds (methyl and ethylmercury, heavy metal phenoxides or other cyclic derivatives with more or less complex molecules) whose formation and persistence mechanisms are topics of current debate and research [29].

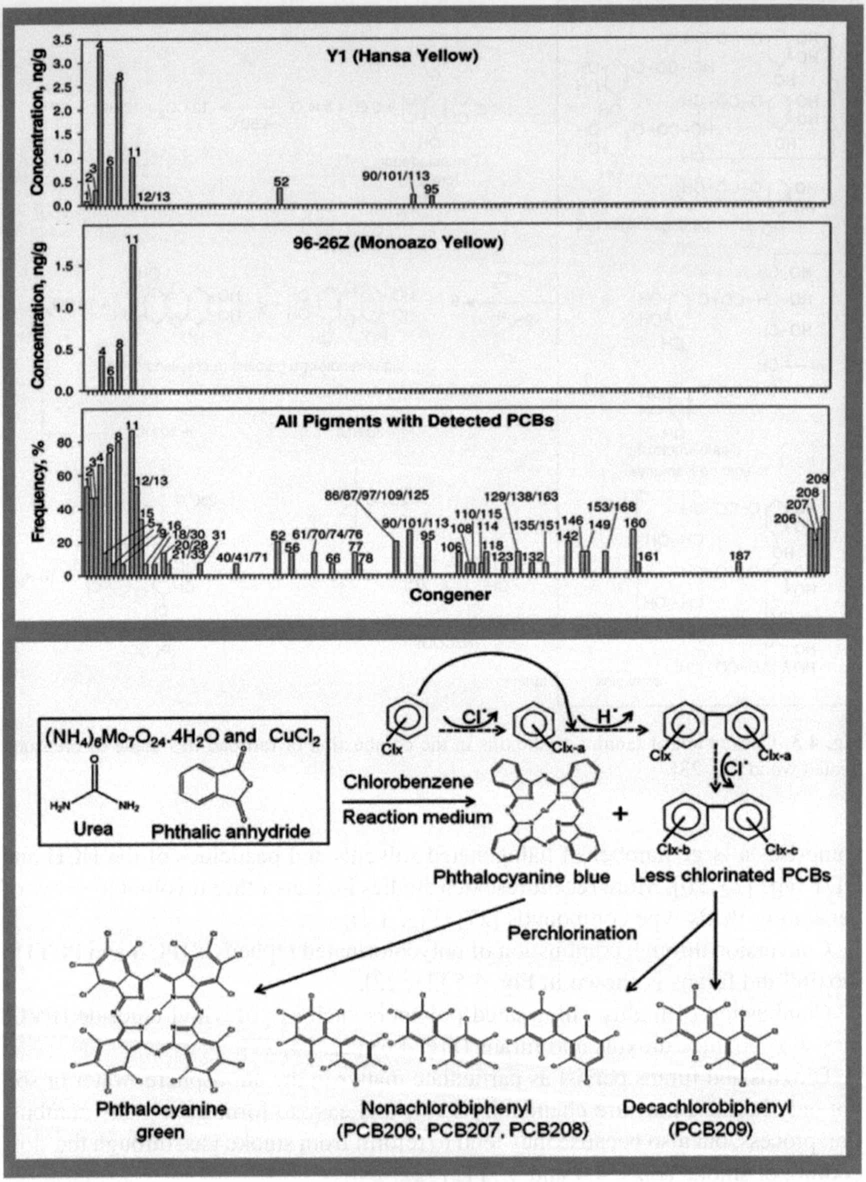

Fig. 4.4 Spectres, concentrations and mechanisms of production/degradation of certain pigments in compounds of the PCB category [27]

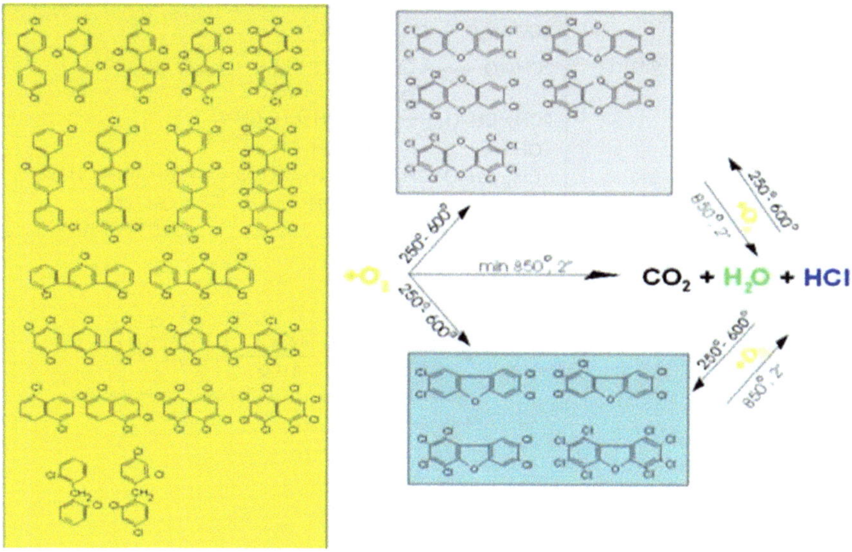

Fig. 4.5 Conversion of PCB and PCT to PCDD and PCDF through incineration. Reconversion of resulting gas mixture to PCDD and PCDF [21, 22]

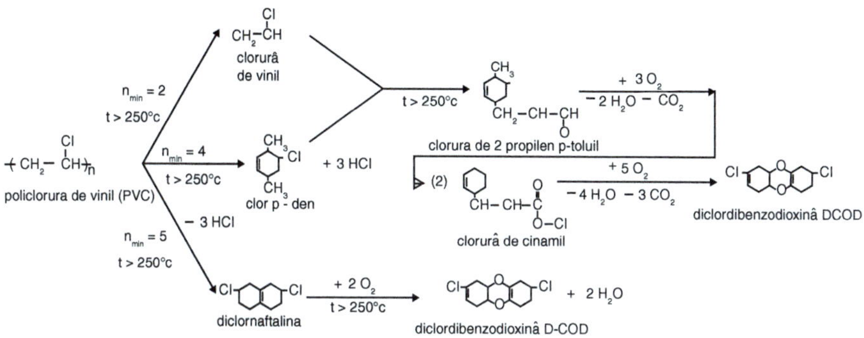

Fig. 4.6 PVC conversion to dioxin in the combustion process [22, 24, 27]

Fig. 4.7 Hypothetical mechanism of dioxin reformation from flue gas

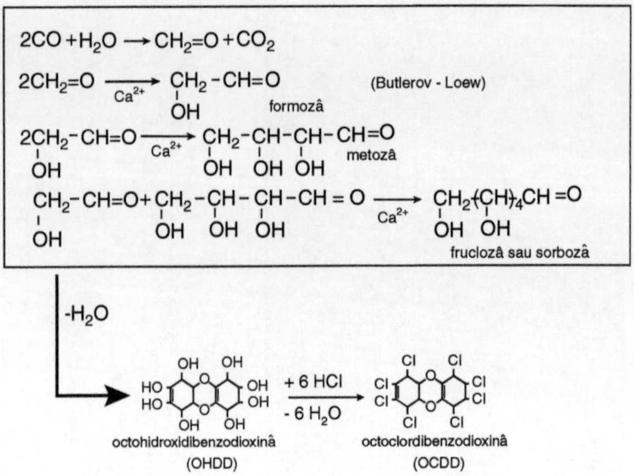

References

1. Smith KR, Veranth JM et al (2006) Acute pulmonary and systemic effects of inhaled coal fly ash in rats: comparison to ambient environmental particles. Toxicol Sci 93:390–399
2. Hurley MD, Ball JC et al (2007) Atmospheric chemistry of a model biodiesel fuel, $CH_3C(O)O(CH_2)_2OC(O)CH_3$: kinetics, mechanisms, and products of Cl atom and OH radical initiated oxidation in the presence and absence of NOx. J Phys Chem 11:2547–2554
3. Ehhalt DH (1987) Free radicals in the atmosphere. Free Radic Res Commun 3(1–5):153–164
4. Monks PS (2005) Gas-phase radical chemistry in the troposphere. Chem Soc Rev 34:376–395
5. McElroy MB, Salawitch RJ et al (1992) The changing stratosphere. Planet Space Sci 40:373–401
6. Bureau H, Keppler H et al (2000) Volcanic degassing of bromine and iodine: experimental fluid/melt partitioning data and applications to stratospheric chemistry. Earth Planet Sci Lett 183:51–60
7. Hara H (2014) Mechanism of acid deposition. EANET Rev. 16 Apr 2014
8. Goodie AS, Middleton NJ (2001) Earth Sci Rev 56:179–204
9. SEPA (2014) Scottish pollutant release inventory rep. Threshold, Edinburg
10. Nielsen OG et al (2007) Atmospheric chemistry of $CF_3CF{=}CH_2$: kinetics and mechanisms of gas-phase reaction with Cl atoms, OH radicals and O_3. Chem Phys Lett 439:18–22
11. Atmospheric ammonia: sources and fate (2000). US EPA AQRS Meet. Rep
12. Mulu AK et al (2013) Photooxidation of ammonia on TiO_2 as a source of NO and NO_2 under atmospheric conditions. J Am Chem Soc 135:8606–8615
13. Schuur EAG (2011) Ecology: nitrogen from the deep. Nature 477:39
14. Galloway JN et al (2004) Nitrogen cycles: past, present, and future. Biogeochemistry 70:153–226
15. Renard JJ, Calidonna SE et al (2004) Fate of ammonia in the atmosphere–a review for applicability to hazardous releases. J Hazard Mater 108:29–60
16. Strâmbeanu N, Demetrovici L et al (2009) Comparative calculation of incineration costs at 3,500 tons/year and 22,000 tons/ years capacity. Int. Symp. Env. Ind, III, 66, Bucharest, 28–30 October 2009
17. Lungu M, Strâmbeanu N et al (2011) Some consideration on the nanoparticles manipulation in fluid media using dielectrophoresis. Rom J Phys 11:103
18. Lungu M, Strâmbeanu N et al (2012) Reduction of nanoparticle emission by electrohydrodynamic filtering of residual combustion gases. International conference ANMBES 2012, 107. Transilvania University of Brasov, Romania, 24–27 May 2012

19. Strâmbeanu N (2007) Criterii de evaluare a potenţialului energetic şi de risc ecologic în procesul incinerării deşeurilor speciale şi toxice. Bul AGIR 3:67
20. Csunderlik, C, Medeleanu, M (2014) Reac□iile compuşilor organici, Facultatea de Chimie Industrială si Inginerie a Mediului din Timişoara (PP)
21. Strâmbeanu N, Pode V (2006) sa, Mecanisme probabile de formare a dioxinelor în procesul arderilor controlate şi necontrolate. Revista Română de Chimie 8: 36
22. Strâmbeanu N (2006) Modele ipotetice ale formării-reformării dioxinelor în procesul incinerării deşeurilor speciale. Bul AGIR 1–2:177
23. Strâmbeanu N, Demetrovici L (2008) sa, Evaluarea energetică a deşeurilor toxice supuse incinerarii, Simp ARS, Sinaia 14–15 Apr 2008
24. Santoleri E et al (2000) Introduction to Hazardous Waste Incineration, 2nd edn. Wiley, New York
25. Convenţia de la Stockholm, obiective cheie, măsuri, instrumente şi acţiuni ale Planului Naţional de Implementare în România, Bucureşti, 2003
26. Mc Kay G (2002) Dioxin characterization, formation and minimization during municipal solid waste (MSW) incineration: review. Chem Eng J 86:343–368
27. Winneke G, Ranft U (2014) Behavioral sexual dimorphism in school-age children and early developmental exposure to dioxins and PCBs: a follow-up study of the Duisburg cohort. EHP Rev
28. Fiedler H et al (2003) Persistent organic pollutants. Springer, Berlin
29. Craig PG et al (2003) Organometallic compounds in the environment, 2nd edn. Wiley, Leicester

Part II
Effects of Nanoparticles on Human Health

Chapter 5
Nanopathology: The Nano-bio-interaction of Nanoparticles Inside the Human Body

Antonietta M. Gatti and Stefano Montanari

Abstract This chapter deals with the invasive character of nanoparticles inside the human body and describes the potential risk that the nanosized particulate matter represents. The dispersion of submicronic and nanoparticles through inhalation, ingestion and other, less common, ways of entrance into the organism are discussed. A preliminary discussion on engineered nanoparticles and incidental (involuntarily released mainly by high-temperature combustive processes) ones is presented and their implications examined. The study presents electron-microscopy images of particles dispersed in organs affected by illnesses. The presence of submicronic and nanosized particles in organs like the liver, the kidney, the pancreas, the bladder, the brain affected by a few forms of cancer is shown by Scanning-Electron-Microscopic images and their elemental chemical composition evaluated by an X-ray microprobe. Some investigations on fetal malformations and miscarriages are shown to demonstrate the translocation of tiny particulate matter from mother to fetus. These pieces of evidence demonstrate that the physical presence of particulate matter can modify the embryonic growth, so fetal modifications may be not hereditary but can be induced accidentally during pregnancy. The delay and inadequacy of national and international rules against the industrial use of nanoparticles is also discussed. And the necessity of a new, multidisciplinary approach to Medicine is presented as well.

5.1 Introduction

Men, like any other terrestrial living being, have shared their environment with dust all along. Dust is produced naturally by volcanoes, by the erosion of rocks, by the raising of pulverized ground and by all combustions, normally rare in nature and

---

A.M. Gatti (✉)
ISTEC, National Council of Research of Italy, via Granarolo, 64, 48018 Faenza, RA, Italy
e-mail: gatti@nanodiagnostics.it

S. Montanari
NANODIAGNOSTICS srl, via E.Fermi 1/L, 41057 San Vito, MO, Italy

M. Lungu et al. (eds.), *Nanoparticles' Promises and Risks*,
DOI 10.1007/978-3-319-11728-7_5

generally confined to woods. For thousands of millennia those have been the only important sources of particulate matter and, besides the comparatively rare exceptions represented in the vast majority of cases by some occasional volcanic activities, the most interesting characteristic for the discussion that follows is constituted by the particle size: Natural particles are rather coarse.

Taking into account a great variability of dating depending on many factors, roughly about half a million years ago Man started to be able to light a fire and use it. That was the start of a no-more natural pollution, but caused by an animal that had begun to behave in a completely different way from that inborn of all the other inhabitants of the Planet.

All combustions, without exceptions, produce solid particles whose chemical composition depends on what is being burnt and on what those chemical elements find in the environment at the time when the particle is being formed.

As the different technologies grew more and more sophisticated, an increasing variety of chemicals was introduced in the processing of the products that were catching on and, as a consequence, particles became more and more complex in their composition. If for many centuries the phenomenon had a relevance restricted only to specific environments like, for instance, those of glass blowers or of metal smelters, the so-called First Industrial Revolution of the Eighteenth Century started to give it an important impetus, and even more impetus was given about one century later with the so-called Second Industrial Revolution.

Attaining high temperatures had been a problem for millennia, and that difficulty was one of the greatest limitations to technical progress. With the widespread, massive introduction first of coal and later of petroleum as industrial fuels, and with the technical progresses and inventions of more and more sophisticated machines that replaced manual labor with a much more effective capacity for work, higher temperatures could be touched and producing materials, for instance, glass, became easier and cheaper, and more and more until-then non-existing products could be fabricated.

Now temperatures far exceeding 10,000 °C are easily available at a relatively low cost with all the benefits they can give but also with a few drawbacks.

As stated above, all combustions generate particles and, as a rule of thumb, the higher the temperature, the smaller the particles. The importance of this matter of fact will be described later.

For elementary geometric reasons, if the diameter of a particle is reduced ten times, its volume and, as a consequence, its mass becomes 1,000 times smaller. A reduction of diameter of a hundred times implies a mass reduction of one million times, and so on, since diameter and volumes have a relationship to the third power.

It is anything but common to find in the atmosphere natural solid particles smaller than 1 μm, the usual ones being comparatively bigger. For instance, sand grains can travel thousands of kilometers as can be seen looking at satellite photographs of Sahara sand flying at a relative high altitude on the Atlantic Ocean from West Africa as far as the eastern coast of the United States. Smaller particles like the ones produced by volcanoes can be the cause of important ecological phenomena. A famous one is the famous "year without a summer": The 1815 eruption of Mount

Tambora in the Dutch East Indies generated a huge aerosol veil that capped virtually all the north hemisphere, preventing the sun rays from reaching the Earth with their usual efficiency. The result was an ecological, sanitary, economic and social disaster.

As mentioned just above, a small reduction in diameter size is enough to reduce dramatically the volume of a particle and, as a consequence, its being more prone to stay suspended in the air and be carried by the wind and other atmospheric phenomena. Nowadays industrial plants, waste incinerators, internal-combustion engines and, in general, all systems working by burning something emit dust, and the use of increasingly high temperatures delivers in the atmosphere ever smaller and more mobile particles. Also explosions are sources of particulate matter. New weapons like, among others, depleted-Uranium or Tungsten bombs work because of the very high temperatures they reach when they hit their mark. That heat is enough to have bomb and part of the target vaporize. In a matter of seconds, as soon as the atoms and the small molecules of that gaseous-state substance are cold enough, they consolidate to form particles.

But particles made by human activities are not restricted to high-temperature, incidentally created ones as an unwelcome side effect.

In the late 1950s scientists realized that matter, when its size is in the nanometer order, behaves in a then-unexpected way, a way that could be exploited in many industrial applications. That was the notion that begot Nanotechnology (or Nanotech, as the name is usually shortened), something that can represent a revolution to our society, if only for its being the hugest financial investment in Man's history.

Artificial matter can be created starting from molecules or even from atoms and assembled into a new lattice (crystalline structure) that we call nanoparticles, i.e. entities with a size generally not exceeding 100 nm. Those are the basic constituents, sophisticated bricks of even more sophisticated buildings, of products with extraordinary properties. Applications are as numerous as they are varied and the limit to nanotechnologists seems to be imagination. In most cases utilizations and products are still at an experimental stage, but in other they have already taken the form of everyday products and, as a matter of fact, reached a multitude of unaware users. Cosmetics like, for example, sun screening and beauty creams; clothing, sports equipment, antimicrobial bandages, household appliances (washing machines, air conditioners, refrigerators, etc.), self-cleaning glass and tiles, paints, scratch-resisting coatings, swimming-pool disinfectants, drugs and even food are but some of the nanotech goods already commercially available.

All those things are undoubtedly exciting and promise to make life easier, but the problem, or one of the problems, is that we are not sure they do not have a negative impact on the environment and on living organisms' health. We are not sure simply because no truly exhaustive research has preceded the marketing of most products and what, in a way, is more important, has preceded the investments. The consequence is that the investors seek to recoup their investment as quickly as possible and are not particularly enthusiastic about waiting for the results of researches for which they should spend more money fearing a negative answer.

Worldwide concern is currently shared by experts about the effects of nanoparticles, be they engineered or be they the result of undesirable side effects of other technologies [1, 2].

The concern is mostly due to the fact that, because of their very small size, they can interact directly with proteins, enzymes, the DNA, and other substances present in all organisms, and that "nano-bio-interaction" can induce toxic effects at cell level about which we know very little. And even less is known about the long-term effects that these products can have on human and animal life and on the environment. As to the fate of nanoparticles at the end of their life cycle as products or product components, we know nearly nothing.

As mentioned above, nanoparticles started to be used also in Medicine. Their capability of crossing the blood–brain barrier offers great possibilities to drugs aimed at the brain and brain pathologies. At present the molecules of the neurological drugs are recognized as not-self and rejected by the blood–brain barrier. The concept is rather simple: the drug is embedded/absorbed by the nanoparticles that once injected can cross the physiological barrier and when the target has been reached they start to biodegrade and release the active principle. They act as a "Trojan-horse." The evident problem is the biocompatibility of the particles. Concept that is unknown to the nanotechnologists, namely that the human body has a defense system able to recognize molecules and chemical compounds, but can be invaded by nanoparticles since they represent non-specific stimulus, but once they interact with cell compounds or organelles some reaction of "non-acceptability" can be triggered, since they are "foreign bodies."

Another use that has been attempted, actually with very dubious results, so that almost all products have been phased out, was the injection of iron-oxide nanoparticles with substances that show a particular affinity for tumor cells, internalized. Once the patient is introduced into a strong magnetic field, the magnetic particulate matter is driven to the cancerous tissue while sparing the healthy cells and for their higher X-ray density than the surrounding tissue they help to define with high sensitivity the borders of cancer. These images help the surgeon to better define the contours of the area to be excised. There is the possibility to treat directly the cancer. An external microwave field can be generated inducing a heating of the iron-oxide particles. The heating occurs at a temperature high enough to kill the cells, that the cancer is somehow eliminated. In this case after the treatment the iron-oxide nanoparticles remain in the dead tissue.

Once again, the problem is there of the non-biocompatibility of particles whose removal is something of which no solid evidence exists. Probably the long-term permanence and the non-biodegradability of these types of nanoparticles can induce other bad side-effects.

If the ones just briefly described are just experimental attempts, the oral intake of colloidal Silver is a widespread practice of self-medication. Apart from the many uses of Silver, be they or be they not backed by demonstrations of their true usefulness, those particles, much smaller than 1 μm as they are, are not biodegradable and there is no reason why they should behave in a different way from any other non-biodegradable particles. There is no evidence that their chronic use is the cause of pathologies, but it must be said that they, as particles taken for long periods in small quantities do, could take a long time before their adverse effects grow manifest. Besides, it is at least unusual that the etiology of a disease is attributed to something occurred many years ago and Silver could be considered free from blame simply because no one thinks of it. However, the FDA issued a Final Rule declaring that all over-the-counter drugs containing colloidal Silver or Silver salts are not recognized as safe and effective and are misbranded [3].

But even if national and international regulatory organizations (ISO, OECD, etc.) are actively working with the aim of governing the subject (from definitions to nomenclature, from tests to devising prevention measures), market runs much faster than regulators. Therefore, taking advantage from the virtual absence of rules, nanoproducts are blindly invading the market, blindly because no one really knows the land on which we are all treading. In short, we do not know whether the risks brought about by engineered nanoparticles outweigh their benefits for society [4]. What we are sure of is that confusion is rampant: research needs money and time, two things on which no one seems willing to invest, and the lack of knowledge hinders the regulators who are often forced to act extrapolating from other, non-specific knowledge. On the other hand, end-users are kept unaware not only of the potential dangers they could encounter, but in many cases they do not even know that they are using a nanoproduct, since declaring their presence is not mandatory for manufacturing companies. So, many of them prefer not to let their customers know what they are buying, lest they would refuse the purchase. Even people who work in laboratories where nanoparticles are engineered not always handle them with due care. As to investors and manufacturers, they too are at risk. Should nanoproducts be proven to be really dangerous, the economic loss would be a certitude.

If, from the health point of view, we do not know much about engineered particles, we know that unintentionally made nanoparticles, i.e. those generated by car traffic, waste incinerators, factories working at high temperature, etc. are extremely invasive and can rather easily enter the organism where they are the documented responsible for deleterious reactions. We know that they are inhaled and ingested, that they can cross the lung and digestive-system barriers and enter the blood stream where, in particular conditions, they can have the blood clot. Then, carried by the blood, they can reach virtually any organ and tissue where they are trapped without having the possibility of being expelled. And we also know that they can cross the cell membrane and enter the nucleus. In any case, they are foreign bodies and so they are perceived by the organisms. There is no reason why engineered particles should be different in their behavior.

Like never before, the environment we are living in contains huge quantities of micro- and nanoparticles, and very unusual ones, at that. The images that follow demonstrate their extreme invasiveness, showing engineered and incidental particles embedded in pathological tissues. The images were obtained by means of a novel ultramicroscopic technique developed by using a Field-Emission-Gun Environmental Scanning-Electron Microscope (FEG-ESEM Quanta 250, FEI, the Netherlands) and an X-ray microprobe of an Energy Dispersive Spectroscopy (EDS by

EDAX, USA). That equipment allows spotting inorganic foreign bodies and analyzing their elemental composition in biopsic, autoptic and surgical samples or in the blood and in other body fluids.

5.2 The Nano-bio-interactions

In 2002, a new word was coined: Nanopathology, and that name was used as the title of a Project (Nanopathology FP5-QOL-2002-147) coordinated by Dr. Antonietta Gatti, one of the authors of this chapter. She worked in collaboration with the Universities of Cambridge (UK) and Mainz (Germany), Biomatech (France) and with FEI Company (The Netherlands). The project's goal was to realize the "impact of micro and nanoparticle in inducing pathologies." At the beginning the word had a simple meaning: "The branch of learning that deals with how the organism reacts to the presence of micro- and nanoparticles." Now, after years of study and research, the word has taken on other aspects and implications and probably, in the next future, Nanopathology will mean a new approach to patients and their symptoms as a whole with the environment where he lives: a sort of customized or Personalized Medicine.

The respiratory and gastrointestinal tract have at their disposal several mechanisms to keep the mucosal surfaces free from dead cell debris and particles deposited by respiration [5, 6]. However, the physiological barriers of the pulmonary and gastrointestinal mucosa may prove ineffective to prevent solid micro- and nano-sized particles from entering the organism. Direct observations in pathological samples directly demonstrated the presence of particulate inorganic matter (like dust, burnt oil residues or asbestos, etc.) [7–9].

The images presented in Figs. 5.1, 5.2, 5.3, 5.4, 5.5, 5.6, 5.7, 5.8, 5.9, 5.10, 5.11, and 5.12 show these particulate matter deeply embedded in biological tissues. They concern the presence of debris of different chemistry in the lungs (Fig. 5.1a, c) affected by adenocarcinoma or by mesothelioma (Fig. 5.4a, c), in the liver (Figs. 5.2 and 5.3), in the stomach affected by cancer (Fig. 5.5), in the pancreas affected by cancer (Fig. 5.7), in the colon (Fig. 5.6), in the kidneys (Fig. 5.8), in the brain (Figs. 5.9 and 5.10) and also in the seminal fluid (Fig. 5.11). These images witness the particle invasive character.

Micro- and nanoparticles share some features of their physical behavior with gases and, like gases, can be inhaled, reaching, when they are small enough as is the case with nanoparticles, as deep as the alveoli. According to a research carried out by the University of Leuven (Belgium) [10] on 100-nm-sized particulate matter, it takes it no longer than a few tens of seconds to pass from the alveoli to the blood stream. If those particles are insoluble, they may act as a trigger to the transformation of fibrinogen into fibrin, the insoluble serum protein that is a sort of scaffold on which a thrombus forms. In most cases the organism can counter that process by producing urokinase, the serine protease that works as a plasminogen activator, and quickly dissolve the thrombus; but there are instances in which, for a number of

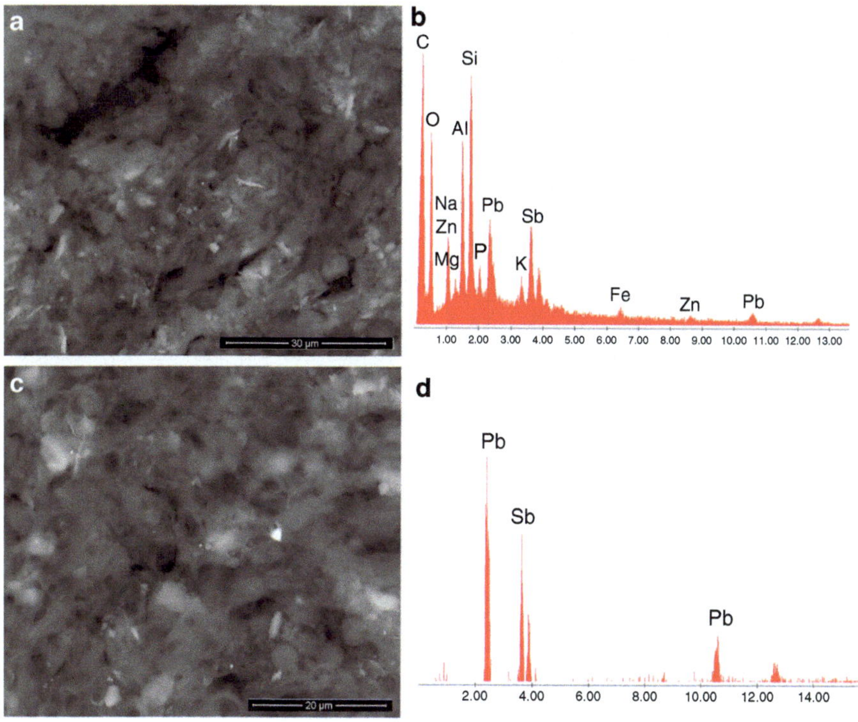

Fig. 5.1 The scanning electron microscope (SEM) image shows debris (*white spots*) of environmental or occupational origin deeply embedded in a lung tissue affected by adenocarcinoma. Among silicon-based particles (**b**) there are also toxic Lead-Antimony-Zinc debris. In the *y*-axis of the EDS spectra the intensity of the signal in counts is shown and in the *x*-axis the energy in keV

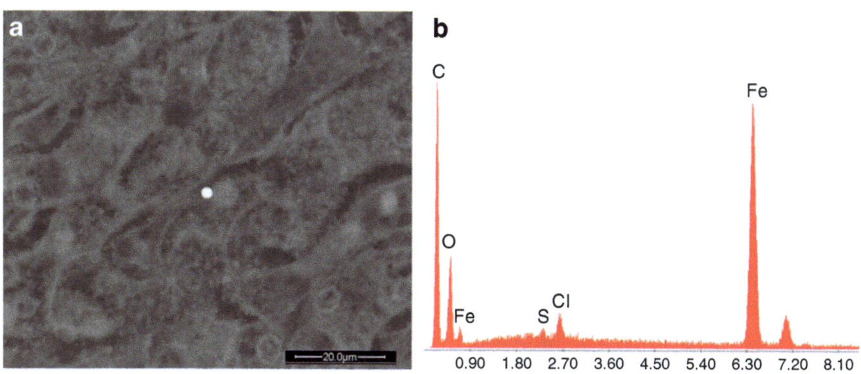

Fig. 5.2 SEM image of an Iron-Sulfur spherule detected in the liver of a patient affected by pleural mesothelioma. The 2 µm-sized Iron-Sulfur-Chlorine round-shaped debris is of combustive origin

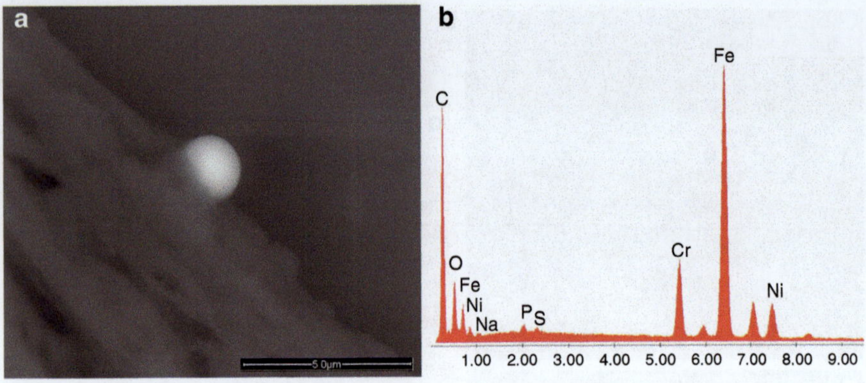

Fig. 5.3 SEM image of a stainless-steel spherule (Iron-Chromium-Nickel) found in a liver biopsy in a patient affected by liver carcinoma

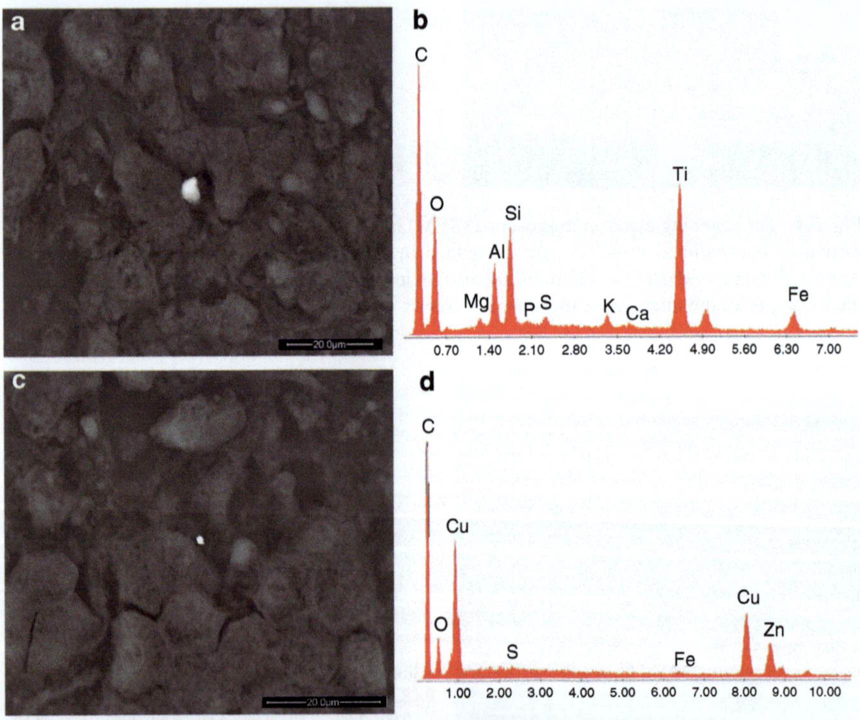

Fig. 5.4 SEM images of liver biopsies (**a**, **c**) of a patient affected by lung mesothelioma. Metallic micron-sized debris, probably inhaled, invaded also other internal organs. They are composed of Titanium, Silicon, Aluminum (**b**), etc, and Copper, Zinc, Iron and Sulfur (**d**)

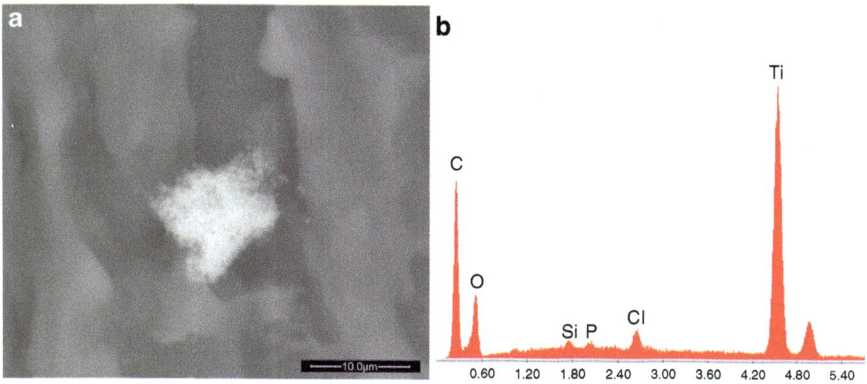

Fig. 5.5 SEM image of a cluster of Titanium-Silicon-Phosphorus-Chlorine nanoparticles found inside a stomach tissue in a patient affected by stomach cancer

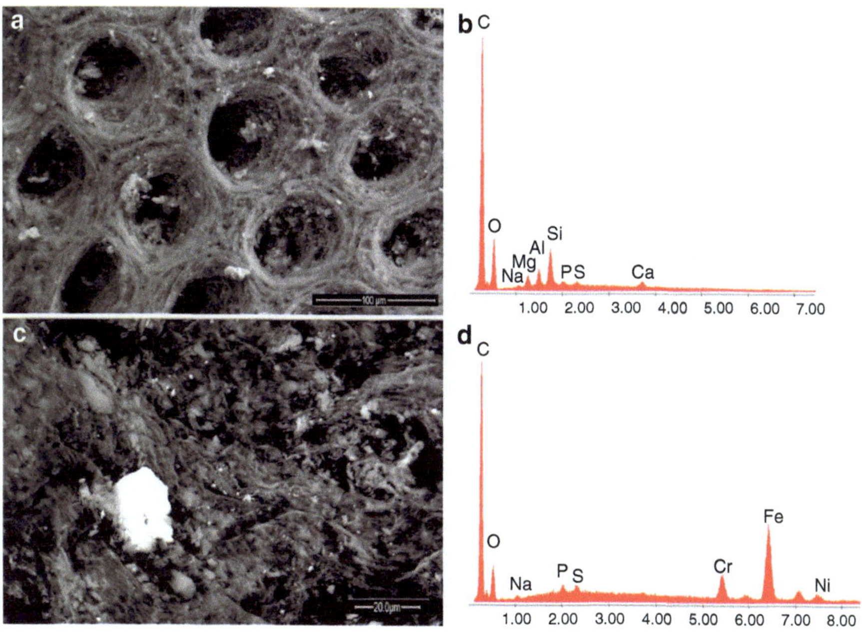

Fig. 5.6 SEM images of a colon tissue affected by cancer. A myriad of micron-sized and nanoparticles are visible. They are silicates (**b**) and stainless-steel (**d**)

reasons, no efficient defense mechanisms are available to the subject. In those cases, a thrombus forms without originating at and being attached to the vascular wall as usually happens. So, being free to float, it necessarily migrates along the veins—if the phenomenon has taken place in the venous system—to reach the lung circulation and be the cause of pulmonary thromboembolism. In case all that occurs in the arteries, the outcome can be a stroke or a myocardial infarction. It is only

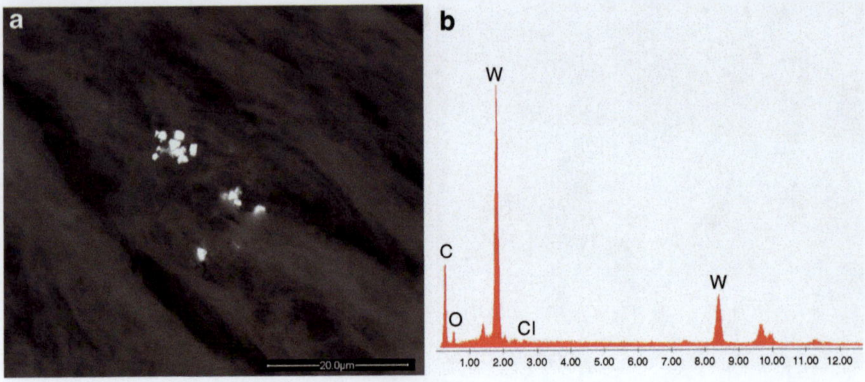

Fig. 5.7 SEM image of submicronic debris composed of Tungsten. They were embedded in a pancreatic tissue affected by cancer.

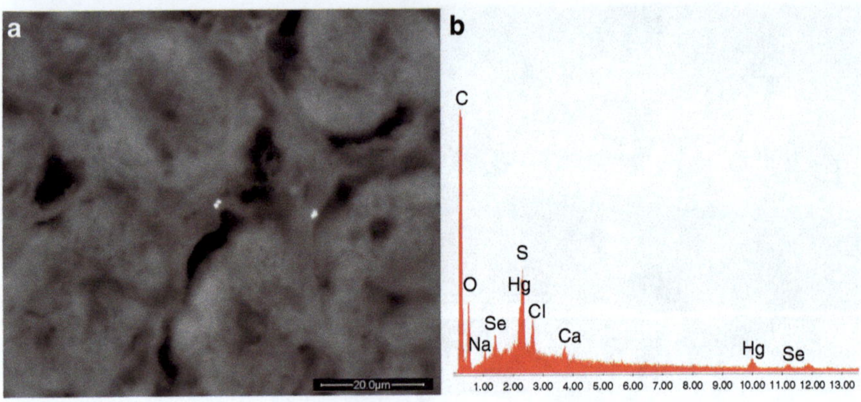

Fig. 5.8 SEM image of single nanoparticles composed of Mercury, Chlorine, Sulfur, Selenium, Sodium and Calcium identified in a kidney tissue in a patient affected by Gulf War Syndrome

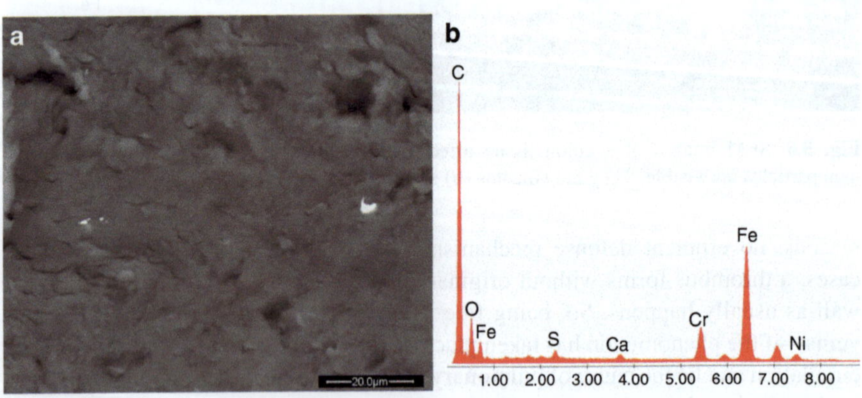

Fig. 5.9 The SEM image shows three stainless-steel submicronic debris identified in a cerebellum affected by neurinoma

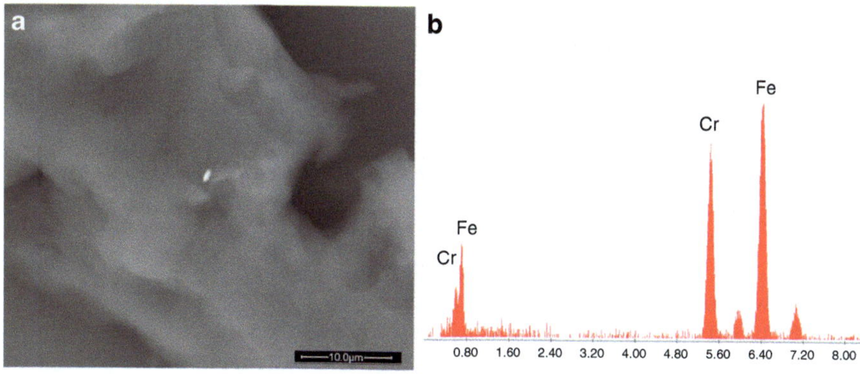

Fig. 5.10 SEM image of an Iron-Chromium nanoparticles embedded in a brain tissue affected by cancer

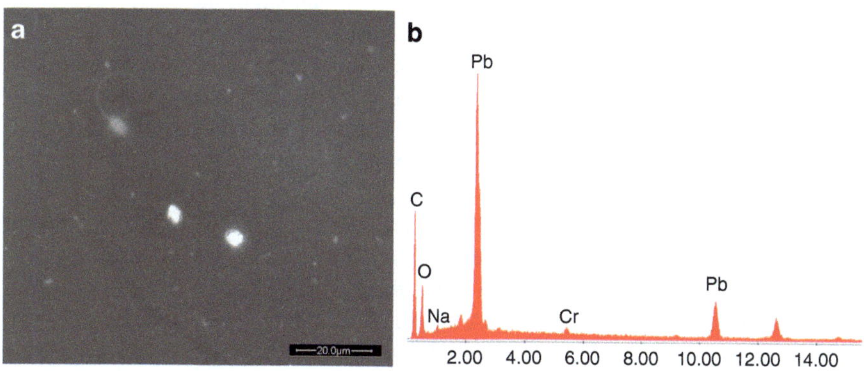

Fig. 5.11 The image shows 2 μm-sized Lead-Chromium particles identified in the seminal fluid of a patient affected by lymphoma

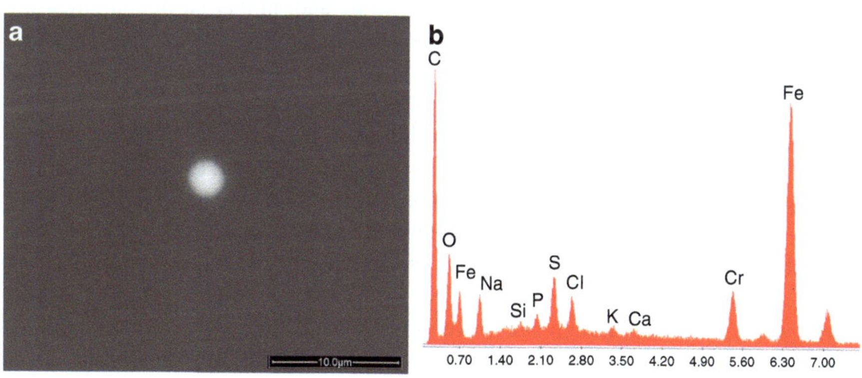

Fig. 5.12 The image shows a 2 μm-sized stainless-steel spherule found in the seminal fluid of a patient affected by Hodgkin's lymphoma. The Sulfur content is too high for a normal steel; that means that the spherule is the product of an controlled combustion in a foundry

obvious that something that behaves that way cannot be considered biocompatible. A few analyses carried out on thrombi extracted from the coronary arteries the first day after a myocardial infarction revealed the presence of inorganic foreign bodies inside the thrombus, and so did the analyses of venous thrombi captured by vena-cava filters. But, in the majority of circumstances, the particles that are carried by the blood reach virtually any organ—impossible, as far as we know, to foretell the butt—and those organs, with very little differences, behave like mechanical filters, capturing that solid matter [11, 12].

Ingestion is another doorway for particles. In that case those that are in touch with the wall of the digestive system long enough can negotiate that barrier and enter the blood circulation, thus sharing their lot with the ones passed through the lungs. According to what we observed, the digestive system allows the passage of particles much bigger than those entered through inhalation.

But inhalation and ingestion are not the only possibility for particles to get into the organism. We found particulate pollutants in injectable preparations and it is only obvious that those will be introduced in the organism of the subject receiving the injection. And we have also found them as additives in cosmetics. In that case, we have no solid-enough evidence to say that they can negotiate the skin barrier, this remaining one of the many aspects of particle behavior to be clarified.

G. Oberdörster [13] demonstrated that nanoparticles can move along the olfactory nerve and it will not be a surprise if also other nerves or even all nerves could host the same phenomenon.

Anyway, no matter how they got there, in the course of our research we keep finding micro- and nano-sized particles in all organs: the liver, the kidney, the spleen, the brain, the gonads, the thyroid, the lymph nodes. We find them even inside cell nuclei in contact with DNA and that can induce DNA damages [14, 15].

Most particles are not biodegradable, i.e. they are bio-persistent and, once they have been imprisoned by a tissue, cannot be expelled, since no such mechanism exists. Their permanence in the tissue can activate physico-chemical reactions that, over time, are likely to induce adverse effects.

In some instances, particles tend to concentrate and, in that particular situation, the tissue that hosts them perceives them as foreign bodies—as they actually are—and starts the classical foreign-body reaction, with the formation of a granulation tissue and the establishment of a chronic inflammatory condition. According to the current medical literature and to our experience, in the long run a chronic inflammation with such characteristics can be the cause of a cancer [16, 17].

How long the phenomenon will take to be completed is hard to foretell. What we can say in keeping with what we have seen in the more than 2,000 human cases we had a chance to analyze so far, is that particularly intense exposures to particles can induce a cancer in a matter of months. Slow, diluted, chronic exposures can take many years or even, though we have no evidence but it is just a matter of speculation, so long as to exceed the natural lifespan of the subject.

It is a fact that cardiovascular accidents and cancer in many different varieties are the most common adverse effects particles can bring about in the organism, but they are not alone.

Particles are suspected to play a part in pathologies like Alzheimer's disease, Parkinson's disease, autism and other pathologies involving the brain. We have no evidence available from our laboratory and the only thing we can say is that further research is necessary to dispel all doubts.

More than once we found particulate matter in the seminal fluid, in most cases samples from soldiers repatriated from missions in war theatres. That undue presence can be the cause of male sterility and of a condition called "Burning Semen Disease" involving the carrier's female sexual partner. In that case, after unprotected sexual intercourse the woman develops bleeding, painful sores in her vaginal channel tissue, and those sores cannot be successfully treated either surgically or pharmacologically. Male sterility had been already observed long ago in workers of some factories, but no satisfactory explanation had ever been given to the phenomenon. The excess of cases of endometriosis in the inhabitants of Taranto, an Italian town particularly polluted by the largest steelworks in Europe, could also find a reasonable answer as to their origin. We add that particles can be artificially removed from the seminal fluid and the couple can resort to assisted reproduction techniques. Of course, this can be done only if the particular condition is known.

We also proved that the passage of particles from mother to fetus is possible, a situation that can give rise to miscarriages when the particulate matter gets to the embryo at an early stage of gestation. When the phenomenon occurs later, these foreign bodies can induce damage compatible with life and the result can be a malformed child.

One of the tasks of the nanopathologists is tracing the exposure to pollutants the patient underwent. So, they must sift through the environment the subject lives or has lived in, including food and drugs. A focused anamnestic work is conducted with the fundamental aid of the patient and, if the collaboration works, it is not rare that the source of pollution can be identified and, if still present, eliminated from the patient's life. Such a detective work is generally far from quick and far from simple, but the combination of the chemical composition of the particles, their size and their shape is often a sort of fingerprint left by a specific type of pollution.

The current Toxicology tests are not always suitable to explain the effects of new pollutants, and that is particularly true with micro- and nanoparticles. The basic concept that Toxicology fails to consider is the nature of particulate matter. Of course particles are made of atoms, each of which can exert its own characteristic effect on the organism, but limiting oneself to just taking into account those effects is in most cases a simplistic approach and a misleading one.

What is very important to understand the consequence that can derive from particles present in the organism is their being foreign bodies, and that is largely independent of their elemental composition. Also size and shape influence their behavior, both being determining factors for particle mobility. And the surface/volume ratio of each particle is also of the utmost importance, as the reactivity increases with the increase of the ratio, a property amply exploited by nanotechnologists. A further aspect to be considered is the ability of particles below a certain size to enter cell nuclei and interfere with the DNA, with organelles, with proteins, etc. The chemical composition of particles becomes important when corrosion occurs. In that case oxidized elements can exert their toxic effects. So, the concept of Toxicology needs to be extended, new tests must be devised and a new expression must be invented to describe specific behaviors over time.

Conclusions

To sum up what we have tried to clarify in this chapter, Nanopathology is a key to help diagnose old and new diseases. Considering the organism a collection of organs with just faint and hazy mutual relationships is proving a more and more unsatisfactory approach to Medicine, particularly when so-far unknown or little-known pathologies are concerned. This is even more true when syndromes that bring together diseases that apparently have nothing to do with each other is the case.

There is nothing new in saying that the organism works as a harmony among cells, tissues and organs and that each component interacts more or less strictly with the rest of the body. The obvious consequence is that anything troubling that harmony can disturb its equilibrium with effects that may be not immediately understandable or not understandable at all if the sectional view of physiology and pathology is the one adopted. The concept can easily be amplified observing that the organism pursues its metabolism immersed in an environment by which is obviously influenced.

Nanopathology deals with the often complex interactions between the organism and what surrounds it. More in particular, what enters the organism: air, food, and drugs. As already mentioned above, besides the detection and the analysis of the particulate pollutants in the patient's samples, applying Nanopathologies demands the search of the origin of the exposure the subject underwent; of how the pollutants entered; of how they moved throughout the organism; of the symptoms they induced. All that, as a first step, aimed at eliminating the pollutants from the patient's life. The further steps are necessarily those of finding proper therapies and techniques to get rid of them from the tissues. For the time being, prevention is the most powerful—the only, in truth—weapon we have available [18, 19].

If we want to be effective, the first thing to do is recognize the problem and define its components. Then we must realize that this is a new kind of Medicine that requires necessarily a multidisciplinary approach. Medical doctors expert in particular fields (oncologists, endocrinologists, hematologists, gynecologists, toxicologists, etc.) must work together with materialists, chemists, physicists and biologists. But, given the complexity of the subject, engineers, agronomists, pharmacists, and many other specialists could be necessary to understand particular cases.

Acknowledgements The authors wish to thank Dr. Federico Capitani and Ms. Lavinia Nitu for their valuable help in the analytical work.

References

1. Gatti A, Puntes V (2009) While scientists are still wondering, industry is already selling. NanoNow 12:28–30
2. Report of joint FAO/WHO expert meeting on the application of nanotechnologies in the food and agriculture sectors: potential food safety implications 1–5 June 2009. pp. 1–133. http://www.who.int/foodsafety/fs_management/meetings/nano_june09/en/index.html
3. http://www.anh-usa.org/fda-declares-silver-has-no-therapeutic-value/
4. Colvin V-L (2003) The potential environmental impact of engineered nanomaterials. Nat Biotechnol 21:1166–1170
5. Kreyling W-G, Scheuch G (2000) In: Gehr P, Heyder J (eds) Clearance of particles deposited in the lungs. Marcel Dekker Inc., New York. pp. 323–376
6. Schlesinger RB, Ben-Jebria A, Dahl AR, Snipes MB, Ultman J (1997) In: Massaro EJ (ed) Disposition of inhaled toxicants. CRC, New York. pp. 493–550
7. Ballestri M, Baraldi A, Gatti AM, Furci L, Bagni A, Loria P, Rapana RM, Carulli N, Albertazzi A (2001) Liver and kidney foreign bodies granulomatosis in a patient with malocclusion, bruxism, and worn dental prostheses. Gastroenterology 121:1234–1238
8. Gatti A, Montanari S, Gambarelli A, Capitani F, Salvatori R (2005) In-vivo short- and long-term evaluation of the interaction material-blood. J Mater Sci Mater Med 16:1213–1219
9. Gatti A, Montanari S (2008) Nanopathology: The health impact of nanoparticles. PanStanford Publishing Pte. Ltd, Singapore, pp 1–298
10. Nemmar A, Hoet P-H, Vanquickenborne B, Dinsdale D, Thomeer M, Hoylaerts M-F, Vanbilloen H, Mortelmans L, Nemery B (2002) Passage of inhaled particles into the blood circulation in humans. Circulation 105(4):411–4
11. Gatti A, Rivasi F (2002) Biocompatibility of micro- and nanoparticles. Part I: in liver and kidney. Biomaterials 23:2381–2387
12. Gatti A (2004) Biocompatibility of micro- and nano-particles in the colon. Part II. Biomaterials 25:385–392
13. Oberdorster G, Oberdorster E, Oberdorster J (2005) Nanotoxicology: An emerging discipline evolving from studies of ultrafine particles. Environ Health Perspect 113:823–839
14. Gatti A, Quaglino D, Sighinolfi GL (2009) A morphological approach to monitor the nanoparticle-cell interaction. Int J Imaging 2(S09):2–21 (Editorial 1)
15. AshaRani PV, Low Kah Mun G, Prakash Hande M, Valiyaveettil S (2009) Cytotoxicity and genotoxicity of silver nanoparticles in human cells. ACS Nano 3(2):279–290
16. Shacter E, Weitzman SA (2002) Chronic inflammation and cancer oncology (review article). January http://www.cancernetwork.com/review-article/chronic-inflammation-and-cancer
17. Coussens LM, Werb Z (2002) Inflammation and cancer. Nature 420:860–867
18. Protecting the nanotechnology workforce: NIOSH nanotechnology research and guidance strategic plan (2013–2016)
19. http://www.nanowerk.com/nanotechnology-news/newsid=35536.php

Chapter 6
Nanotoxicity: Must Consider Aspect of Nanoparticle Development

Ranjita Shegokar

Abstract Early years are proof of nanoparticle developments and understanding basic mechanisms. With no doubt, nanotechnology based drug delivery systems are very effective to achieve improved therapeutic effects in vivo and properties in vitro. Drug discovery programs are generating pharmacologically promising drug candidates but they suffer from solubility, stability, and toxicity. Nanoparticle engineering could find various solutions to solve these issues but unknowingly it contributes to the toxicity at cellular and organ levels. New ways to treat diseases are being researched include approaches like gene therapy and immunotherapy, which suffer from other formulation disadvantages and toxicity issues. Overall, many exciting results are obtained using nanotechnology research in various infectious and non-infectious diseases. However, the main challenge is to pass the safety testing during preclinical and clinical trial. This chapter illustrates various aspects of nanoparticle toxicity, its evolution and presents the status on the regulatory and industry discussions about it.

6.1 Introduction

6.1.1 Background on Nanotoxicity

Nanotechnology has already reached successfully from lab to market with a lot of scientific interest from pharma industry. Nanoparticles to sub-micron sized particles are used in food, cosmetics and pharma as drug delivery carriers and imaging agents. Increased use and expected growth raised the eyes on other side of nanoparticles, i.e., adverse effects. Large numbers of government and non-government sectors are working on the increasing public awareness on nanomaterials (particularly non-pharma), identification of improved characterization techniques and robust toxicity screening tools. Nanotechnology for drug

R. Shegokar, Ph.D. (✉)
Department of Pharmaceutics, Biopharmaceutics and NutriCosmetics, Free University of Berlin, Kelchstr. 31, 12169 Berlin, Germany
e-mail: ranjita@arcsindia.com

M. Lungu et al. (eds.), *Nanoparticles' Promises and Risks*,
DOI 10.1007/978-3-319-11728-7_6

delivery in pharma industry is expected to grow from a current value of $2.3 billion to $136 billion by the year 2021 which is accounted for 15 % of global nanotechnology market lead by Asian country with compound annual growth rate (CAGR) of 32.5 % [1]. Injectables are mainly to contribute to this growth in areas of liposomes, microspheres and nanoparticles in diabetes and oncology followed by a steady succession of imaging and diagnostic products. In summary, nanotechnology is a growing field in pharmaceuticals employed via injectable, oral and dermal route and offers great potential. To make optimum use of nanoparticles is equally important to understand the correlations between their properties and the related biological responses in vitro and in vivo.

Potential higher risk is mainly for particles below 100 nm which is defined as the "nanoparticle" by the FDA defined as National Nanotechnology Initiative (NNI). Depending upon the nature of nanoparticle, e.g. metal, magnetic it show unique mechanism of breakdown in body which might lead to unpredictable and exceptional toxic effect, that is difficult to predict as they are involved in many catalytic and oxidative reactions in vivo. Reliable systematic quantitative predictions of pharmacokinetics and pharmacodynamics of nanoparticles are essential to develop or to design nanoparticles for therapeutic and diagnostic applications. This information will help to understand the behavior of non-specificity of nanoparticle toward tissues and different cell types which later can serve as the basis in determination of related toxicity and future investigative directions.

Many leading organization worldwide, e.g. Organization for Economic Co-operation and Development (OECD), European Commission (Registration, Evaluation, Authorisation and Restriction of Chemicals legislation—REACH, NANO Safety Cluster), European Food Safety Authority (EFSA), Scientific Committee on Consumer Products (SCCP), and Environmental Protection Agency (EPA) are trying to determine and regulate the suitable endpoints in toxicity assessment testing in addition to the legislated tests and experimental procedures. The main purpose of this chapter is to discuss toxicity aspects of nanoparticles and understand the basic factors that might affect the toxicity of nanoparticle. This chapter reviews techniques, assays and other prediction tools, which can be used in assessment of toxicity behavior of nanoparticles.

6.1.2 Currently Used Nano-based Drug Delivery Systems and the Stress They Generate In Vivo

Recently, the U.S. Food and Drug Administration had issued three final guidances and one draft guidance aimed at providing greater regulatory clarity on the use of nanotechnology in FDA-regulated products to help ongoing research activities and developments by industries [2]. FDA has approved more than 25 nanoparticles based on various drug delivery systems. This clearly indicates potential efficacy of nanoparticles in the treatment of infectious or non-infectious diseases. In pharma, liposomes, nanolipidsomes, lipid nanoparticles, nanostructured lipid carriers, nanoemulsions, polymeric nanoparticles, lipid nanocapsules, lipid–polymer complexes, lipid–polymer nanoparticles, nanocrystals, dendrimers, fullerences, and nanoshell are widely studied as drug delivery systems by academia and industries. On the other side, carbon nanotubes, graphene nanoparticles, metal nanoparticles and quantum dots are emerging nanomaterials used in imaging, diagnosis and as drug delivery systems. These systems provide efficient organ and cellular targeting. Functionalization of the surface offers additional manipulation to the behavior of nanoparticles in vivo. Overall, nanoparticles or nanomedicine is the future of 21st century.

Nanoparticle-induced toxicity involves generation of oxidative stress (free radical or reactive oxygen species generation, ROS), inflammation, genetic damage, and the inhibition of cell division and cell death. ROS stress further causes genotoxicity, inflammation, fibrosis, and carcinogenesis by release of some harmful cytokines (Fig. 6.1).

Each nanoparticle has unique mechanism of ROS generation which is till date not fully understood, e.g. metal nanoparticle induce ROS via Fenton-type reactions while carbon nanotubes by secretion of harmful cytokines. The damage of mitochondria and some cellular organelle is induced by nanoparticles without generation of any ROS. This could also be the case with stents and scaffold implanted.

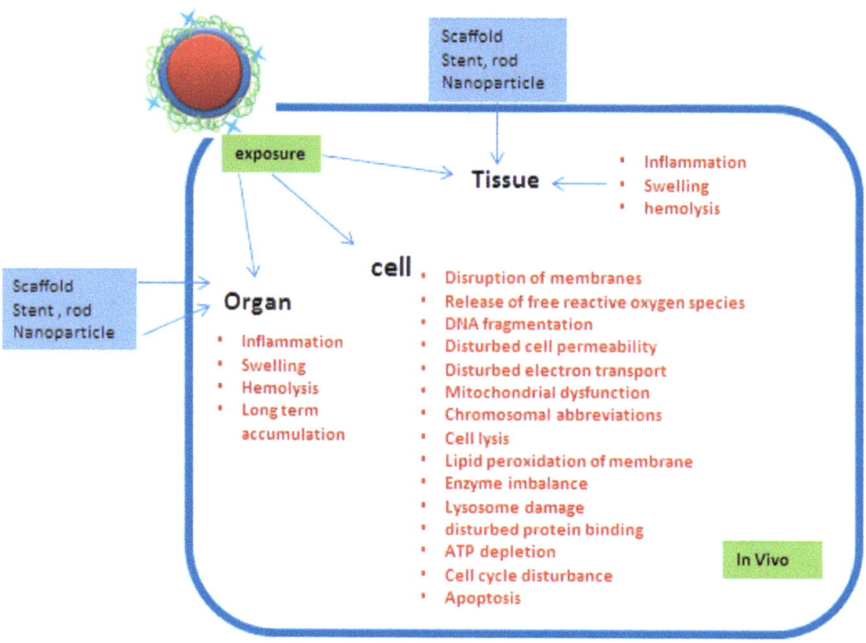

Fig. 6.1 Interactions of nanoparticles and cellular components and induction of related toxicological effects

6.2 Attention to Nanotoxicity Is Required: Why? What Are the Challenges?

Nanotechnology is an emerging technology and has applications in pharma (drug delivery, imaging, etc.), food, and cosmetics. It offers wide range of advances but attention to their side effects or toxicity effects is equally important. These nanomaterials interact at cellular levels and while exerting advantageous effects they also lead to some toxicological reactions if not formulated well. There are several challenges in understanding their toxicological behavior. The main challenge in assessing toxicity mainly includes availability of limited analytical techniques. It is very important to have validated, predictable materials characterization tool and protocols to foresee the toxicity and hazard potential based on pharmacology and toxicology response related to dose (acute and chronic exposure) alongside with the determination of physiological reactivity (e.g. molecular, cellular, tissue, organ toxicity). Currently, multiple techniques are available for macroscale biomaterials and for soluble pharmaceutical products; however for nanoparticle, development of such systems is still in preliminary phase. For implants, the situation might be different to understand toxicity at local tissue. Another challenge is to correlate results from literature and in vitro and in vitro experiments done. There is a strong need to develop such a database where scientist can learn from experiences, observation from others in developing own protocol to predict toxicity. Other challenges are right choice of excipients, dose selection, type of cell line and animal model selection and toxicity assessment protocol suitability. In the following sections, these parameters are discussed one by one. These parameters are related to the properties of nanoparticles as well as processing techniques.

6.3 Aspects of Nanotoxicity to Be Understood

6.3.1 Parameters Related to Nanoparticles

Toxic effects are sensitive, varied and complex and very much dependent on the molecular interaction of nanoparticle at the cellular level which is directly affected by the physico-chemical properties of the particles. Factors that might affect toxicity depend on the type of nanoparticles and their composition, e.g. metal, magnetic or polymeric nanoparticles. Factors such as composition, capillary forces, optical effects, color, melting points, conductivity, ionization potential, electron affinity, magnetism, surface energy, and reactivity contribute to the toxicity. Surface topology, porosity, texture, and high-density fabrication are must considered factors (Fig. 6.2).

Other nanoparticle properties that affect toxicity and induce bioreactions in body are intrinsic colloidal instability, aggregation phenomena in aqueous milieu, bio-accumulation in the tissues, adsorption and transport phenomenon. Currently,

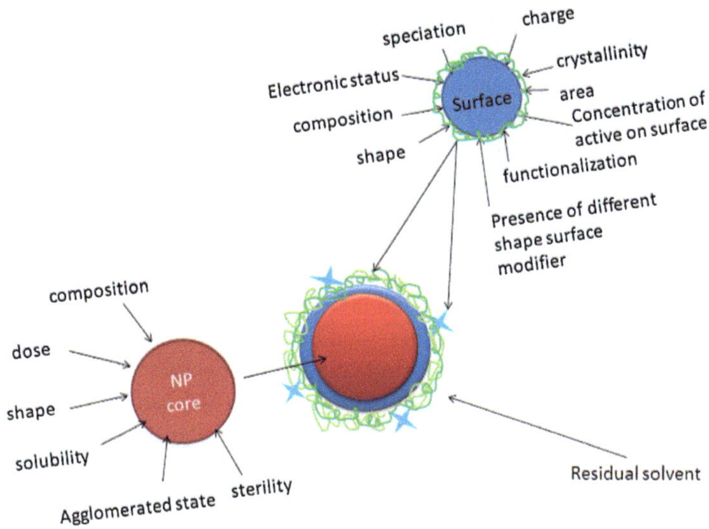

Fig. 6.2 Nanoparticle surface and core-related parameters which might catalyze toxicological interactions

there are limited analytical tools to predict about the intrinsic risks, tolerance, hazards, toxicity, or dose–response relationships for almost all classes of nanoparticles. One of the emerging fields in this direction is "Nanotolerance and Nanotoxicology" which deals with constraints placed by the size and surface properties toward toxicity by tracking properties of nanoparticle and their reactivity in biological or physiological systems mainly by using in vitro assays. Attempts to develop advance analytical tools and assays are still in primitive stages. These critical determinants, discussed briefly here, include size, shape, particle surface, surface positive charges, surface-containing groups, particle dissolution, metal ion release from nanometals and nanometal oxides, UV light activation, aggregation, mode of interaction with cells, inflammation, and pH of the medium.

These physico-chemical properties mainly affect the ultimate accumulation at target site and also translocation of nanoparticles in body, e.g. nanoparticles administered to the lung can translocate to the liver, pulmonary accumulation of crystalline silica with accumulations in the liver, spleen, peritoneal cavity and bone marrow. Surface contamination of biomaterials may range from adventitious adsorption to surface oxidation, charge generation or initiation of electron transfer reactions. The adventitious adsorption of bacteria-derived ubiquitous endotoxins—pyrogens or lipopolysaccharides (LPS)—or polyaromatic hydrocarbons (PAHs—known carcinogens) is one concern. Surface composition and its control are requisite to understand nanomaterial interactions with cells. Correct calculations of biologically effective dose of nanoparticles based on mass or concentration of particles per unit tissue or per number of cells or surface area of cells in cell culture is very essential to get right dose–response relation and for toxicity predictions. The toxicity of nanoparticles has been reported as in vitro assays (in cell line) and in vivo (rodents, humans, rabbits and aquatic species, such as zebrafish, catfish, algae). Giannaccini et al. studied the non-mammalian vertebrate embryos such as chicken, zebrafish and Xenopus as models for biomedical applications [3].

The selection of nanoparticle dose to get accurate therapeutic measure is very challenging [4]. Preclinical, stability assessment in aqueous media, plasma, and protein adsorption must be analyzed along with purity (indicated by the absence of lingering contaminants such as antioxidants or homopolymers), reproducibility of manufacture, and drug release and biodegradability profiles. In the case of metal nanoparticle, size is mainly determined in dry state while cell line studies are done in wet condition. It is important that conditions of experiment are the same for in vivo or in vitro studies and characterization of nanoparticles [5].

Quantitative Structure–Activity Relationships (QSAR) can be applied to small organic compounds with limited applications to nanomaterials. Scientists can take help of computational molecular methods to predict properties of small molecule up to large biomolecules, their reactivity and mechanisms of actions. Quantum chemical calculations and molecular dynamics simulations can help to further understand toxicological aspect. In addition to computational modeling, physiologically based pharmacokinetic (PBPK) models and modeling dose–response. Mesoscale dynamics (MesoD) techniques can be applied to assess the properties of NPs like molecular volume, surface charges, dipole moment, band gap, transition energies, and ionization potentials. For structure encoding, various commercial programs such as DRAGON, CODESSA, CAChe, etc. are used [6].

6.3.2 Assays Selection

Toxicity protocol depends very much on the routes of biological exposure. It is equally important to aware of batch to batch variability of nanoparticles and the decision on number of batches to be tested. The factors also include reagents and chemicals used during cell assay, sterility of nanoparticles and cell cultures, treatment given to cells (exposure time, sonication, temperature (25 °C vs. 37 °C)). Safety pharmacology assessment of dextran-coated nanoparticle graphene at single dose injected intravenously could be predicted in extended acute toxicity study [7]. Mechanism of toxicity is either by accumulation of nanoparticle at cellular or organ level, resulting in excessive ROS causing oxidative stress resulting in differences in physiology of cell, cytotoxicity, apoptosis, and cancer initiation. DNA is a direct target of ROS, results in base and sugar lesions, DNA eprotein crosslinks and breaks in single- and double-strand. Highly reactive radicals, such as hydroxyl radicals, can damage DNA quickly in the vicinity. Superoxide dismutase, peroxidases, and catalases are some of the prominent antioxidant enzymes that efficiently protect against these harmful biological events. Cell proliferation assay (SRB assay) on optical nanoparticles like gold nanospheres, gold nanorods, silver nanopheres, silver triangular nanoplates and quantum dots composed with different material and surface chemical can be correlated to the physical factors like material composition, physical size/surface area, shape, and surface chemical property [8].

6.3.3 Excipient Selection

It has been shown that genotoxity and oxidative stress of quantum dots made up of core/shell CdSe/ZnS can be attenuated by PEG encapsulation when studied by neutral comet assay and H2AX foci formation [9]. Diagnotic iron oxide nanoparticles proved its effectiveness to determine mechanistic interactions at the nanobiointerface with cells of the immune, neuroglial and neurovascular units in vivo [10]. Selection of surfactant, surfactant concentration, chemical nature of polymer, % polymer used, solvent used in nanoparticle manufacturing process are very important. Some surfactant has inherited toxicity profile. In case of lipid nanoparticles, nature of lipid, lipid concentration, and type of technique, e.g. solvent evaporation are key parameters. Excipient compatibility, excipient stability after nanonization, and freeze drying are other areas to take into consideration. Purity of excipient and nanoparticle is equally important in toxicity determination.

The outer structure of nanoparticles determines the protein adsorption patterns in vivo which ultimately determine the fate of nanoparticle [11]. Excipient choice determines the protein adsorption on nanoparticle and its ultimate fate, e.g. polysorbate 80 is known to induce adsorption of ApoE on the surface of nanoparticle which led to brain targeting. Similarly, nanoparticles surface modified with disaccharide could led to macrophage targeting and ultimately to spleen [12]. This kind of excipient directed, drug delivery systems can be used for organ-specific targeting for various diseases, e.g. splenic targeting for HIV, diagnostic or imaging for polysorbate 80 decorate nanoparticles. Readers are advised to refer a chapter on application of multidimensional gel electrophoresis in this book to know the basics of in vitro protein characterization.

6.3.4 Selection of Cell Lines

In vitro assays are simple, faster, and more cost-efficient to that of animal studies. Assessment of toxicity endpoints beforehand, complexity of animal model, and metabolic activity are determining factors in obtaining in vivo biodistribution and are mostly neglected. Cytotoxicty reactions involve effects on cell signaling, membrane perturbations, influence on the cellular electron transfer cascades, production of cytokines, chemokines and ROS, transcytosis and intercellular transport, gene regulation, overt toxic reactivity, no observable toxicity, and cell necrosis or apoptosis. A major challenge with in vitro assay is determination of correct and meaningful endpoints that exactly reflects physiological stress, toxicity, or therapy or some other phenomenon detectable in vivo. This endpoint detection is very important to form a link between in vitro and in vivo results. The assurance of cell assay predictability, biocompatibility with biomaterial parameters needs wisely judged by quality control standards for cell behavior and biological relevance. These protocols or procedures are needed to be implemented, compared and enforced during study.

In vitro cell line studies are used to predict a human response at the administered dose and to determine relative toxicity in a human or related cell line [13]. These in vitro assays are vital in screening nanoparticles or formulations for cytotoxicity or adverse biological effects before in vivo testing. These in vitro cell assay methods could significantly reduce the number of in vivo experiments by giving flexibility to chose formulation and thereby reduction in use of research animals. Various in vitro toxicity assays have been utilized to determine the viability and/or cytotoxicity in cultured cells based on colorimetric measurements or by use of fluorescent dyes as markers to determine cell viability assessing membrane integrity (i.e. neutral red, calcein AM) or cell metabolism (i.e. MTT, alamar Blue). These types of assays are commonly used for small molecule, and to limited extent for nanoparticles. However, there is strong need to use sophisticated techniques like high-throughput screening techniques to understand in detail the mechanism of cellular interactions. One of the study done on four carbon-based nanomaterials viz. CB, SWCNT, C60, and nC60 and one non-carbon-based quantum dots were evaluated in nanomaterials by nine viability assays and flow cytometry in human skin epithelial cell line (keratinocytes) supports the concept of high-throughput screening and multiple assay determination. These cell-based assays must be validated by imaging techniques like transmission electron microscopy or fluorescence microscopy or by chemical marker-based viability assays [13]. ESR spin trapping techniques are employed to identify the formation of ROS.

Many commercial sectors offer thousands of transformed cell lines to choose from. Many cell lines are deposited within commercial registries (i.e. American Tissue Type Culture Collection, ATCC) (Fig. 6.3).

Most commonly, commercial available cell lines are genetically altered or tissue-harvested (primary cells) either with or without serum. Leukocytes and immune modulatory cells (e.g. neutrophils, macrophages, and dendritic cells), phagocytic cells (e.g. monocyte and macrophage phenotypes) have been used as experimental models in uptake studies. In primary phagocytes, special need to take care of cell differentiation state (i.e. to terminally differentiated macrophages) which could significantly alter phagocytic and proliferative tendencies of cell. Primary cells obtained from tumor biopsy may provide a closer representation of the tumor cell type. However, 2-D monocultures are generally not representative of the in vivo case because the culture lacks the spatial interactions and phenotypic/cell type cross-talk of the actual tumor. Additionally, primary cell culture often requires animal resources, maintenance fees, and post-harvest animal sacrifice, raising costs, ethical controversies and significant paperwork for the investigator (e.g. IACUC and IRB applications).

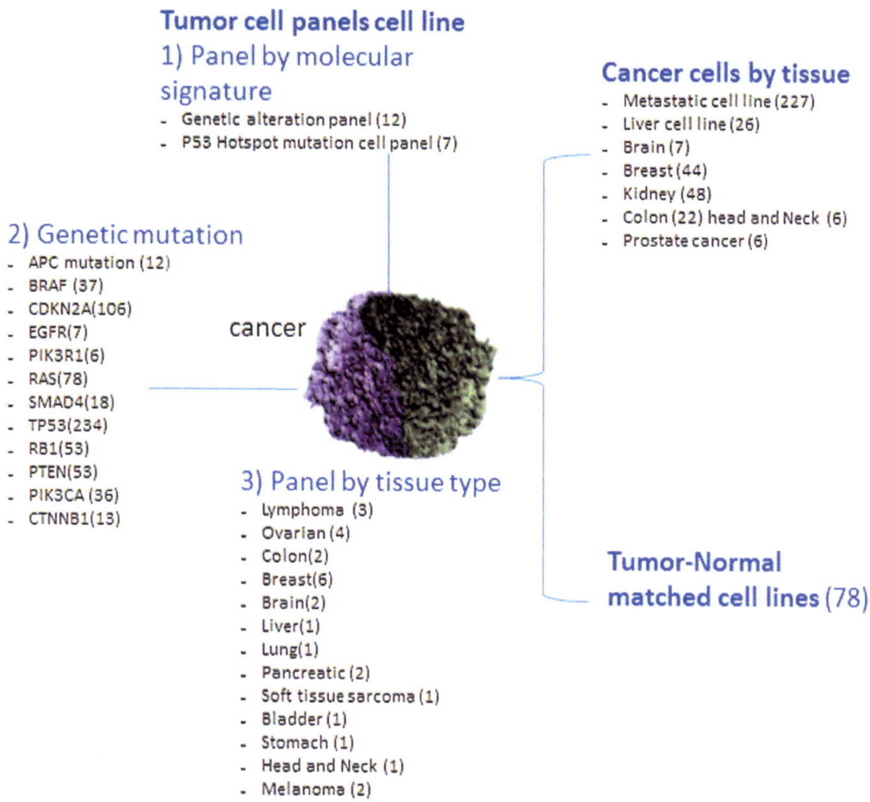

Fig. 6.3 Different cell lines (and numbers of other cell line available) employed in cellular assay related to cancer. The number indicated in the bracket is the number of subtype (available) cell line. The data collection is from ATCC

Cell models can be selected from dermal, oral-gastric, colonic, pulmonary, nasal, kidney and vaginal epithelial depending upon the need and the desired endpoint. Other common transformed epithelial lines include MDCK (kidney), HT-1080, HT29-18-C1 (colonic), and HeLa (cervical). Other relevant cell types include endothelial cells (BAECs and HUVECs) and red blood cells (erythrocytes), platelets, and leukocytes in cultures. Secondary transformed endothelial-derived cell lines include COS-7 (kidney) and MS-1 (pancreas). Human aortic endothelial cells (HAECs), human umbilical vascular endothelial cells (HUVECs), and human microvascular endothelial cells can be used as models of endothelial inflammatory response and resultant arthrosclerosis. Beside from common precancerous cells, numerous oncogene-transformed cancerous cell lines (MSTO211H, HL60,WTK1, 1321N1, and HeLa) are employed in vitro for materials testing, pulmonary cells are represented by immortalized secondary cell lines phenotype, e.g. THP-1, RAW, J7,

Mono mac 6 cells or IC-21 lines or as primary cultures, e.g. alveolar macrophages. Gastric, nasal, oral, and urinary tract epithelium are all characterized in vivo by viscous mucin production and tissue coatings that is absent in cell-based assays, which does not represent in vivo condition. The condition maintained in vitro must be as close as possible to the in vivo scenario. Long et al. used BV2 cells (immortalized mouse microglia), rat dopaminergic (DA) neurons (N27), and primary cultures of embryonic rat striatum to model the characteristic ROS-burst from particle insult to these macrophage-like cells in the context of the brain [14].

The changes in cells are indication of modulation of protein or gene expression, phagocytic ability, and/or inflammation reactions that alter normal phenotype. Short-lived ROS species are main cause of these changes and can be determined in cell media by fluorescein-compound-based tests (fluorescein probes, 2′,7′-difluorescein-diacetate dichlorodihydrofluorescein diacetate) or by Electron Paramagnetic Resonance (EPR). ROS in cell may be quantified by the well-known glutathione (GSH) assay. Other assays are based on the analysis of key oxidized species damage cell membranes or DNA mainly by immunocytochemistry which allows detection of specific DNA lesions and directly measures the ROS involvement in DNA damage. BODIPY-C11 is a fluorescent dye which adapts different colors when oxidized and unoxidized into lipid bilayers, is another way to determine stress or lipid peroxidation at cell membrane. Figure 6.4 numbers the cell lines available for various infectious and non-infectious diseases according to ATCC.

Cell viability assay gives indication of live cells as a response to dose and also answers question of biocompatibility. Various dye-based assays are used widely which mainly employ neutral red, trypan blue. Other assays which determine cell viability are LIVE/DEAD™ kits by Life Technologies, lactate dehydrogenase (LDH), formazan-based assays (MTT), MTS, WST, Alamar blue (resazurin), Coomassie blue, ATP-luciferin luminescence, adenylate kinase (AK) release assay, determination of mitochondrial membrane potential (MMP) and thiobarbituric (TBA) assays. Fluorescence-activated cell sorting (FACS) is an efficient way of categorizing and quantifying cells based on the status as healthy, dead, apoptotic, or necrotic. It can distinguish alterations in the cell cycle by measuring parameters like hypodiploid DNA which is indirect indicative of apoptosis.

Other pathway-sensitive cell viability assays are enzyme-linked immunosorbent assay (ELISA) to quantify fragmented DNA, comet assay to detect DNA fragmentation by gel electrophoresis, Caspase Glo3/7 indication of mitochondrial apoptotic pathway, Hoechst-DNA, and TUNEL assays (TdT-mediated dUTP-biotin nick end labeling) measure fragmented DNA. Old to new ELISA assay version allows direct detection of hundreds of cytokine protein in multiplexed assay formats including 8 (LINCOplex). Change patterns in gene/protein expression can be detected by polymerase chain reaction (PCR), or real-time PCR (qPCR) for quantitative determination, western blotting, and total protein assay (i.e. BCA or Bradford). This type of assay can give indication on the modulation of CDKN1A (cell-cycle arrest gene), GADD45β (DNA damage-dependent), IL-6 and NFκBIA (inflammatory response),

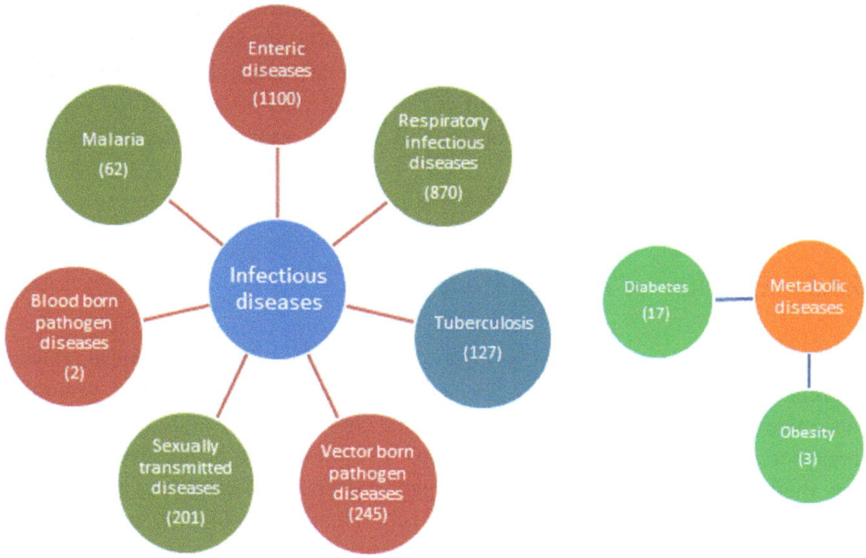

Fig. 6.4 Infectious and non-infectious diseases and number of cell lines available according to ATCC

EGFP (reporter gene), and NF-κB (inflammatory growth factors), cytokines, and adhesion molecules.

Phagocytic activity, internalization, and organelle interaction of nanoparticles of particles larger than 500 nm is mainly triggered by protein–particle adsorption and subsequent receptor-mediated activation of F-actin-driven pseudopods. A number of cell-particle uptake assay kits are available in market, however it is advisable to combine them together with microscopy techniques. Optical techniques mainly include TEM, high-resolution (HRTEM) and energy-filtering (EFTEM) TEM, Energy Loss Spectroscopy (EELS) and confocal Laser Scanning Microscopy and fluorescence microscopy. Advance LSM can capture pseudo-3D images by the use of reconstruction algorithms combining several axial and lateral images [15]. Another technique called multiple particle tracking (MPT) assays can help to understand particle interactions with various cell components in real-time using video-interfaced phase microscopy by real-time information on pore sizes, particle adhesion, intracellular particle transport mechanism, and barriers to particle transport or localization. Additionally, two-point micro-rheology, a variation of MPT, allows visualization of cell–particle interactions which is not accessible by any other technique.

Other mammalian cell-based assays include use of single-celled ciliated protozoa and tetrahymena thermophila. In short, high-throughput screening methods using multiple concentrations and multiple cells model ensures accuracy of results. The ultra-small 384-well cell culture plates or nanodrop sample assay allows multiple assays with suitable detection features (e.g. fluorescence and luminescence) [15]. Advanced kit like Nano-toolkit allows determination of concentration, size, chemical composition, density, route of exposure, time of exposure, and the point of embryonic development at which the test material is administered by using scoring system of 0 (normal phenotype); 1 (minor phenotypic changes); 2 (moderate alterations); 3 (severe embryo deformation); and 4 (embryo death).

6.3.5 Selection of Animal Model

Assessment of nanotoxicity can be done at (1) sublethal doses, (2) chronic doses, (3) repeated doses, and (4) long-term doses. In animal experimentation, generally rats, mice, rabbits and dogs are employed. The choice of correlative and predictive models to assess acute and chronic toxicities is very essential in the development of new therapeutics. However, there is still gap in deeper understanding and correlation of physiochemical factors and pharmacological, toxicological screens, pharmacokinetic and biodistribution pattern obtained in vivo. The causes of autophage in animals after administration of nanoparticles are still unknown and literature report involvement of Akt/mTOR pathway [16]. The toxicology of nanoparticles when administered by non-inhalation route is very complex and needs interdisciplinary efforts to investigate pharmacological risk assessment at cellular and biochemical levels. While respiratory tract has more than 40 different cell types, type of cells, polarity, tight junctions, and other characteristics contribute to the differences in biological/toxicological responses against inhaled particles besides understanding of complicated pulmonary pathways.

Efficacy studies are mainly performed in mice and pharmakinetic and toxicity studies in rats and dogs. To reduce time and investment involved in establishing mammalian animal model for toxicity, Zebrafish *(Danio rerio)* model has been involved in recent years because of its close homology with human genome and immunologic response behaviorr [17]. Optimal clarity of zebrafish allows close experimental watch on pathological processes in real time, such as organogenesis, vasculogenesis, and deleterious effects during anatomical development during nanoparticle treatment. Cost of maintenance and husbandry (200–300 eggs/day for every 5–7 days) is very much less compared to the other animal model (3–6 months).

Another model, Xenopus, allows microsurgery at very minute level and can be used as model for gene function testing. There is a database developed to assimilate information on resources on use of Xenopus as model (www.xenbase.org). For in vivo angionesis testing, chicken embryo is widely used as model. Schweitzer A rated animal models at scale of 1–5 depending upon their use in pharmacological experiments: (1) low sensibility/consciousness: mollusks; (2) some sensibility: cephalopods, fishes, amphibians; (3) sentient, but potentially limited in consciousness: reptiles; (4) sentient and conscious: mammals, birds; (5) sentient, highly intelligent and precognitive: primates, carnivores, cetaceans [3].

3-D kidney organoid proximal tubule culture was found to be suitable to detect the toxicity of hydroxylated generation-5 PAMAM dendrimer and gold nanoparticles [18]. While biochemistry assay, pathological examination and metal content measurement in tissue might be interesting to understand in vivo fate of metal nanoparticles. Intracellular stress response is often investigated with H2DCF-DA (20,70-dichlorodihydrofluoresc(e) in diacetate) probe or mitochondrial membrane potential (MMP).

6.4 Analytical Techniques Used In Vitro and In Vivo Studies

Deep understanding of factors like amount of nanoparticles reaching to the cells, experimental conditions, and amount of nanoparticles/volume cell culture medium containing the cell can improve the correlations between dose and response. Equally important is technique selection.

In addition to particle sizing and aggregation stability assays performed in vitro at different pH conditions and in relevant biological simulated fluid is an important to predict their interaction in body. Many techniques are used to characterize particle size, e.g. TEM, scanning electron microscopy (SEM), optical spectroscopy, dynamic light scattering, laser diffractometry, and fluorescence polarization. Mainly conclusions are based on results obtained from two techniques. Additionally, each method possesses its own advantages and limitations. Large angle X-ray diffraction (XRD) is mainly used to differentiate between crystalline and amorphous nature material while small angle X-ray scattering (SAXS) and small angle neutron scattering (SANS) can be used to analyze particle size and shape of core and shell structure. In many cases, multi-angle laser light scattering (MALLS) is used together with UV–visible spectroscopy in field flow fractionation (FFF) for size determination and separation.

Specific surface analysis methods are limited in number and can be employed to analyze surface contaminants of nanoparticles by using one or many surface analysis techniques like time-of-flight secondary ion mass spectrometry (ToF-SIMS), X-ray photoelectron spectroscopy (XPS), X-ray fluorescence (XRF), energy dispersive X-ray analysis (EDX), time-resolved fluorescence polarization anisotropy (TRFPA) can size 1–10 nm particles may be sized with 0.1 nm resolution and surface-enhanced Raman spectroscopy (SERS). These techniques can also be used to identify impurities in products or determination of mass coverage by functionalization. Metal content, metal based impurities can be detected by coupled plasma-mass spectroscopy (ICP-MS), inductively coupled plasma-atomic emission spectroscopy (ICPAES) but these methods have limitations over detection of size and shape.

Electrochemistry is widely used as an alternative characterization technique to microscopy and spectroscopy. New electrochemical detection methods like “nano-impacts” or “single nanoparticle collisions” offer in-situ detection of metal nanoparticles. Sometimes, toxicity causes are inherited from techniques used to produce nanoparticles, e.g. milling. Wet media milling is widely used to produce nanocrystals. The milling time and rotation speed are key parameters which might affect traces of metal in samples. This can be controlled by optimizing milling time and type of ball material [19]. Gamma scintigraphy is used for determination of location of nanoparticles in vivo.

6.5 Industry and Regulatory Approach and Patent on Toxicity Determination

Nanoparticles have application as fluorescent biological markers detection of proteins, probing of DNA structures, separation and purification of biological molecules and cells, MRI contrast enhancement, tumor destruction via heating, tissue engineering and drug and gene delivery. FDA nanotechnology products are main in the area of medical devices, biotechnology, tissue engineering, vaccines, cosmetics, drug delivery systems, and combination products (drug-device, drug-biologic, device-biologic or drug-device-biologic that are physically or chemically combined, co-packaged in a kit or separate cross-labeled products). Multi-component system that may consist of carrier/delivery system (drug or device), therapeutic agent (drug or biologic), imaging agent, targeting agent, implantable microchip-based delivery systems that deliver different drugs under controlled conditions and injectable delivery systems (transdermal microneedles). Imaging agents are mainly for MRI, ultrasound or for radioactive imagine, while therapeutic agents could be small molecule, nucleic acid or proteins. Receptor ligands, antibody serves as targeting moiety. FDA closely keep watch on nanoparticles research advancement and is working with Center of Drug Evaluation and Research (CDER), Center for Biologic Evaluation and Research (CBER), National Center for Toxicological Research (NCTR), and Center for Food Safety and Applied Nutrition (CFSAN).

Currently, approved nanoscale therapeutics include gadolinium chelate for MRI imaging (Gd-DTPA Dimeglumine), iron oxide particles for MRI imaging (Feridex), Products using NanoCrystaltechnology (Rapamune, Emend), Liposomes (Doxil, DaunoXome), Microemulsions (Cyclosporine) and Albumin-bound nanoparticles (Abraxane), Silver nanoparticles (antibacterial wound dressing), engineered Calcium Phosphate (NanOssTM, duplicates microstructure, composition and performance of human bone), nanoparticle dental restorative (3M ESPE Filtek), Cosmetics (containing lipid nanoparticles or "nanosomes" used as delivery systems, for controlled release of active ingredients; L'Oreal, Estee Lauder), sunscreens (containing titanium dioxide and zinc oxide nanoparticles which make the product appear transparent) (Fig. 6.5).

As particle size gets smaller, then the particle shows size-specific effects on biological activity, then it becomes very critical to evaluate to what extent these small particles get access to tissues and cells at normal conditions. Another question is how long is the residence time, what is the clearance mechanism, what type of

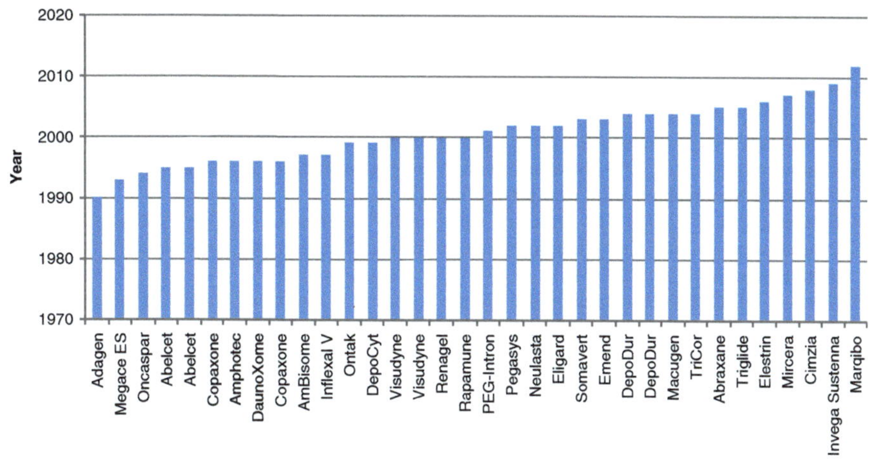

Fig. 6.5 FDA-approved nanotechnology-based products in pharma and their launch year (The Name can be read as: brand name and type of drug delivery system)

effects these small particles can exert are these effects are transient or permanent, to what extent and what type of cell might be a target for these small particles are the main safety concern for industry and for FDA.

These effects are route-specific, e.g. for oral increased bioavailability, subcutaneous-sensitization, dermal penetration/phototoxicity, occcular-intravetreal retention, inhalation-systemic bioavailability, intravenous-sterility; this information can give alter on safety. Currently, CDER/FDA's guidelines on safety requirements are rigorous. However, if required the demands from research in academia and industry might result in additional regulation on toxicological testing. It is equally important to educate public on use of nanoparticle specification at which concentration or at which particle size they might cause adverse effects.

Conclusion and Perspective

Nanotechnology in health care system is growing rapidly and needs attention from scientists in finding suitable toxicological assays and analysis methods. To date, used techniques and assays are well established, however sophisticated assays combined with new techniques might provide more advance data and understanding of toxicity behavior. Toxicogenomics endeavors to understand the biological effects of cellular products controlled by the genomes such as RNA, proteins, and metabolites. Compared with a conventional assay, new combinations such as assay-electrochemical analysis, assay-cell ROS generation, assay-visualization techniques might be useful to get more insight in toxicology. Altogether, nanotoxicity research needs multidisciplinary collaboration and debate on needs. The role of regulatory bodies is very crucial in getting inputs for new assays.

References

1. http://www.cientifica.com/research/market-reports/nanotechnology-in-drug-delivery-global-markets
2. http://www.fda.gov/ScienceResearch/SpecialTopics/Nanotechnology/default.htm, 2014
3. Giannaccini M et al (2014) Non-mammalian vertebrate embryos as models in nanomedicine. Nanomed Nanotechnol Biol Med 10(4):703–719
4. Soenen SJ et al (2014) The effect of nanoparticle degradation on poly(methacrylic acid)-coated quantum dot toxicity: the importance of particle functionality assessment in toxicology. Acta Biomater 10(2):732–741
5. Warheit DB (2008) How meaningful are the results of nanotoxicity studies in the absence of adequate material characterization? Toxicol Sci 101(2):183–185
6. Gajewicz A et al (2012) Advancing risk assessment of engineered nanomaterials: application of computational approaches. Adv Drug Deliv Rev 64(15):1663–1693
7. Kanakia S et al (2014) Dose ranging, expanded acute toxicity and safety pharmacology studies for intravenously administered functionalized graphene nanoparticle formulations. Biomaterials 35(25):7022–7031
8. Qinghui Z et al (2014) The nanotoxicity investigation of optical nanoparticles to cultured cells in vitro. Toxicol Rep 1:137–144
9. Ju L et al (2013) Quantum dot-related genotoxicity perturbation can be attenuated by PEG encapsulation. Mutat Res 753(1):54–64
10. Kim Y et al (2013) In vivo nanoneurotoxicity screening using oxidative stress and neuroinflammation paradigms. Nanomed Nanotechnol Biol Med 9(7):1057–1066
11. Shegokar R et al (2011) In vitro protein adsorption studies on nevirapine nanosuspensions for HIV/AIDS chemotherapy. Nanomed Nanotechnol Biol Med 7(3):333–340
12. Shegokar R, Singh KK (2012) Preparation characterization and cell based delivery of stavudine surface modified lipid nanoparticles. J Nanomed Biotherapeutic Discov 1:192
13. Monteiro-Riviere NA, Inman AO, Zhang LW (2009) Limitations and relative utility of screening assays to assess engineered nanoparticle toxicity in a human cell line. Toxicol Appl Pharmacol 234(2):222–235
14. Long TC et al (2007) Nanosize titanium dioxide stimulates reactive oxygen species in brain microglia and damages neurons in vitro. Environ Health Perspect 115(11):1631–1637
15. Jones CF, Grainger DW (2009) In vitro assessments of nanomaterial toxicity. Adv Drug Deliv Rev 61(6):438–456
16. Soenen SJ et al (2013) Turning a frown upside down: Exploiting nanoparticle toxicity for anticancer therapy. Nano Today 8:121–125
17. Fako VE, Furgeson DY (2009) Zebrafish as a correlative and predictive model for assessing biomaterial nanotoxicity. Adv Drug Deliv Rev 61(6):478–486
18. http://dx.doi.org/10.1016/j.biomaterials.2014.04.060, 2014
19. Shegokar R, Müller RH (2010) Nanocrystals: industrially feasible multifunctional formulation technology for poorly soluble actives. Int J Pharm 399(1–2):129–139

Chapter 7
Pros and Cons on Magnetic Nanoparticles Use in Biomedicine and Biotechnologies Applications

Florina M. Bojin and Virgil Paunescu

Abstract In recent years, the design and synthesis of colloidal magnetic suspensions have attracted an increased interest especially in the fields of biotechnology and biomedicine because they have many applications including targeted drug delivery, cell labeling and magnetic cell separation, hyperthermia, tissue repairing, magnetic resonance imaging (MRI) contrast enhancement, enzyme immobilization, immunoassays, protein purification, etc.

7.1 Introduction

Magnetic nanoparticles (MNPs) used in biomedicine must meet several requirements. They have to be non-toxic, chemically stabile, uniform in size, well stabilized under physiological conditions, biocompatible and to present high magnetization. Magnetite (Fe_3O_4) and maghemite (γ-Fe_2O_3) are the most suitable iron oxide nanoparticles employed for biomedical applications because they are biocompatible, have low toxicity in the human body and show a superparamagnetic behavior.

Therefore, synthesis of magnetic iron oxide nanoparticles with tailored properties has attracted considerable scientific and technological interest. Various synthesis routes were developed for producing magnetic particles, such as co-precipitation, microemulsion method, thermal decomposition of different organic precursors, spray pyrolysis, and sol-gel method.

For biomedical applications, magnetic iron oxide nanoparticles must be dispersed in biocompatible media in order to obtain stable colloidal suspensions. In order to prevent the particle aggregation and to improve the biocompatibility and stability, nanoparticles are coated with various surfactants: poly(ethylene glycol),

F.M. Bojin (✉) • V. Paunescu
Department of Functional Sciences, "Victor Babes" University of Medicine and Pharmacy Timisoara, Etimie Murgu Square No. 2A, 300041 Timisoara, Romania
e-mail: florinabojin@umft.ro

M. Lungu et al. (eds.), *Nanoparticles' Promises and Risks*,
DOI 10.1007/978-3-319-11728-7_7

oleic acid, poly(acrylic acid), gluconic acid, other polymers, liposome and fatty acids.

We investigated the influence of colloidal suspensions of magnetic iron oxide nanoparticles on tumor and normal (adult bone marrow-derived mesenchymal stem cells (MSCs)) cell lines cultivated in vitro conditions. Magnetite nanoparticles were prepared by combustion synthesis as well by co-precipitation route. Combustion synthesized Fe_3O_4 nanoparticles proved to be superior to the co-precipitated magnetite nanoparticles in terms of cell viability. The influence of different concentrations on MNPs on cellular morphology, ultrastructure expression of phenotypical markers, and viability were examined in order to establish whether the combustion synthesized nanoparticles may be used for in vitro and in vivo applications. Tumor cells treated with combustion synthesized nanoparticles had a lower viability when compared to similar cells treated with co-precipitated nanoparticles. When treated with combustion synthesized nanoparticles, tumor cells were enucleated and lost their adhesion abilities. Normal cells treated with combustion synthesized nanoparticles developed anchorage structures, which made them more resistant to the chemical stress. This remarkable behavior of combustion synthesized Fe_3O_4 nanoparticles opens a whole new perspective on using the combustion synthesized magnetite nanoparticles in cancer therapy due to their selective intrinsic behavior.

7.2 Theoretical Study

7.2.1 Magnetic Nanoparticles. General Considerations

Magnetic materials can be considered indispensable for modern technology. They can enter in composition of different electronic devices and appliances and are intensely used in industrial and medical equipment.

The ferromagnetic and ferrimagnetic materials with nanometric dimensions present a magnetism form called superparamagnetism. This involves smaller than 20 nm dimensions of the particles. Superparamagnetic materials present a single magnetic domain and do not present hysteresis [1–4]. In case that the particle dimension is increasing above one critical size, which depends on the nature of the material, it becomes multi-domain. The most used magnetic material is magnetite, Fe_3O_4 and or maghemite—γ-Fe_2O_3 [5, 6].

Magnetite is a natural mineral that presents a crystalline structure of inversed spinel (Fig. 7.1) with a cubic cellular unit with the central sides constituted from 56 atoms: 32 O^{2-} anions, 16 Fe^{3+} cations, and 8 Fe^{2+} cations.

In the spinel inverse structure of Fe_3O_4, half of the F^{3+} ions are tetrahedrally coordinated being surrounded by four oxygen atoms, and the other half of the F^{3+} ions and all Fe^{2+} ions are octahedral coordinated [8, 9].

Magnetite is oxidized at maghemite rapidly in air which is also ferromagnetic but has a slightly reduced magnetic response. This process occurs only at the

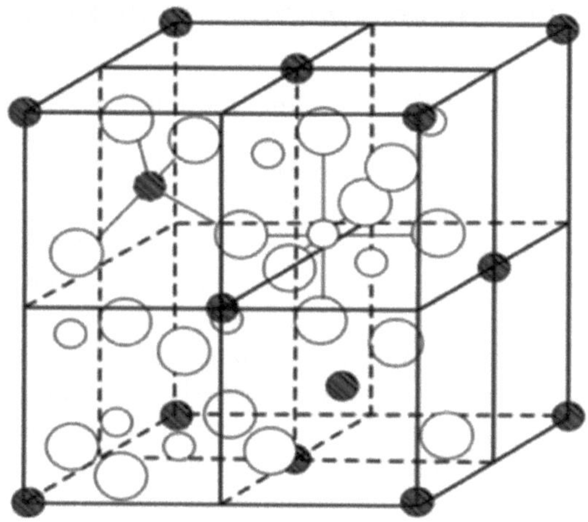

Fig. 7.1 Inversed spinel structure of magnetite [7]

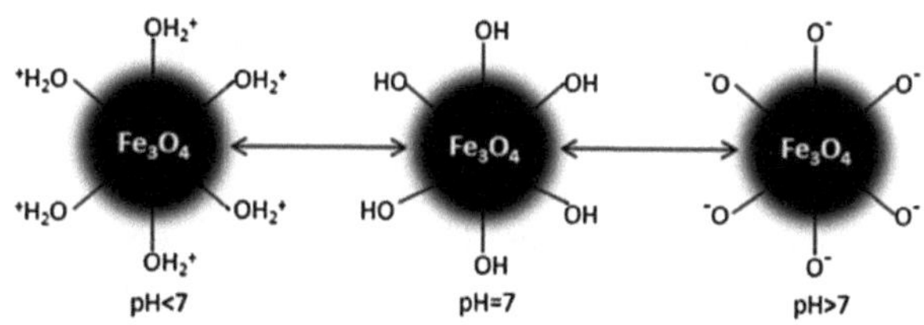

Fig. 7.2 Behavior of the magnetite particle according to the acid or basic pH [11]

surface of the crystals. The center of the crystals is also oxidized, by the diffusion of the Fe^{2+} ions at the surface were converted to Fe^{3+}. The speed at which oxidation occurs is determined by the diffusion speed of the Fe^{2+} ions and by the distance to the surface of the particle. This is why the particles that are bigger remain unaffected by the oxidation phenomena, while the small particles can be oxidized even at room temperature.

At temperature higher than 300 °C, magnetite is oxidized at hematite (α-Fe_2O_3) [10]. This is antiferromagnetic, as a consequence this conversion can damage if is used in specific applications.

The chemistry of the surface and its proprieties are especially important for specific applications. The iron atoms from the surface of the magnetite which are not bound by the oxygen act as Lewis acids, and coordinate the molecules that can donate a pair of electrons. In aqueous systems, these atoms coordinate the water molecules that dissociate quickly resulting in surface functionalized magnetite with hydroxyl groupings like Fe-OH. In this way the chemistry of the magnetite particles is strongly dependent on the pH values; at low pH values the surface of the magnetite particles is charged positively, and at high pH values it is charged negatively (Fig. 7.2). The hydroxyl groupings formed at the surface of the magnetite have an amphoteric character.

7.2.2 *Methods for Synthesis of Magnetic Nanoparticles*

In the last years numerous research were focused on the synthesis of magnetic nanoparticles. Many scientific publications described efficient methods for synthesis that allows the obtaining of single dispersed magnetic nanoparticles stable in time and with controllable shape.

The synthesis of magnetic nanoparticles experienced substantial progress in the last years but with all of these the high-quality magnetic nanoparticles with controllable proprieties represent a continuous challenge.

The superparamagnetic nanoparticles synthesis is a complex process. First of all an appropriate method of synthesis, that does not involve complicated purification steps and that can be used on industrial scale, must be selected. The most important stage of this process is represented by the establishment of experimental conditions that can assure the obtaining of nanoparticles with proprieties specific for the application domain.

From the iron oxides, magnetite (Fe_3O_4) and maghemite (γ-Fe_2O_3) are the most used in a variety of fields and can be considered superparamagnetic nanoparticles—in certain synthesis conditions.

In Table 7.1, some of the proprieties of the two oxides are presented [10, 12, 13]. Synthetic iron oxides can be obtained by a number of methods, such as precipitation of iron salts [14–22], thermic decomposing of the organometallic precursors [23–28], sol-gel method [29–32], microemulsion [33–36] laser pyrolysis [37–40], combustion method [41–46], hydrothermal method [47–52], sonochemical method [53–56], etc.; from all these methods the most common and used is Fe^{2+} and Fe^{3+} salts precipitation method.

7.2.2.1 Iron Salts Precipitation Method

The iron salts precipitation method is the most simple and efficient method for obtaining magnetic particles, due to the large quantity of particles that can be synthetized.

This method consists of mixing two salts Fe^{3+} and Fe^{2+} in 2:1 molar ratio, in aqueous media, followed by precipitation of these salts using a precipitation agent (a base) [57]. The chemical reaction equation for the formation of magnetite can be described as follows:

$$Fe^{2+} + 2Fe^{3+} + 8OH^- \rightarrow Fe_3O_4 + 4H_2O \tag{7.1}$$

Table 7.1 Magnetite and maghemite physical and magnetic characteristics

Characteristics	Magnetite	Maghemite
Chemical formula	Fe_3O_4	γ-Fe_2O_3
Color	Black	Brown-red
Density (g/cm^3)	5.18	4.87
Melting temperature (°C)	1,583–1,597	–
Dourness	5.5	5
Magnetism type	Ferromagnetic	Ferromagnetic
Curie temperature (K)	858	820–986
Saturation magnetization (emu/g)	92–100	60–80
Crystallization system	Cubic	Cubic or tetragonal
Structural type	Inverse spinel	Spinel with flows
Network parameter (nm)	$a = 0.8396$	$a = 0.83474$ (cubic)
		$a = 0.8347$, $c = 2.501$ (tetragonal)

The complete precipitation of the magnetite should be produced at pH values from 8 to 14, in a non-oxidative media, since magnetite is sensitive to oxidation resulting in maghemite according to the following reaction:

$$2Fe_3O_4 + \frac{1}{2}O_2 \rightarrow 3\gamma Fe_2O_3 \tag{7.2}$$

The precipitation process consists of two steps [10, 11, 58–61]:

- Nucleation, when the species concentration reaches supersaturation;
- Slow increase of the nucleus.

7.2.2.2 Sol-Gel Method

This method consists of hydrolysis and condensation of some precursors in solution that results in obtaining a sol of nanometric particles. Organic condensation leads to the formation of a tridimensional network of metal oxides, called wet gel. Since this reaction occurs at room temperature, thermic treatments are required in order to obtain the final crystalline structure [62, 63].

Gamarra et al. [64] precipitated iron oxyhydroxide (FeOOH) in water in the presence of a surfactant, after which they reduced the Fe^{3+} at Fe^{2+} partially, by a light drying in N_2 atmosphere, obtaining at the end magnetite particles.

This method has certain advantages like [65]:

- Depending on experimental conditions materials with pre-established structure can be obtained;
- The possibility of obtaining some amorphous pure phases and the possibility of obtaining single dispersed particles with a good dimension control;
- The possibility of encapsulating of the iron oxide in different matrices, maintaining the properties and stability of the particles.

7.2.2.3 Microemulsion Method

A microemulsion is an isotopic thermodynamically stable dispersion of two liquids that are immiscible (water and oil), in the presence of surfactant, which forms a film at the interface between oil and water, having the hydrocarbon chain, non-polar, dissolved in oil and the polar group in the aqueous phase [66].

Vidal-Vidal et al. [67] obtained spherical single dispersed maghemite particles covered with oleilamine or oleic acid, which presents a narrow distribution of the dimension between 0.6 and 3.5 nm, with magnetization saturation very high (76.3 Am^2/kg, for uncoated nanoparticles; 35.2 Am^2/kg, for magnetic nanoparticles coated with oleic acid, and 33.2 Am^2/kg for magnetic nanoparticles coated with oleilamine). The results demonstrated that oleilamine acts as a precipitation and bonding agent. Chin and Yaacob [68] demonstrated that by the synthesis of the nanoparticles of iron oxide using microemulsion method, nanoparticles under 10 nm can be obtained.

This method is difficult to control, and the efficiency is low compared with other synthesis methods for magnetic nanoparticles, and the obtained particles are polydispersed.

7.2.2.4 Laser Pyrolysis

Laser pyrolysis from gas phase was demonstrated to be an essential procedure in the synthesis of uniformed iron nanoparticles, where the reactants are diluted, which results in the obtaining of fine particles with narrow distributions for the dimension of the particles and the controlled purity, without congesting them [38, 39].

The method for obtaining the magnetic particles by laser pyrolysis was initiated by Haggerty in 1981 for preparing ultrafine silicon powder [69].

The nanoparticles obtained by laser pyrolysis presents a core-shell structure of Fe/FeO, in which the thickness of the shell is not dependent on initial conditions.

The mean diameter of the particles depends on the laser power and it varies between 4 and 11 nm. In the case of FeO particles with the diameter between 9 and 11 nm, the saturation magnetization of the nanopowder reaches 80 Am^2/kg [70].

Core-shell nanoparticles covered with carbon layer present a special interest for biomedical utilization.

The laser pyrolysis synthesis method is one of the most promising with respect to the industrial scale production of nanopowders with the diameter between 5 and 20 nm.

7.2.2.5 Combustion Method

Combustion method involves a strong exothermal redox reaction, between an oxidant agent and diverse organic reducing agents. Initiation of combustion process develops through rapid heating of the mixture, at relatively decreased temperature, below 500 °C.

Compared to other methods, combustion method has certain advantages, such as: is a simple method, short reaction time, low energy consumption, and friendly with the environment. Moreover, the end product is directly obtained as a result of combustion, without further calcination, so that without supplementary energy consumption.

Increased interest for this unconventional synthesis method is due to diverse range of variables, through which the combustion processes can be conducted and directed so that the reaction products can be fitted within large limits.

Factors influencing self-propagated combustion reaction are [71, 72]:

- Nature of oxidant agent and combustion agent;
- Molar ratio between combustion agent/oxidant;
- Presence of additives with auxiliary roles;
- Initiation temperature and heating velocity;
- Volume of raw matter mixtures;
- Water amount (water) from raw matters;
- Pressure.

Mukasyan et al. [73] showed that maximum temperature reached during the reaction and reaction duration are two main elements controlling the properties of the resulting powder, mainly the amorphous or crystalline characteristics and size of particles from the reaction product. The difference between particles size obtained by using different combustion agents is explained by different combustion gases amount which is released by different exothermal reactions, and different temperature from the reactant system [74].

Li et al. [75] showed that for different metallic cations, organic combustion agents with different functional groups have different complexion power. This will influence both formation and morphology of the desired product.

McKittrick [71] and Jung [76] declare that utilization of larger amount of combustion agent will result in increased combustion temperature developed during the reaction.

Reduced size of particles is due to increased gases volume, which will induce sample expansion, thus impairing development of particles synthesis and growing in size. Considering this aspect, Ozuna et al. [77] formulated the hypothesis that higher the pressure in raw matter is, higher will be the specific surface of the resulting powder, because the oxidation process is strongly exothermal, and the reaction time is very reduced (~1 s); pressure within the system increases even more, and the resulting combustion gases disintegrate the structure of solid material, thus contributing to significant reduction of particle size and spectacular increase of specific surface.

7.2.2.6 Hydrothermal Method

Hydrothermal method includes diverse wet-chemical technologies for crystallization of substances mixture in a close system (autoclave). Hydrothermal method is successfully used when an increase of iron oxide nanoparticle crystals is desired.

Synthesis of iron oxide magnetic nanoparticles requires high temperatures (above 200 °C) and high pressures (0.3–4 MPa), using two main routs: hydrolysis and oxidation or neutralization of metallic hydroxides mixture. The most important parameters of the process are the solvent, temperature, and reaction time [78].

Chen and Xu [78] demonstrated that the size of Fe_3O_4 particles are increasing as the reaction time increases.

Zheng et al. [79] obtained Fe_3O_4 particles using the hydrothermal method, with particle size of 27 nm, using sodium bis(2-ethylhexyl)sulfosuccinate as surfactant. The magnetite nanoparticles obtained are endowed with superparamagnetic behavior at room temperature.

Wang et al. [80] present obtaining of Fe_3O_4 nanoparticles with high crystalline feature. The nanoparticles size obtained using this method at 140 °C for 6 h was 40 nm, saturation magnetization was 85.8 emu/g, hardly decreased compared to bulk Fe_3O_4 bulk (92 emu/g).

Daou et al. [81] reported obtaining monodispersed magnetite nanoparticles of 39 nm in size, synthesized first by co-precipitation method at 70 °C, followed by a second synthesis phase—hydrothermal treatments at 250 °C. Magnetite nanoparticles obtained by co-precipitation method were 12 nm in size, being oxidized in contact with air.

7.2.2.7 Sonochemical Method

Sonochemical method was largely used in generation of new materials with unusual properties. Chemical effects of ultrasounds seem to be acoustic cavitation, which means that the bubbles are forming, grow, and are implosively falling into fluid, which generates a localized hot spot, as gaseous phase. This method was applied in synthesis of nanocomposites, and its versatility was demonstrated in case of preparing the iron oxide nanoparticles [55]. Vijayakumar et al. [82] reported obtaining of magnetite nanoparticles using the sonochemical method by sonication of iron acetate in water, in absence of air. Particles obtained by this method were 10 nm in size, presenting superparamagnetic behavior and very low saturation magnetization at room temperature (below 1.25 emu/g).

Pinkas et al. [83] obtained amorphous iron oxide with very large specific surface using sonochemical method. They sonicated $Fe(acac)_3$ in small amount of water, in argon atmosphere. Organic content and specific surface of Fe_2O_3 can be controlled by water amount within the reaction mixture. Thus, nanoparticles of 48 m^2/g specific surfaces were obtained for using a solvent and 260 m^2/g when working within humid argon atmosphere.

7.2.3 Stabilization/Functioning of Magnetic Nanoparticles

Although within iron oxide magnetic nanoparticles synthesis there are significant advances, maintenance of their stability over a longer time period, without their agglomeration or precipitation, is a very important issue. Iron oxide magnetic nanoparticles stability is required for almost any application, so that development of effective strategies related to improvement of chemical stability is mandatory.

Agglomeration of magnetic nanoparticles is related to van der Waal's and magnetic forces. Van der Waal's interaction occurs due to fluctuations of electron orbitals from one particle, which induces oscillatory dipoles in a neighbor particle. The simplest and direct method seems to be coating of Fe_3O_4 magnetic nanoparticles with an impenetrable cover, so that the oxygen cannot reach the surface of magnetic particle.

Coating strategies applied to magnetic nanoparticles can be divided into two main groups:

- Cover of magnetic nanoparticles with organic compounds, including surfactants [84–87] and polymers [88–91];
- Coating of magnetic nanoparticles with inorganic compounds, including silica gel [92–94], carbon [95, 96], and precious metals (Au [97, 98], Ag [99]).

Surfactants should have functional chemical moieties capable of interacting with hydroxyl groups on the surface of preformed magnetite particles (hydrogen or covalent bonds) and should be stable within media imposed by application fields.

In case of ferrofluids, the main factors providing their stability are: shape, particle size, and chemical structure of the coating layer, responsible of compatibility with dispersion environment. There are three methods for impairing the contact between magnetic nanoparticles and reduction of dipole–dipole interaction: steric stabilization, electrostatic stabilization, and mix stabilization [100].

Generally, surfactants or polymers can be chemically anchored of physically adsorbed on the surface of magnetic nanoparticles, in single layer or double layer, thus creating repulsive forces in order to balance the van der Waal's and magnetic forces which are acting on the surface of magnetic nanoparticles.

Most usual functional groups which can bind on the surface of magnetite are phosphates, sulfates, and carboxylates [10]. Carboxyl group of oleic acid ($CH_3(CH_2)_7CH{=}CH(CH_2)_7COOH$) is involved in formation of hydrogen bonds with hydroxyl groups from the surface of magnetite particles, and thus the coating of particles and their stabilization against agglomeration. Magnetite particles covered with oleic acid are used for obtaining of magnetic fluids based on hydrocarbons [101, 102].

Willis et al. [103] showed that degradation of oleic acid during thermal decomposition, method used for obtaining iron oxide nano-crystallite, will result in formation of high-quality γ-Fe_2O_3 nano-crystals.

Fauconnier et al. [104] investigated adsorption of citric and gluconic acid on surface of maghemite particles for further use in biomedical application.

Polyethylene glycol (PEG) is a hydrophilic polymer, water-soluble, biocompatible, which can be used in synthesis of biocompatible nanoparticles with increased resistance within blood circulation [105].

Another alternative for magnetite particles coating is represented by use of co-polymers, which conduct to particles core-shell, which possible applications in drug transport and delivery (drug vector).

Kumagai et al. [106] developed a simple method for nanoparticles synthesis, followed by their treatment with copolymer block polyethylene glycole-polyaspartic acid. The nanoparticles obtained by this method are endowed with increased stability and solubility in aqueous solutions and biological media.

Koneracka et al. [107] synthesized double-layer coated magnetite nanoparticles for further dispersion in water. Primary surfactant was sodium oleate ($C_{17}H_{33}COONa$), while secondary surfactant was polyethylene glycol which is a biocompatible surfactant. Magnetic nanofluid is used in generation of polymeric nanospheres containing cytostatic drugs.

Moeser et al. [108] prepared magnetite nanoparticles covered with a bifunctional polymer composed of polypropylene oxide, as primary surfactant, and polyacrylic acid anchored with polyethylene oxide chains, as secondary surfactant. Nanoparticles obtained by this method were used in separation of organic compounds from aqueous media.

Utilization of inorganic compounds, such as gold, silver, silica gel, carbon, as surfactants, not only that provides a good stability of the particles, but also allows functionalization of their surface due to engraftment of certain biological ligands.

Silica gel is the most used compound for preparation of iron oxide nanoparticles with functionalized surface, because has few advantages: excellent biocompatibility, hydrophilic ability, integration of other functional groups on its surface, stabilization of iron oxide magnetic nanoparticles in solutions, prevents interactions between particles and their agglomeration, thus providing a better encapsulation [109, 110].

The most common method for synthesis in obtaining magnetic nanoparticles covered with silica is Stöber method [111].

Generally, silica layer increases particle size, so that magnetic properties will be changed. However, the thickness of silica layer can be adjusted by changing TEOS: water ratio, ammonium concentration, hydrolysis time [111].

Many studies described the role of functional groups, which control reactivity and colloidal properties of magnetic suspension, as well as the influence of alkaline reagents, concentration of coated nanoparticles, water/alcohol ratio, or concentration of TEOS on final morphological aspect of these nanostructures [92, 93, 112, 113].

Magnetic nanoparticles using gold as surfactant seem to be ideal, due to its decreased reactivity; however, direct coating of magnetic nanoparticles with gold is very difficult, due to different characteristics of the two surfaces [114–116].

Good coating with carbon layers provides an effective barrier against oxidation and acid erosion of magnetic nanoparticles. So that, it is possible to synthesize magnetic nanoparticles covered with carbon, which are stable from the thermal and biocompatibility point of view, and which are presenting an increased stability against oxidation, which is crucial for certain application fields [117].

7.2.4 *Applications of Magnetic Nanoparticles*

Increased interest for magnetic nanoparticles can be explained due to diverse applications of these compounds. The fields in which these particles are used are: biomedical, catalysis, and industrial field. Magnetite particles (Fe_3O_4) dispersed in a fluid were largely used as ferrofluids [118, 119] in diverse applications, such as: electric transformers, pressure transducers, inertial sensors (acceleration, slope, gravity) [120], position sensors [121], devices for information storage [122], heat transfer [123], optics [124, 125], electronics [126], and biomedical engineering [22, 127–130].

7.2.4.1 Applications in Biomedical Field

In case of biomedical applications, colloidal suspensions, obtained because of magnetic nanoparticles dispersion in biological media, should have colloidal stability over a long period. The magnetic core of the particle should respond to an external magnetic field, so that it could be directed and positioned in a certain location, thus facilitating Magnetic Resonance Imaging (MRI) for medical diagnosis, as well as antitumor therapies assisted by alternative magnetic field.

Magnetite and/or maghemite nanoparticles are the most desired for biomedical applications due to their strong ferromagnetic behavior, relatively decreased toxicity, decreased sensitivity to oxidation, as well as increased values of saturation magnetization, compared to other materials (cobalt, nickel, more susceptible to oxidation, with increased toxicity).

Regarding the use of nanoparticles in medical applications, these can be grouped in two main categories: in vivo and in vitro applications. In vivo applications are especially based on diagnostic procedures (MRI) and therapeutic applications (hyperthermia, targeted drug delivery).

Main utilization of magnetic nanoparticles for in vitro experiments is related to diagnosis (cellular separation and selection [131–134], and magnetic relaxometry [135, 136]).

Magnetic resonance imaging (MRI) is a non-invasive technique, without exposure to radiations, which can give transversal imaging within solid material and living organisms [137, 138]. Development of MRI as clinical diagnostic tool largely contributed to pharmaceutical products advance, generating so-called magneto-pharmaceutical products. The purpose of these magneto-pharmaceutical products, in case of clinical use, is to increase the contrast between damaged and healthy tissue, and/or to indicate the function of an organ or blood vessels [139].

These magneto-pharmaceutical products were introduced for the first time as contrast agents, used in magnetic resonance imaging, for localization and diagnostic of brain injuries, myocardial infarction or liver lesions/tumors, where the magnetic nanoparticles have a tendency to highly accumulate, due to differences between tissue composition and endocytotoxic cellular processes [140–143].

Hyperthermia is a therapeutically procedure used for increasing body temperature in a certain region (between 41 and 46 °C, especially in cancer therapy) [144–149]. This technique can be used together with other antitumor treatments (chemotherapy, radiotherapy, and immunotherapy).

Increasing the temperature required for hyperthermia can be accomplished using fine iron oxide magnetic particles. Using an external magnetic field, these particles can be transported and can target the tumor cells. The advantage of hyperthermia is that induces heating of localized magnetic particles and the surrounding tissue, which is enough for destroying the tumor cells, more sensitive to temperature variations.

The first attempt for targeted drug therapy on humans was reported by Lübbe et al. [150], when they used magnetic nanoparticles coated with epirubicin [151] for treatment of solid tumor. Treatment procedure consisted of intravenous infusion of this mixture (magnetic nanoparticles coated with drug) followed by one cycle of chemotherapy. During perfusion and 45 min after, a magnetic field was constructed as close as possible to the tumor site, and they demonstrated that epirubicin-carrying magnetic nanoparticles were successfully directed and transported toward the tumor area.

Effectiveness of this therapy is dependent upon the intensity of magnetic field and the properties of magnetic nanoparticles which are used. Dependent on administration route of drug-carrying magnetic nanoparticles (intravenous, intra-arterial), a series of important parameters should be considered: blood flow, infusion pathway, circulation time, distance from the magnetic source, strength of bound between magnetic nanoparticle-drug, tumor volume, etc. [128, 152].

7.3 Experimental Data

7.3.1 Introduction

In recent years, the design and synthesis of colloidal magnetic suspensions have attracted an increased interest especially in the fields of biotechnology and biomedicine because they have many applications including targeted drug delivery, cell labeling and magnetic cell separation, hyperthermia, tissue repairing, magnetic resonance imaging (MRI) contrast enhancement, enzyme immobilization, immunoassays, protein purification, etc.

Magnetic nanoparticles (MNPs) used in biomedicine must meet several requirements. They have to be non-toxic, chemically stabile, uniform in size, well stabilized under physiological conditions, biocompatible and to present high magnetization. Magnetite (Fe_3O_4) and maghemite (γ-Fe_2O_3) are the most suitable iron oxide nanoparticles employed for biomedical applications because they are biocompatible, have low toxicity in the human body and show a superparamagnetic behavior.

Therefore, synthesis of magnetic iron oxide nanoparticles with tailored properties has attracted considerable scientific and technological interest. Various synthesis routes were developed for producing magnetic particles, such as co-precipitation, microemulsion method, thermal decomposition of different organic precursors, spray pyrolysis, and sol-gel method.

In the recent years, combustion synthesis has been often reported as a useful method for the preparation of metal oxide nanopowders. Still, there are very few papers dealing with combustion synthesis of Fe_3O_4 nanopowders and virtually there are no studies on the in vitro toxicity of the as-prepared iron oxide nanoparticles. In 2012, Ianoş et al. [46] reported a simple combustion technique for the preparation of Fe_3O_4 nanopowders, which is based on conducting the combustion reaction inside a round bottom flask and evolving gases are bubbled in a beaker filled with water.

For biomedical applications, magnetic iron oxide nanoparticles must be dispersed in biocompatible media in order to obtain stable colloidal suspensions. In order to prevent the particle aggregation and to improve the biocompatibility and stability, nanoparticles are coated with various surfactants: poly(ethylene glycol), oleic acid, poly(acrylic acid), gluconic acid, other polymers, liposome and fatty acids.

In this study, we investigated the influence of colloidal suspensions of magnetic iron oxide nanoparticles on tumor (breast cancer cell line SK-BR-3) and normal (adult bone marrow-derived mesenchymal stem cells (MSCs)) cell lines cultivated in vitro conditions. The MNPs used for the preparation of colloidal suspensions were synthesized using a new version of the combustion method.

As far as we know, this could be the first evaluation of the toxic effects of combustion synthesized magnetic iron oxide nanoparticles. For comparison, magnetite nanoparticles were also prepared by the well-known co-precipitation route. The influence of different concentrations on MNPs on cellular morphology, ultrastructure expression of phenotypical markers, and viability were examined in order to establish whether the combustion synthesized nanoparticles may be used for in vitro and in vivo applications.

7.3.2 *Results and Discussion*

7.3.2.1 Characterization of Fe_3O_4 Nanoparticles

The main characteristics of iron oxide nanopowders prepared by combustion synthesis (powder 1) and co-precipitation (powder 2) are shown in Table 7.2.

Table 7.2 Characteristics of Fe_3O_4 nanoparticles prepared by solution combustion synthesis [25] (powder 1) and co-precipitation (powder 2), respectively

Powder no.	XRD phase composition	D_{XRD} (nm)	S_{BET} (m^2/g)	D_{TEM} (nm)	M_s (emu/g)	M_r (emu/g)	H_c (kA/m)
1	Fe_3O_4	10	106	15–20	55.3	3.3	3.0
2	Fe_3O_4	9	98	10–15	61.3	0.8	0.5

D_{XRD} crystallite size, S_{BET} specific surface area, D_{TEM} particle size from TEM, M_s saturation magnetization, M_r remanent magnetization, H_c coercivity

One can easily notice that that the two powders are very similar in terms of phase composition (single-phase Fe_3O_4), crystallite size, and specific surface area and particle size.

Considering the magnetic properties of the resulted powders one may notice that the saturation magnetization of the sample prepared by combustion synthesis is slightly smaller than the sample prepared by co-precipitation (Table 7.2). At the same time, the remnant magnetization and the coercivity of combustion synthesized Fe_3O_4 are a little bit larger than the co-precipitated Fe_3O_4, yet very close to the superparamagnetic behavior.

7.3.2.2 Structural Characteristics of Colloidal Suspensions Based on Fe_3O_4 Nanoparticles

After coating the nanoparticles with a double layer of oleic acid, the stabilized nanoparticles were dispersed in phosphate buffer saline (PBS), leading to two sets of stable colloidal fluids, termed 1F (deriving from powder 1) and 2F (deriving from powder 2). TEM analysis conducted on samples 1F and 2F, evidenced that the actual size of combustion synthesized Fe_3O_4 nanoparticles is around 15–20 nm (Fig. 7.3), whilst the co-precipitated magnetite particles have 10–15 nm. Both types of nanoparticles have a sphere-like shape.

The intensity distribution of particle size revealed that the colloidal suspension deriving from combustion synthesized magnetite (sample 1F) has virtually a single family of particles and an average hydrodynamic diameter of 107 nm (Fig. 7.2). On the other hand, the colloidal suspension deriving from the co-precipitated magnetite (sample 2F) has a bimodal particle size distribution, which suggests the presence of two populations of particles (Fig. 7.4).

The average hydrodynamic diameter of sample 2F (61 nm) is smaller when compared to sample 1F. However, in both cases, the hydrodynamic diameter is larger than the core size of Fe_3O_4 nanoparticles observed in TEM images (Fig. 7.3), which is probably due to the presence of oleic acid double layer as well as the possible aggregate formation.

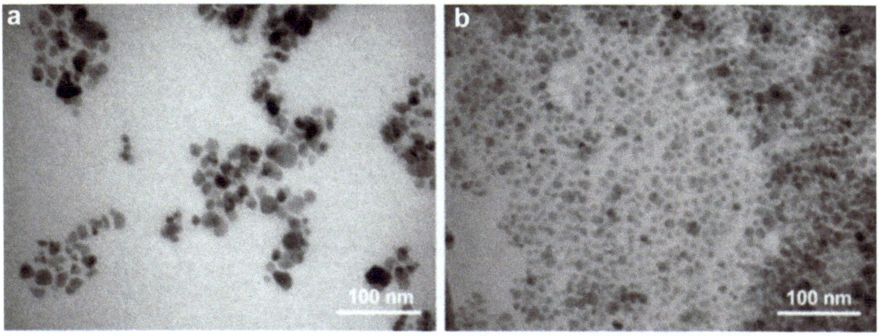

Fig. 7.3 TEM images of colloidal suspensions 1F (**a**) and 2F (**b**)

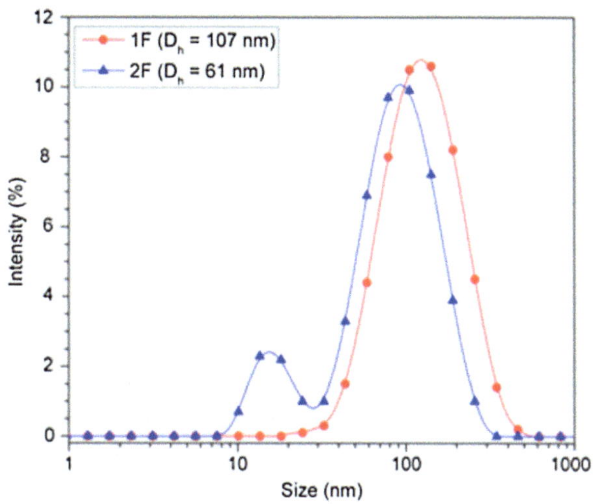

Fig. 7.4 The intensity distribution of particle size (DLS) of suspensions 1F and 2F

7.3.2.3 Ultrastructural and Morphological Cell Changes Induced by Fe_3O_4-Based Colloids

Optic microscopy did not reveal any morphological changes of adult bone marrow-derived mesenchymal stem cells (MSCs) but scanning electron microscopy showed several changes in cellular morphology, of both normal (MSCs) and tumor cells (breast cancer cell line SK-BR-3).

Bone marrow-derived mesenchymal stem cells (MSCs) are fibroblast-like, presenting thin (filopodia) and thick (lamellipodia) elongations for inter-cellular contact and adhesion to substrate (Fig. 7.5a). When fluid 1F was added in culture media for 48 h, MSCs developed cellular protrusions—microtentacles, which is usually associated with increased adhesion capacity (Fig. 7.5b). When fluid 2F was added in culture, MSCs seem to lose their filopodia, become flattened and irregular

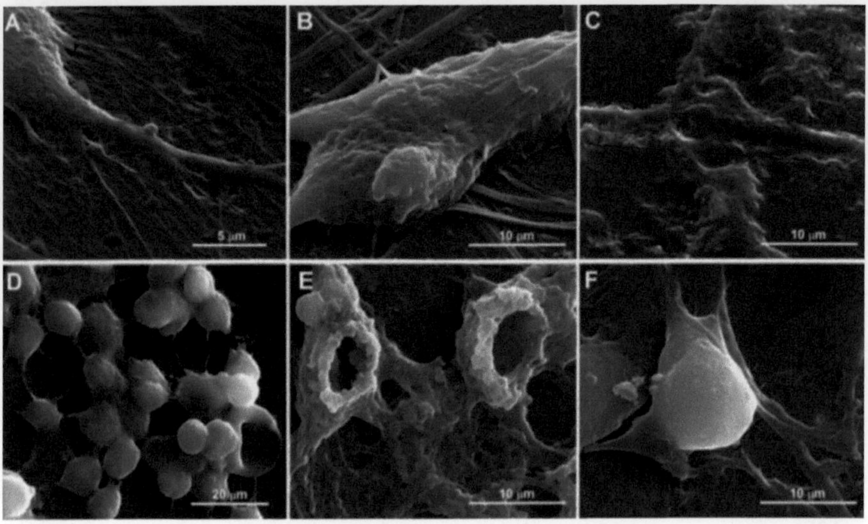

Fig. 7.5 SEM images of MSCs and SK-BR-3 cells before and after the treatment with colloidal suspensions: (**a**) control (untreated) MSCs, (**b**) MSCs treated with suspension 1F, (**c**) MSCs treated with suspension 2F, (**d**) control (untreated) SK-BR-3 cells, (**e**) SK-BR-3 cells treated with suspension 1F, (**f**) SK-BR-3 cells treated with suspension 2F

in morphological appearance, while the magnetite nanoparticles are revealed covering the cell surface (Fig. 7.5c).

Morphological characteristics of SK-BR-3 cells are depicted in Fig. 7.5d, showing the round shape, small diameter and cluster-like growth of this cellular type in vitro, which are forming more like a network of cellular elongations and contact points, with increased deposition of extracellular matrix within the resulted mesh. SK-BR-3 tumor cells presented a very unusual behavior when left in contact with fluid 1F (deriving from combustion synthesized nanoparticles) for 48 h, extruding the nucleus, so that the cells were enucleated (Fig. 7.5e)—which is a very rare phenomenon.

When fluid 2F (deriving from co-precipitated nanoparticles) was added on these tumor cells, they developed thicker elongations and anchorage structures, long, and less interconnected with the surrounding cells (Fig. 7.5f).

Transmission electron microscopy was performed on both cellular types, adherent on cell culture inserts and suspension cells. Ultrastructural description of MSCs focuses on segmented nucleus, multiple elongations, numerous mitochondria and lysosomes, rare endoplasmic reticulum and intracytoplasmic vacuoles (Fig. 7.6a, b).

SK-BR-3 tumor cells are characterized by large nucleus, multiple lipid vacuoles, well represented endoplasmic reticulum distributed along the entire cytoplasm, polyribosomes and mitochondria, showing intense metabolic processes these cells are involved in (Fig. 7.6c, d). Addition of fluid 2F (deriving from co-precipitated nanoparticles) on SK-BR-3 cells determined some morphological changes, inducing accumulation of MNPs at the cytoplasmic level, in large clusters (Fig. 7.7a, b),

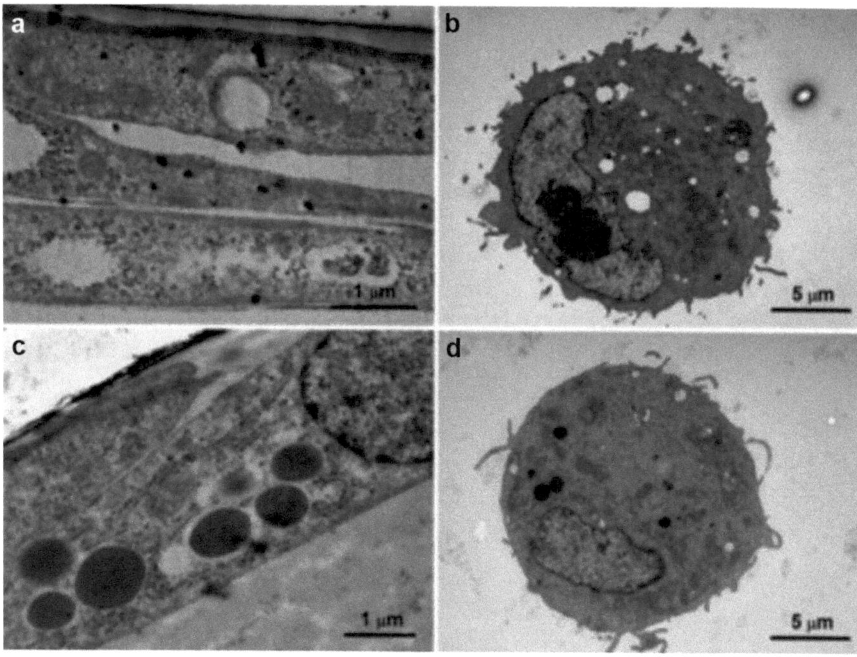

Fig. 7.6 TEM images of untreated control cells: (**a**) adherent MSCs grown in cell culture inserts, (**b**) suspension of MSCs, (**c**) adherent SK-BR-3 cells cultured on cell culture inserts, (**d**) suspension of SK-BR-3 cells

decreased number of mitochondria and occurrence of large vacuoles within the lysosomal structures.

When fluid 1F (deriving from combustion synthesized nanoparticles) was added in culture media, most of the SK-BR-3 (>90 % of the cells, counted on five different TEM images) cells appeared without nucleus (as seen on the SEM image—Fig. 7.5e) and less of the cytoplasmic structures were evidenced. A spiral-like concentrically oriented pattern was distinguished within the entire cytoplasm, revealing the stress fibers inducing the morphological change of cytoskeleton (Fig. 7.7c, d).

Interestingly, the enucleation phenomenon occurred only in the case of SK-BR-3 tumor cells treated with the fluid deriving from combustion synthesized Fe_3O_4 nanoparticles, which might indicate a certain degree of toxic selectivity of these nanoparticles with respect to SK-BR-3 tumor cells. Although the use of superparamagnetic iron oxide nanoparticles (SPIONs) in cancer therapy (hyperthermia) has been widely investigated [1–6, 8, 28–36, 153], no reports were published about the potential use of combustion synthesized Fe_3O_4 nanoparticles in cancer therapy due to their intrinsic selective action on SK-BR-3 tumor cells.

Suspension fixed MSCs treated with the two colloidal suspensions showed the entire inner architecture disturbed, with large lysosomes accumulating the nanoparticles, suggesting an accentuated endocytosis process for removal of iron

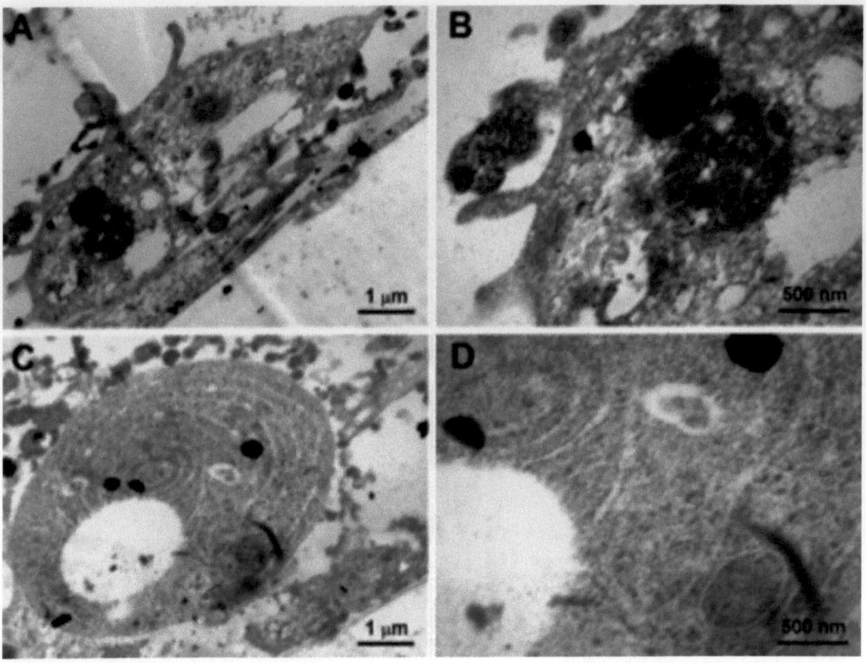

Fig. 7.7 TEM of SK-BR-3 cells treated with colloidal suspensions: (**a**, **b**) SK-BR-3 cells treated with fluid 2F (deriving from co-precipitated nanoparticles), (**c**, **d**) SK-BR-3 cells treated with fluid 1F (deriving from combustion synthesized nanoparticles)

oxide nanoparticles (Fig. 7.8a–d). There is no evidence of stress fibers within the cytoplasm, and the polyribosomes are well represented for both treatments used on these cells (Fig. 7.8a–d).

7.3.2.4 Immunophenotypical Cell Changes Induced by the Addition of Fe_3O_4-Based Colloids

MSCs were stained for cytoskeleton protein vimentin, which does not change the expression pattern within the cytoplasm of MSCs control (Fig. 7.9a), MSCs treated with fluid 1F (Fig. 7.9b) or MSCs treated with fluid 2F (Fig. 7.9c). Morphological analysis shows that MSCs in contact with fluid 1F became more elongated, stretched and fusiform (Fig. 7.9b). SK-BR-3 cells presented initially an increased expression of Her2 oncoprotein (Fig. 7.9d). The expression of this marker was absent/diminished in the case of SK-BR-3 cells treated with suspensions 1F (Fig. 7.9e) and 2F (Fig. 7.9f).

Flowcytometric evaluation of characteristic stem cell markers CD90 and CD73 showed that expression of these molecules on MSCs surface was maintained on cells treated with the suspension 2F deriving from co-precipitated nanoparticles, and decreased significantly on MSCs treated with fluid 1F, prepared from

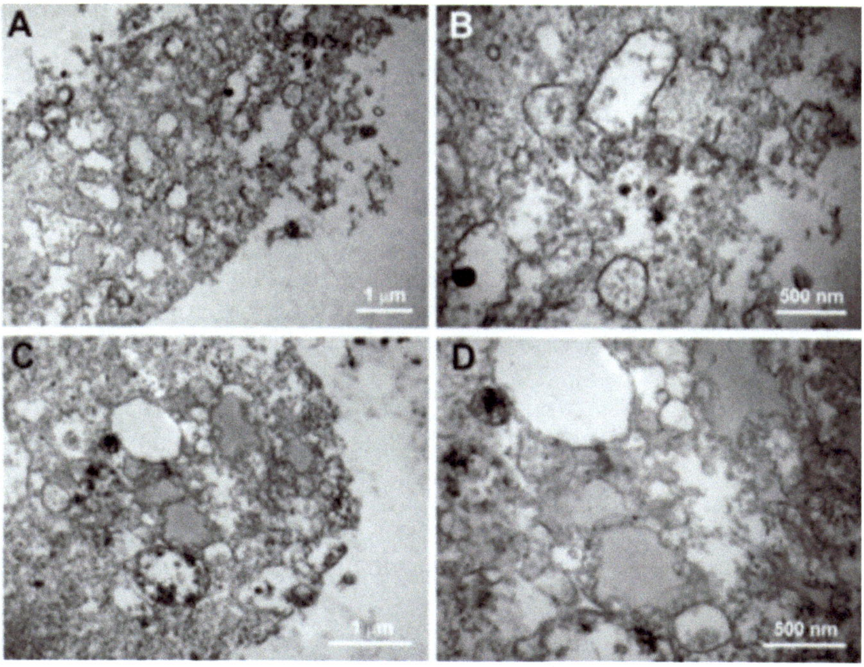

Fig. 7.8 TEM of MSCs treated with colloidal suspensions: (**a**, **b**) MSCs treated with fluid 2F (deriving from co-precipitated nanoparticles), (**c**, **d**) MSCs treated with fluid 1F (deriving from combustion synthesized nanoparticles)

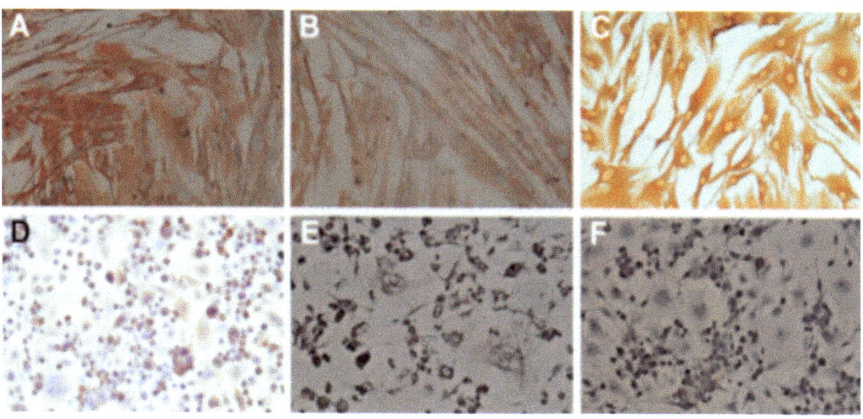

Fig. 7.9 Immunocytochemistry on MSCs and SK-BR3 cells stained for Vimentin and Her2

combustion synthesized nanoparticles. Similar pattern was obtained in case of the other molecule investigated—CD29, which is an adhesion molecule, involved in migratory capacity of these cells (Fig. 7.10).

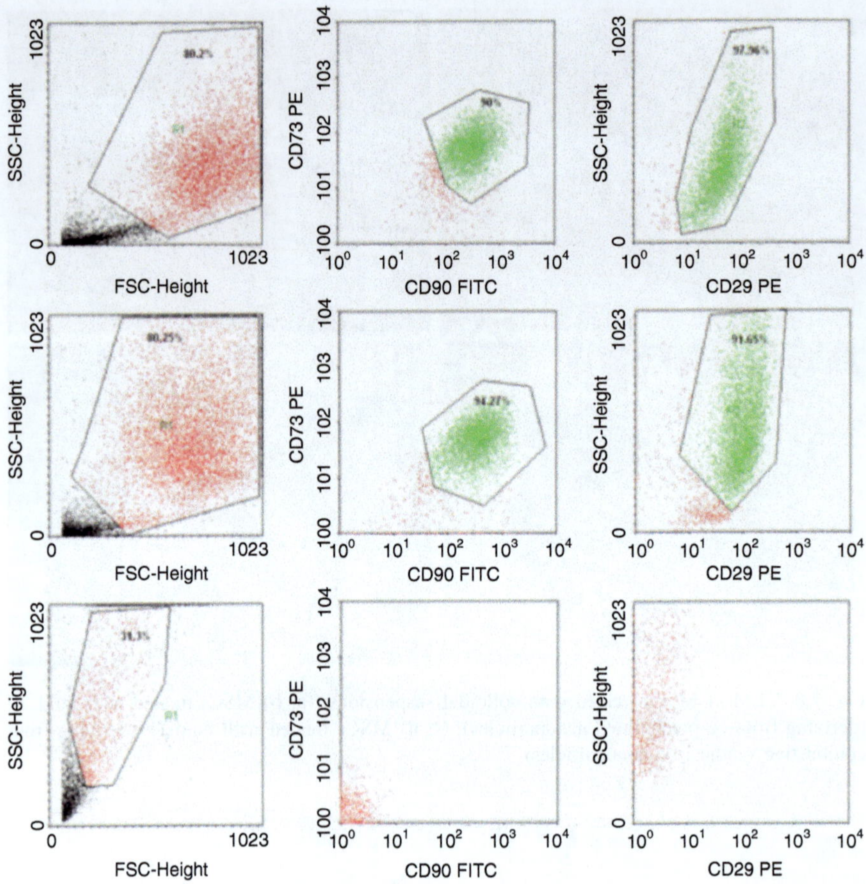

Fig. 7.10 Flowcytometric analysis of MSCs before and after treatment with colloidal suspensions: MSCs control (untreated cells), MSCs treated with fluid 2F, and MSCs treated with fluid 1F

SK-BR-3 tumor cells treated with each of the two colloidal suspensions increased their aggressive potential by highly expressing Her2 oncoprotein on the cellular surface (Fig. 7.11). Another interesting aspect revealed by the flowcytometric analysis is the translation of the SK-BR-3 cells to the left panel of FCS axis and upper part of the SSC axis, suggestive for decreased size and increased granularity of these cells treated with magnetite-based colloids. Data are correlated with TEM images showing accumulation of nanoparticles within the cytoplasm.

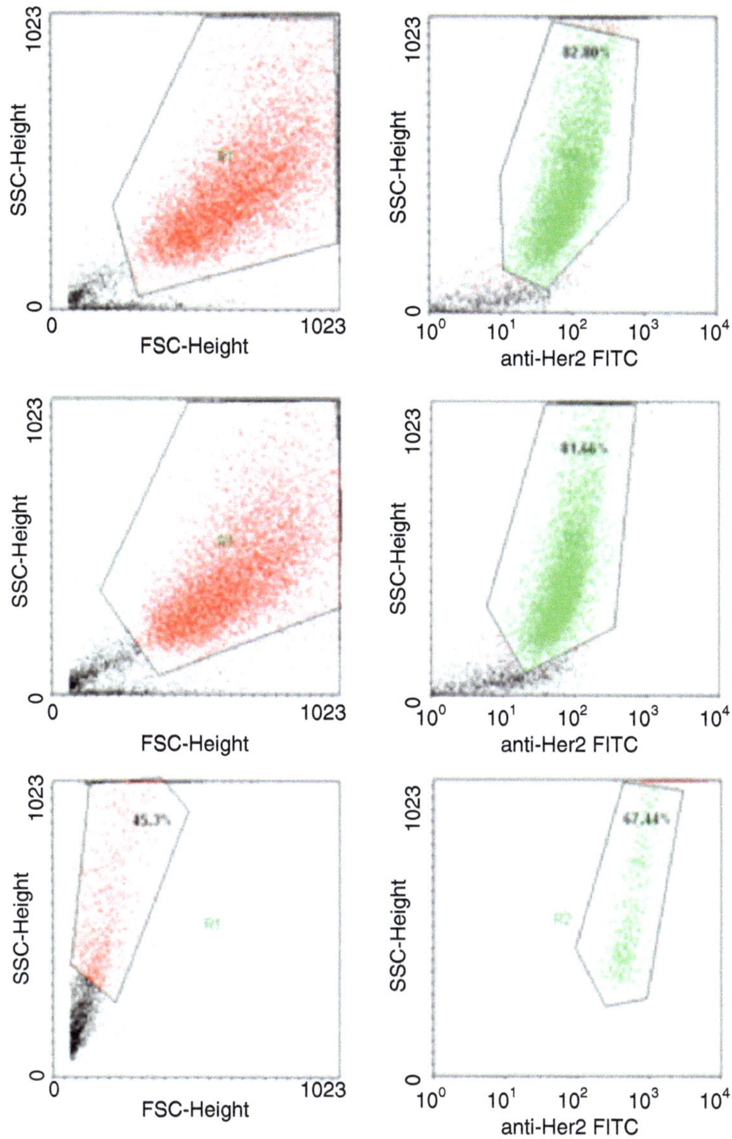

Fig. 7.11 Flowcytometric analysis of SK-BR-3 cells before and after treatment with colloidal suspensions: SK-BR-3 control (untreated cells), SK-BR-3 cells treated with fluid 2F, and SK-BR-3 cells treated with fluid 1F

7.3.2.5 Toxic Effects of Fe_3O_4-Based Colloids

Annexin V-FITC/PI assay was employed to compare the viability of normal MSCs and SK-BR-3 tumor cells after 48 h of co-culture with the two colloidal suspensions

Table 7.3 Results of Annexin V-FITC/PI assay—viability of MSCs and SK-BR-3 tumor cells treated with magnetite-based colloidal suspensions

Cellular type	Substance	Live cells (%)	Apoptotic cells (%)	Dead cells (%)
MSC	Control (untreated)	90.63	7.93	1.44
MSC	Fluid 1F	66.94	7.87	25.20
MSC	Fluid 2F	61.28	30.62	8.10
SK-BR-3	Control (untreated)	73.10	23.35	3.55
SK-BR-3	Fluid 1F	61.87	32.41	5.72
SK-BR-3	Fluid 2F	65.57	27.38	7.05

having a similar concentration of magnetite nanoparticles, ~5×10^{-2} mg/mL (Table 7.3).

The viability of MSCs treated with the colloidal suspension deriving from combustion synthesized magnetite (fluid 1F) is higher than that deriving from co-precipitated magnetite (fluid 2F). The viability of SK-BR-3 tumor cells treated with the colloidal suspension deriving from combustion synthesized Fe_3O_4 (fluid 1F) is smaller than that deriving from co-precipitated Fe_3O_4 (fluid 2F).

The results of Annexin V-FITC/PI assay—viability of MSCs and SK-BR-3 tumor cells treated with magnetite-based colloidal suspensions—support the idea that combustion synthesized nanoparticles are more active in reducing the proliferation rate of SK-BR-3 tumor cells. This behavior is in excellent agreement with the enucleation process of SK-BR-3 tumor cells observed on SEM (Fig. 7.5e) and TEM (Fig. 7.7c, d) images.

7.3.3 Experimental Procedures

Synthesis of Fe_3O_4 nanoparticles: magnetite nanoparticles were prepared by two different routes, namely solution combustion synthesis and co-precipitation respectively.

(a) Combustion synthesis: an aqueous solution containing $Fe(NO_3)_3 \times 9H_2O$ (Roth, pro analysis) and $C_6H_{12}O_6$ (Riedel de Haën, pro analysis) was heated to 400 °C in the absence of air, in a round bottom flask. As the water evaporates, a smoldering combustion reaction occurs, leading to the formation of a black powder (powder 1) which consists of nanocrystalline Fe_3O_4. A more detailed description of combustion synthesis of magnetite nanoparticles can be found in one of our papers published in 2012 [25]. The resulted black powder was hand crushed, washed with warm distilled water and dried at 80 °C.

(b) Co-precipitation: excess NH_4OH (Chimopar, pro analysis) was added under vigorous stirring to a hot (80 °C) aqueous solution of $FeSO_4 \times 7H_2O$ (Chimopar, pro analysis) and $FeCl_3 \times 6H_2O$ (Merck, pro analysis), according to the procedure reported by Bica [11, 37]. A black precipitate, of single-phase magnetite was obtained (powder 2).

Fe_3O_4 nanoparticle characterization: iron oxide nanoparticles have been characterized in terms of XRD phase composition (Rigaku Ultima IV, $Cu_{K\alpha}$, Tokyo, Japan), specific surface area (nitrogen adsorption–desorption, Micromeritics ASAP 2020 (Micromeritics Instrument Corporation, Norcross, USA)) and magnetic properties (VSM 880 ADE/DMS magnetometer (DMS/ADE Technologies, Massachusetts, USA)).

Preparation of stable colloidal suspensions: Fe_3O_4 nanoparticles prepared by the two synthesis routes have been used to prepare stable colloidal suspensions. For this purpose, the iron oxide nanoparticles prepared by combustion synthesis were sonicated for several hours and then covered with a first layer of oleic acid (Merck, 65–88 %) [11, 37]. In the case of the co-precipitation synthesized particles, the nanoparticles were coated with a first layer of oleic acid as soon as the precipitation reaction started. This step was followed by a washing process with warm fresh distilled water. Subsequently, the pH of the solution containing the magnetic nanoparticles was adjusted to 9 with NH_4OH (Chimopar, pro analysis) and a second layer of oleic acid was achieved by adding some more oleic acid. The oleic acid double layer-coated nanoparticles were dispersed in phosphate buffered saline (PBS—Sigma Aldrich), leading to stable colloidal suspensions termed sample 1F (derived from powder 1, prepared by combustion synthesis) and sample 2F (derived from powder 2, prepared by co-precipitation). The two colloidal suspensions contained a similar amount of Fe_3O_4 nanoparticles ($\sim 5 \times 10^{-2}$ mg/mL).

Colloidal suspension characterization: the resulted colloidal suspensions based on PBS were characterized by dynamic laser scattering (DLS) using a ZetaSizer NanoZS Malvern Instrument (Worcestershire, UK). A drop of diluted aqueous suspension of magnetic nanoparticles (MNPs) was placed on the grids and allowed to air-dry, and transmission electron microscopy (TEM) was performed for diluted MNPs with FEI Tecnai 12 transmission electron microscope (FEI Company, Eindhoven, NL). The in vitro toxicity of colloidal suspensions based on magnetic iron oxide nanoparticles on tumor and normal cell lines was investigated as follows.

Cell lines: unprocessed bone marrow (10 mL) obtained from 10 human adult subjects free of hematological disorders was used for isolation of mesenchymal stem cells (MSCs). Bone marrow-derived human mesenchymal stem cells (MSCs) were isolated following 2 cellular passages based on plastic adherence, fibroblast-like morphology, and were further used in our experiments. The MSCs culture and expansion media contained *alpha*-minimum essential medium (MEM; Gibco BRL, Invitrogen, Carlsbad, CA, USA), supplemented with 10 % fetal calf serum (FCS; PromoCell, Heidelberg, Germany) and 2 % Penicillin/Streptomycin mixture (Pen/Strep, 10,000 IU/mL; PromoCell), and cells were grown at 37 °C in 5 % CO_2 atmosphere. Medium replacement was performed every third days and when reaching 80–90 % confluence, the cells were passed using 0.25 % Trypsin-EDTA solution (Sigma Aldrich Company, Ayrshire, UK) followed by centrifugation (10 min, 300 g) and replated in T75 culture flasks at a density of 10,000 cells/cm^2. SK-BR-3 (breast cancer) cells were procured from American Type Culture Collection (ATCC, Manassas, VA, USA) and further maintained and expanded in McCoy's 5A medium (Gibco) supplemented with 10 % FCS (PromoCell) and 1 %

Pen/Strep solution 10,000 IU/mL (PromoCell) in an incubator, at 37 °C in a humidified and 5 % CO_2 atmosphere. Cells were seeded at 10,000 cells/cm^2 in appropriate well plates and allowed to attach for 24 h previous to MNPs addition. Bone marrow samples were obtained after signing the written informed consent elaborated under an approved protocol by the Ethics Committee of "Victor Babeş" University of Medicine and Pharmacy Timişoara, according to the World Medical Association Declaration of Helsinki.

Cells scanning electron microscopy (*SEM*): scanning electron microscopy was performed for identification of morphological changes in MSCs and SK-BR-3 cell lines induced by the colloidal suspensions. Cells were cultured at 7,000 cells/cm^2 in 24-well format cell culture inserts (BD Labware Europe, Le Pont De Claix, France) and colloidal suspensions were added 48 h after colloidal suspensions addition, cells were pre-fixed for 1 h with 2.5 % buffered glutaraldehyde (in PBS), rinsed three times in PBS, and the 0.4 μm pore-sized membranes were detached from the culture inserts. For better image quality, cells fixed on the membranes were sputter-coated with platinum-palladium and examined with a FEI Quanta 3D FEG electron microscope (FEI Company, Eindhoven, NL) generating digital electron micrographs.

Cells transmission electron microscopy (*TEM*): MSCs and SK-BR-3 tumor cells were employed for investigation of ultrastructural details changes, 72 h after addition of colloidal suspensions in different concentrations. Cells were prefixed for 1 h with glutaraldehyde (2.5 % in PBS), rinsed three times in PBS, and post-fixed for 1 h in osmium acid (2 % in PBS). Dehydration was done in graded acetone in distilled water dilutions, followed by infiltration with Epon resin. Sections of about 100 nm, obtained on a diamond knife (Diatome) with Leica UC6 ultramicrotome were post-stained with lead citrate and uranyl acetate. The grids were examined with a FEI Tecnai 12 transmission electron microscope (FEI Company).

Annexin V/PI assay: because MTT viability assay is based on spectrophotometric detection of color change due to active substrate metabolization in live cells, and the presence of suspended Fe_3O_4 nanoparticles can influence color detection, we used Annexin V-FITC/PI viability assay, which is based on fluorescence emission and is performed on flowcytometer, thus resulting in less ambiguities of results. Annexin V-FITC (Miltenyi Biotec, Gladbach, Germany) was used in cell death flowcytometric studies (apoptosis) combined with Propidium Iodide Staining Solution (BD Biosciences, San Jose, CA, USA) following the manufacturer's protocol. Shortly, 10^6 cells were washed in 1× Annexin V Binding Buffer (BD Pharmigen) and centrifuged at $300 \times g$ for 10 min, resuspended in the same solution and incubated with 10 μL of Annexin V-FITC for 15 min in the dark. After washing the cells with 1 mL specific binding buffer and centrifugation, the cell pellet was resuspended in 500 μL binding buffer and 1 μg/mL of PI solution was added immediately prior to analysis by flowcytometry.

Flowcytometric analysis: MSCs and SK-BR-3 cells in culture 80–90 % confluence were detached using 0.25 % Trypsin-EDTA (Sigma), washed two times with PBS, resuspended in 100 μL PBS at a concentration of 10^5 cells/mL and incubated in the dark at room temperature for 30 min with mouse anti-human fluorochrome-conjugated antibody at a dilution specified in the manufacturer's protocol. Cells were then washed twice with 1 mL Cell Wash Solution (BD Biosciences, San Jose, CA, USA) each and resuspended in 500 μL of the same solution for further analysis on a four-color capable FACSCalibur (Becton-Dickinson) flow-cytometer. Conjugated antibodies utilized included FITC-conjugated CD90 (BD Pharmingen™), CD44, and Her2 (Abcam), and PE-conjugated CD29 and CD73. Fluorochrome-conjugated antibodies were purchased from BD Pharmingen™ unless otherwise specified. Acquisition and data analyses were performed using CellQuest Pro software (Becton-Dickinson).

Immunocytochemistry/Immunofluorescence: immunohistochemistry was performed for MSCs and SK-BR-3 tumor cell lines. Cells prepared for these analyses were grown in 4-well glass chamber slides, were left to adhere for 24 h, and colloidal suspensions of different concentrations were added. Untreated cells were used as control. 72 h after addition of colloidal suspensions, medium was removed, cells were washed, fixed with 4 % paraformaldehyde and permeabilized with 0.1 % Triton X-100 and then investigated for expression of the proteins of interest, using for labeling the following antibodies: monoclonal mouse anti-swine Vimentin (clone V9), anti-human smooth muscle actin/HRP (clone 1A4), monoclonal anti-human endoglin, CD105 (clone SN6h), and polyclonal rabbit anti-human c-erb-2 oncoprotein, respectively. All primary antibodies were provided by DakoCytomation (Glostrup, Denmark) and tested for human specificity and cross-reactivity. Staining protocol continued with secondary biotinylated antibody binding, substrate addition (EnVision+® System-HRP DAB/AEC for use with primary antibodies; Dako), and hematoxylin counterstaining of the nuclei following the manufacturer's procedures. Microscopy analysis was performed on a Nikon Eclipse E800 microscope.

Conclusions

The influence of stable colloidal suspensions of magnetite nanoparticles (prepared by combustion synthesis as well by co-precipitation route) on tumor (SK-BR-3 breast cancer cell line) and normal (MSCs adult bone marrow-derived mesenchymal stem cells) cell lines cultivated in vitro conditions was investigated. As far as one can tell, this is the first evaluation of the toxic effects of combustion synthesized magnetite nanoparticles.

Experimental results evidenced that combustion synthesized Fe_3O_4 nanoparticles can be used and are compatible to in vitro culture conditions of both normal and tumor cells. After 48 h in culture media, combustion synthesized nanoparticles and co-precipitated nanoparticles decreased the proliferation rate of normal MSCs and SK-BR-3 tumor cells. Viability of normal MSCs treated with combustion synthesized nanoparticles was higher than the viability of MSCs treated with co-precipitated nanoparticles. At the same time, SK-BR-3 tumor cells treated with combustion synthesized

(continued)

nanoparticles had a lower viability when compared to similar cells treated with co-precipitated nanoparticles. Therefore, in terms of cell viability, the combustion synthesized Fe_3O_4 proved to be superior to the co-precipitated magnetite nanoparticles. A very unusual and rare phenomenon was observed in the case of SK-BR-3 tumor cells treated with combustion synthesized nanoparticles, as the SK-BR-3 tumor cells were enucleated and lost their adhesion abilities. On the other hand, in the presence of combustion synthesized nanoparticles normal MSCs developed anchorage structures, which made them more resistant to the chemical stress.

This remarkable behavior of combustion synthesized magnetite nanoparticles opens a whole new perspective on the potential use of combustion synthesized Fe_3O_4 nanoparticles in cancer therapy due to their selective intrinsic behavior, not only due to their superparamagnetic properties (hyperthermia)—as currently reported.

Acknowledgements We would like to thank Alina Tăculescu, PhD; Robert Ianoş, PhD; and Prof. Cornelia Păcurariu, PhD, Politehnica University Timisoara, for their remarkable contribution in characterization and synthesis of MNPs.

References

1. Leslie Pelecky DL, Rieke RD (1996) Magnetic properties of nanostructured materials. Chem Mater 8(8):1770–1783
2. Cullity BD (1972) Introduction to magnetic materials. Addison-Wesley, New-York
3. Hadjipanayis GC, Prinz GA (1991) Science and technology of nanostructured magnetic materials. Plenum Press, New-York
4. Shylesh S, Schünemann V, Thiel WR (2010) Magnetically separable nanocatalysts: bridges between homogeneous and heterogeneous catalysis. Angew Chem Int Ed 49:3428–3459
5. Huber DL (2005) Synthesis, properties, and applications of iron nanoparticles. Small 1 (5):482–501
6. Jun YW, Choi JS, Cheon J (2007) Heterostructured magnetic nanoparticles: their versatility and high performance capabilities. Chem Commun 12:1203–1214
7. Tebble RS, Craik DJ (1969) Magnetic materials. Wiley-Interscience, London
8. West AR (1988) Basic solid state chemistry. Wiley, New York
9. O'Handley RC (2000) Modern magnetic materials—principles and applications. Wiley, New York
10. Cornell RM, Schwertmann U (1996) The iron oxides: structure, properties, reactions, occurrence and uses. VCH, Weinheim, Germany
11. Cornell RM, Schwertmann U (1991) Iron oxides in the laboratory: preparation and characterization. Wiley-VCH Verlag GmbH, Weinheim
12. Jiles D (1998) Introduction to magnetism and magnetic materials, 2nd edn. Chapman & Hall, New York
13. Frankel RB, Moskowitz BM (2003) In: Miller JS, Drillon M (eds) Magnetism: molecules to materials IV: nanosized magnetic materials. Wiley-VCH Verlag GmbH, Weinheim

14. Bee A, Massart R, Neveu S (1995) Synthesis of very fine maghemite particles. J Magn Magn Mater 149(1–2):6–9
15. Kang YS, Risbud S, Rabolt JF, Stroeve P (1996) Synthesis and characterization of nanometer-size Fe_3O_4 and gamma-Fe_2O_3 particles. Chem Mater 8(9):2209–2211
16. Lee JW, Isobe T, Senna M (1996) Magnetic properties of ultrafine magnetite particles and their slurries prepared via in-situ precipitation. Colloids Surf A Physicochem Eng Asp 109:121–127
17. Kim DK, Zhang Y, Voit W, Rao KV, Muhammed M (2001) Synthesis and characterization of surfactant-coated superparamagnetic monodispersed iron oxide nanoparticles. J Magn Magn Mater 225(1–2):30–36
18. Jolivet JP, Chaneac C, Tronc E (2004) Iron oxide chemistry. From molecular clusters to extended solid networks. Chem Commun 5:481–487
19. Si S, Kotal A, Mandal TK, Giri S, Nakamura H, Kohara T (2004) Size-controlled synthesis of magnetite nanoparticles in the presence of polyelectrolytes. Chem Mater 16(18):3489–3496
20. Tartaj P, Morales MP, Gonzalez-Carreno T, Veintemillas-Verdaguer S, Serna CJ (2005) Advances in magnetic nanoparticles for biotechnology applications. J Magn Magn Mater 290:28–34
21. Wu W, He Q, Hu R, Huang J, Chen H (2007) Preparation and characterization of magnetite Fe_3O_4 nanopowders. Rare Metal Mat Eng 36(3):238–243
22. Laurent S, Forge D, Port M, Roch A, Robic C, Vander Elst L, Muller R (2008) Magnetic iron oxide nanoparticles: synthesis, stabilization, vectorization, physicochemical characterizations and biological applications. Chem Rev 108(6):2064–2110
23. Rockenberger J, Scher EC, Alivisatos AP (1999) A new nonhydrolytic single-precursor approach to surfactant-capped nanocrystals of transition metal oxides. J Am Chem Soc 121 (49):11595–11596
24. Hyeon T, Lee SS, Park J, Chung Y, Bin NH (2001) Synthesis of highly crystalline and monodisperse maghemite nanocrystallites without a size-selection process. J Am Chem Soc 123(51):12798–12801
25. Sun SH, Zeng H (2002) Size-controlled synthesis of magnetite nanoparticles. J Am Chem Soc 124(28):8204–8205
26. Hyeon T (2003) Chemical synthesis of magnetic nanoparticles. Chem Commun 8:927–934
27. Sun SH, Zeng H, Robinson DB, Raoux S, Rice PM, Wang SX, Li GX (2004) Monodisperse MFe_2O_4 (M = Fe, Co, Mn) nanoparticles. J Am Chem Soc 126(1):273–279
28. Woo K, Hong J, Choi S, Lee HW, Ahn JP, Kim CS, Lee SW (2004) Easy synthesis and magnetic properties of iron oxide nanoparticles. Chem Mater 16(14):2814–2818
29. Răileanu M, Crişan M, Petrache C, Crişan D, Zaharescu M (2003) Fe_2O_3-SiO_2 nanocomposites obtained by different sol-gel routes. J Optoelectron Adv Mater 5(3):693–698
30. Ismail AA (2005) Synthesis and characterization of $Y_2O_3/Fe_2O_3/TiO_2$ nanoparticles by sol-gel method. Appl Catal B Environ 58(1–2):115–121
31. Durães L, Costa BFO, Vasques J, Campos J, Portugal A (2005) Phase investigation of as-prepared iron oxide/hydroxide produced by sol-gel synthesis. Mater Lett 59(7):859–863
32. Dai ZF, Meiser F, Mohwald H (2005) Nanoengineering of iron oxide and iron oxide/silica hollow spheres by sequential layering combined with a sol-gel process. J Colloid Interface Sci 288(1):298–300
33. Hirai T, Mizumoto JY, Shiojiri S, Komasawa I (1997) Preparation of Fe oxide and composite Ti-Fe oxide ultrafine particles in reverse micellar systems. J Chem Eng Jpn 30(5):938–943
34. Liu C, Zou BS, Rondinone AJ, Zhang ZJ (2000) Reverse micelle synthesis and characterization of superparamagnetic $MnFe_2O_4$ spinel ferrite nanocrystallites. J Phys Chem B 104 (6):1141–1145
35. Santra S, Tapec R, Theodoropoulou N, Dobson J, Hebard A, Tan WH (2001) Synthesis and characterization of silica-coated iron oxide nanoparticles in microemulsion: the effect of nonionic surfactants. Langmuir 17(10):2900–2906

36. Yang HH, Zhang SQ, Chen XL, Zhuang ZX, Xu JG, Wang XR (2004) Magnetite-containing spherical silica nanoparticles for biocatalysis and bioseparations. Anal Chem 76(5):1316–1321
37. Bi XX, Ganguly B, Huffman GP, Huggins FE, Endo M, Eklund PC (1993) Nanocrystalline α-Fe, Fe_3C, and Fe_7C_3 produced by CO_2-laser pyrolysis. J Mater Res 8(7):1666–1674
38. Hofmeister H, Huisken F, Kohn B, Alexandrescu R, Cojocaru S, Crunteanu A, Morjan I, Diamandescu L (2001) Filamentary iron nanostructures from laser-induced pyrolysis of iron pentacarbonyl and ethylene mixtures. Appl Phys Mater Sci Process 72(1):7–11
39. He YQ, Li XG, Swihart MT (2005) Laser-driven aerosol synthesis of nickel nanoparticles. Chem Mater 17(5):1017–1026
40. Leconte Y, Veintemillas-Verdaguer S, Morales MP, Costo R, Rodriguez I, Bonville P, Bouchet-Fabre B, Herlin-Boime N (2007) Continuous production of water dispersible carbon-iron nanocomposites by laser pyrolysis: application as MRI contrasts. J Colloid Interface Sci 313(2):511–518
41. Varma A, Lebrat JP (1992) Combustion synthesis of advanced materials. Chem Eng Sci 47 (9–11):2179–2194
42. Patil KC, Aruna ST, Ekambaram S (1997) Combustion synthesis. Curr Opin Solid State Mater Sci 2:158–165
43. Mukasyan AS, Epstein P, Dinka P (2007) Solution combustion synthesis of nanomaterials. Proc Combust Inst 31(2):1789–1795
44. Patil KC, Hegdeg MS, Ratan T, Aruna ST (2008) Chemistry of nanocrystalline oxide materials. Combustion synthesis. Properties and applications. World Scientific, Singapore
45. Aruna ST, Mukasyan AS (2008) Combustion synthesis and nanomaterials. Curr Opin Solid State Mater Sci 12(3–4):44–50
46. Ianoş R (2009) An efficient solution for the single step synthesis of $4CaO{\cdot}Al_2O_3{\cdot}Fe_2O_3$ powders. J Mater Res 24(1):245–252
47. Murakami S, Hosono T, Jezadevan B, Kamitakahara M, Iouku K (2008) Hydrothermal synthesis of magnetite/hydroxyapatite composite material for hyperthermia therapy for bone cancer. J Ceram Soc Jpn 116:950–954
48. Giri S, Samanta S, Maji S, Ganguli S, Bhaumik A (2005) Magnetic properties of alpha-Fe_2O_3 nanoparticle synthesized by a new hydrothermal method. J Magn Magn Mater 285(1–2):296–302
49. Mao B, Kang Z, Wang E, Lian S, Gao L, Tian C, Wang C (2006) Synthesis of magnetite octahedrons from iron powders through a mild hydrothermal method. Mater Res Bull 41 (12):2226–2231
50. Liu X, Qiu G, Yan A, Wang Z, Li X (2007) Hydrothermal synthesis and characterization of alpha-FeOOH and alpha-Fe_2O_3 uniform nanocrystallines. J Alloys Compd 433(1–2):216–220
51. Zhu H, Yang D, Zhu L (2007) Hydrothermal growth and characterization of magnetite (Fe_3O_4) thin films. Surf Coating Tech 201(12):5870–5874
52. Yang X, Jiang W, Liu L, Chen B, Wu S, Sun D, Li F (2012) One-step hydrothermal synthesis of highly water-soluble secondary structural Fe_3O_4 nanoparticles. J Magn Magn Mater 324:2249–2257
53. Kim EH, Lee HS, Kwak BK, Kim BK (2005) Synthesis of ferrofluid with magnetic nanoparticles by sonochemical method for MRI contrast agent. J Magn Magn Mater 289:328–330
54. Bang JH, Suslick KS (2007) Sonochemical synthesis of nanosized hollow hematite. J Am Chem Soc 129(8):2242–2243
55. Teo BM, Chen F, Hatton AT, Grieser F, Ashokkumar M (2009) Novel one-pot synthesis of magnetite latex nanoparticles by ultrasound irradiation. Langmuir 25(5):2593–2595
56. Feng J, Mao J, Wen XG, Tu MJ (2011) Ultrasonic-assisted in situ synthesis and characterization of superparamagnetic Fe_3O_4 nanoparticles. J Alloys Compd 509:9093–9097
57. Khalafalla SE, Reimers GW (1973) Magnetofluids and their manufacture, US Patent 3764540

58. Boistelle R, Astier JP (1988) Crystallization mechanisms in solution. J Cryst Growth 90:14–30
59. Gribanov NM, Bibik EE, Buzunov OV, Naumov VN (1990) Physicochemical regularities of obtaining highly dispersed magnetite by the method of chemical condensation. J Magn Magn Mater 85(1–3):7–10
60. Sugimoto T (2003) Formation of monodispersed nano- and micro-particles controlled in size, shape, and internal structure. Chem Eng Tech 26(3):313–321
61. Schwarzer HC, Peukert W (2004) Tailoring particle size through nanoparticle precipitation. Chem Eng Comm 191(4):580–606
62. Liu XQ, Tao SW, Shen YS (1997) Preparation and characterization of nanocrystalline alpha-Fe_2O_3 by a sol-gel process. Sensor Actuator B Chem 40(2–3):161–165
63. Kojima K, Miyazaki M, Mizukami F, Maeda K (1997) Selective formation of spinel iron oxide in thin films by complexing agent-assisted sol-gel processing. J Sol-Gel Sci Technol 8 (1–3):77–81
64. Gamarra LF, Brito GES, Pontuschka WM, Amaro E, Parma AHC, Goya GF (2005) Biocompatible superparamagnetic iron oxide nanoparticles used for contrast agents: a structural and magnetic study. J Magn Magn Mater 289:439–441
65. Răileanu M, Crişan M, Petrache C, Crişan D, Jitianu A, Zaharescu M, Predoi D, Kuncser V, Filoti G (2005) Sol-Gel Fe_xO_y-SiO_2 nanocomposites. Rom J Phy 50(5–6):595–606
66. Bagwe RP, Kanicky JR, Palla BJ, Patanjali PK, Shah DO (2001) Improved drug delivery using microemulsions: rationale, recent progress, and new horizons. Crit Rev Ther Drug Carrier Syst 18(1):77–140
67. Vidal-Vidal J, Rivas J, Lopez-Quintela MA (2006) Synthesis of monodisperse maghemite nanoparticles by the microemulsion method. Colloids Surf A Physicochem Eng Asp 288(1–3):44–51
68. Chin AB, Yaacob II (2007) Synthesis and characterization of magnetic iron oxide nanoparticles via w/o microemulsion and Massart's procedure. J Mater Process Technol 191(1–3):235–237
69. Haggerty JS (1981) Controlling powder size with collimated light beam—which selectively vaporises larger particles. US Patent Number US4289952-A
70. Morjan I, Alexandrescu R, Dumitrache F, Birjega R, Fleacă C, Soare I et al (2010) Iron oxide-based nanoparticles with different mean sizes obtained by the laser pyrolysis: structural and magnetic properties. J Nanosci Nanotechnol 10(2):1223–1234
71. McKittrick J, Shea LE, Bacalski CF, Bosze EJ (1999) The influence of processing parameters on luminescent oxides produced by combustion synthesis. Displays 19(4):169–172
72. Garcia R, Hirata GA, McKittrick J (2001) New combustion synthesis technique for the production of $(In_xGa_{1-x})_2O_3$ powders: hydrazine/metal nitrate method. J Mater Res 16 (4):1059–1065
73. Mukasyan AS, Costello C, Sherlock KP, Lafarga D, Varma A (2001) Perovskite membranes by aqueous combustion synthesis: synthesis and properties. Sep Purif Technol 25(1–3):117–126
74. Luo XX, Cao WH, Xing MM (2006) Preparation of nano Y_2O_2S:Eu phosphor by ethanol assisted combustion synthesis method. J Rare Earths 24(1):20–24
75. Li F, Hu K, Li JL, Zhang D, Chen G (2002) Combustion synthesis of gamma-lithium aluminate by using various fuels. J Nucl Mater 300(1):82–88
76. Jung CH, Park JY, Oh SJ, Park HK, Kim YS, Kim DK, Kim JH (1998) Synthesis of Li_2TiO_3 ceramic breeder powders by the combustion process. J Nucl Mater 253:203–212
77. Ozuna O, Hirata GA, McKittrick J (2004) Pressure influenced combustion synthesis of gamma- and alpha-Al_2O_3 nanocrystalline powders. J Phys Condens Matter 16(15):2585–2591
78. Chen D, Xu R (1998) Hydrothermal synthesis and characterization of nanocrystalline Fe_3O_4 powders. Mater Res Bull 33(7):1015–1021

79. Zheng YH, Cheng Y, Bao F, Wang YS (2006) Synthesis and magnetic properties of Fe_3O_4 nanoparticles. Mater Res Bull 41(3):525–529
80. Wang J, Sun JJ, Sun Q, Chen QW (2003) One-step hydrothermal process to prepare highly crystalline Fe_3O_4 nanoparticles with improved magnetic properties. Mater Res Bull 38 (7):1113–1118
81. Daou TJ, Pourroy G, Begin-Colin S, Greneche JM, Ulhaq-Bouillet C, Legare P, Bernhardt P, Leuvrey C, Rogez G (2006) Hydrothermal synthesis of monodisperse magnetite nanoparticles. Chem Mater 18(18):4399–4404
82. Vijayakumar R, Koltypin Y, Felner I, Gedanken A (2000) Sonochemical synthesis and characterization of pure nanometer-sized Fe_3O_4 particles. Materials Science and Engineering A-Structural Materials Properties Microstructure and Processing 286(1):101–105
83. Pinkas J, Reichlova V, Zboril R, Moravec Z, Bezdicka P, Matejkova J (2008) Sonochemical synthesis of amorphous nanoscopic iron(III) oxide from $Fe(acac)_3$. Ultrason Sonochem 15 (3):257–264
84. Sahoo Y, Pizem H, Fried T, Golodnitsky D, Burstein L, Sukenik CN, Markovich G (2001) Alkyl phosphonate/phosphate coating on magnetite nanoparticles: a comparison with fatty acids. Langmuir 17(25):7907–7911
85. Vékás L, Bica D, Marinică O (2006) Magnetic nanofluids stabilized with various chain length surfactants. Rom Rep Phys 58(3):257–267
86. Jiang W, Wu Y, He B, Zeng X, Lai K, Gu Z (2010) Effect of sodium oleate as a buffer on the synthesis of superparamagnetic magnetite colloids. J Colloid Interface Sci 347:1–7
87. Mourdikoudis S, Liz-Marzán LM (2013) Oleylamine in nanoparticle synthesis. Chem Mater 25:1465–1476
88. Euliss LE, Grancharov SG, O'Brien S, Deming TJ, Stucky GD, Murray CB, Held GA (2003) Cooperative assembly of magnetic nanoparticles and block copolypeptides in aqueous media. Nano Lett 3(11):1489–1493
89. Liu XQ, Guan YP, Ma ZY, Liu ZH (2004) Surface modification and characterization of magnetic polymer nanospheres prepared by miniemulsion polymerization. Langmuir 20 (23):10278–10282
90. Hong R, Fischer NO, Emrick T, Rotello VM (2005) Surface PEGylation and ligand exchange chemistry of FePt nanoparticles for biological applications. Chem Mater 17(18):4617–4621
91. Alsmadi NA, Wadajkar AS, Cui W, Nguyen KT (2011) Effects of surfactants on properties of polymer-coated magnetic nanoparticles for drug delivery application. J Nanopart Res 13 (12):7177–7186
92. de Almeida MPS, Caiado KL, Sartoratto PPC, Cintra e Silva DO, Rereira AR, Morais PC (2010) Preparation and size-modulation of silica-coated maghemite nanoparticles. J Alloy Comp 500:149–152
93. Roca AG, Carmona D, Miguel-Sancho N, Bomati-Miguel O, Balas F, Piquer C, Santamaria J (2012) Surface functionalization for tailoring the aggregation and magnetic behaviour of silica-coated iron oxide nanostructures. Nanotechnology 23(15):155603
94. Singh RK, Kim TH, Patel KD, Knowles JC, Kim HW (2012) Biocompatible magnetite nanoparticles with varying silica-coating layer for use in biomedicine: physicochemical and magnetic properties, and cellular compatibility. J Biomed Mater Res A 100(7):1734–1742
95. Lu AH, Schmidt W, Matoussevitch N, Bönnemann H, Spliethoff B, Tesche B, Bill E, Kiefer W, Schüth F (2004) Nanoengineering of a magnetically separable hydrogenation catalyst. Angew Chem Int Ed 43:4303–4306
96. Luo N, Liu KX, Liu ZY, Li XJ, Chen SY, Shen Y, Chen TW (2012) Controllable synthesis of carbon coated iron-based composite nanoparticles. Nanotechnology 23(47):475603
97. Lin J, Zhou WL, Kumbhar A, Wiemann J, Fang JY, Carpenter EE, O'Connor CJ (2001) Gold-coated iron (Fe@Au) nanoparticles: synthesis, characterization, and magnetic field-induced self-assembly. J Solid State Chem 159(1):26–31

98. Mohammad F, Balaji G, Weber A, Uppu RM, Kumar CSSR (2010) Influence of gold nanoshell on hyperthermia of superparamagnetic iron oxide nanoparticles. J Phys Chem C 114(45):19194–19201
99. Sobal NS, Hilgendorff M, Mohwald H, Giersig M, Spasova M, Radetic T, Farle M (2002) Synthesis and structure of colloidal bimetallic nanocrystals: the non-alloying system Ag/Co. Nano Lett 2(6):621–624
100. Vékás L (2013) Magnetic nanofluids. Synthesis, stabilization, properties, applications. Romanian Academy Publ. House, Bucharest
101. Bica D (1995) Preparation of magnetic fluids for various applications. Rom Rep Phys 47(3–5):265–272
102. Bica D, Vékás L, Avdeev MV, Marinică O, Socoliuc V, Bălăşoiu M, Garamus VM (2007) Sterically stabilized water based magnetic fluids: synthesis, structure and properties. J Magn Magn Mater 311:17–21
103. Willis AL, Turro NJ, O'Brien S (2005) Spectroscopic characterization of the surface of iron oxide nanocrystals. Chem Mater 17(24):5970–5975
104. Fauconnier N, Bee A, Roger J, Pons JN (1996) Adsorption of gluconic and citric acids on maghemite particles in aqueous medium. Progr Colloid Polymer Sci 100:212–216
105. Moghimi SM, Hunter AC, Murray JC (2001) Long-circulating and target-specific nanoparticles: theory to practice. Pharmacol Rev 53(2):283–318
106. Kumagai M, Imai Y, Nakamura T, Yamasaki Y, Sekino M, Ueno S, Hanaoka K, Kikuchi K, Nagano T, Kaneko E, Shimokado K, Kataoka K (2007) Iron hydroxide nanoparticles coated with poly(ethylene glycol)-poly(aspartic acid) block copolymer as novel magnetic resonance contrast agents for in vivo cancer imaging. Colloids Surf B Biointerfaces 56(1–2):174–181
107. Koneracka M, Muckova M, Zavisova V, Tomasovicova N, Kopcansky P, Timko M, Jurikova A, Csach K, Kavecansky V, Lancz G (2008) Encapsulation of anticancer drug and magnetic particles in biodegradable polymer nanospheres. J Phys Condens Matter 20 (20):204151
108. Moeser GD, Green WH, Laibinis PE, Linse P, Hatton TA (2004) Structure of polymer-stabilized magnetic fluids: small-angle neutron scattering and mean-field lattice modeling. Langmuir 20(13):5223–5234
109. Bruce IJ, Taylor J, Todd M, Davies MJ, Borioni E, Sangregorio C, Sen T (2004) Synthesis, characterisation and application of silica-magnetite nanocomposites. J Magn Magn Mater 284:145–160
110. Alcala MD, Real C (2006) Synthesis based on the wet impregnation method and characterization of iron and iron oxide-silica nanocomposites. Solid State Ion 177(9–10):955–960
111. Stöber W, Fink A, Bohn E (1968) Controlled growth of monodisperse silica spheres in the micron size range. J Colloid Interface Sci 26(1):62–69
112. van Blaaderen A, Kentgens APM (1992) Particle morphology and chemical microstructure of colloidal silica spheres made from alkoxysilanes. J Non Cryst Solids 149(3):161–178
113. Wang H, Nakamura H, Yao K, Maeda H, Abe E (2001) Effect of solvents on the preparation of silica-coated magnetic particles. Chem Lett 11:1168–1169
114. Cho SJ, Idrobo JC, Olamit J, Liu K, Browning ND, Kauzlarich SM (2005) Growth mechanisms and oxidation resistance of gold-coated iron nanoparticles. Chem Mater 17(12):3181–3186
115. Wang LY, Luo J, Maye MM, Fan Q, Qiang RD, Engelhard MH, Wang CM, Lin YH, Zhong CJ (2005) Iron oxide-gold core-shell nanoparticles and thin film assembly. J Mater Chem 15 (18):1821–1832
116. Căruntu D, Cushing BL, Căruntu G, O'Connor CJ (2005) Attachment of gold nanograins onto colloidal magnetite nanocrystals. Chem Mater 17(13):3398–3402
117. Chan HBS, Ellis BL, Sharma HL, Frost W, Caps V, Shields RA, Tsang SC (2004) Carbon-encapsulated radioactive Tc-99m nanoparticles. Adv Mater 16(2):144–149
118. Rosensweig RE (1989) Magnetic fluids: phenomena and process applications. Chem Eng Progr 85(4):53–61

119. Raj K, Moskowitz B, Casciari R (1995) Advances in ferrofluid technology. J Magn Magn Mater 149(1–2):174–180
120. Vékás L (2009) Ferrofluids and magnetorheological fluids. Adv Sci Technol 54:127–136
121. Raj K, Moskowitz B (1990) Commercial applications of ferrofluids. J Magn Magn Mater 85 (1–3):233–245
122. Todorovic M, Schultz S, Wong J, Scherer A (1999) Writing and reading of single magnetic domain per bit perpendicular patterned media. Appl Phys Lett 74(17):2516–2518
123. Blums E (1995) Some new problems of complex thermomagnetic and diffusion-driven convection in magnetic colloids. J Magn Magn Mater 149(1–2):111–115
124. Philip J, Rao CB, Jayakumar T, Raj B (2000) A new optical technique for detection of defects in ferromagnetic materials and components. NDT Int 33(5):289–295
125. Philip J, Jaykumar T, Kalyanasundaram P, Raj B (2003) A tunable optical filter. Meas Sci Tech 14(8):1289–1294
126. Chiba D, Yamanouchi M, Matsukura F, Ohno H (2003) Electrical manipulation of magnetization reversal in a ferromagnetic semiconductor. Science 301(5635):943–945
127. Tartaj P, Morales MP, Veintemillas-Verdaguer S, González-Carreño T, Serna CJ (2003) The preparation of magnetic nanoparticles for applications in biomedicine. J Phys D Appl Phys 36 (13):R182–R197
128. Pankhurst QA (2006) Nanomagnetic medical sensors and treatment methodologies. BT Technol J 24(3):33–38
129. Villanueva A, Cañete M, Roca AG, Calero M, Veintemillas-Verdaguer S, Serna CJ, Morales MP, Miranda R (2009) The influence of surface functionalization on the enhanced internalization of magnetic nanoparticles in cancer cells. Nanotechnology 20(11):115103
130. Mahmoudi M, Sant S, Wang B, Laurent S, Sen T (2011) Superparamagnetic iron oxide nanoparticles (SPIONs): development, surface modification and applications in chemotherapy. Adv Drug Deliv Rev 63(1–2):24–46
131. Safarikova M, Safarik I (1999) Magnetic solid-phase extraction. J Magn Magn Mater 194(1–3):108–112
132. Shinkai M (2002) Functional magnetic particles for medical application. J Biosci Bioeng 94 (6):606–613
133. Yoza B, Matsumoto M, Matsunaga T (2002) DNA extraction using modified bacterial magnetic particles in the presence of amino silane compound. J Biotechnol 94(3):217–224
134. Nam JM, Thaxton CS, Mirkin CA (2003) Nanoparticle-based bio-bar codes for the ultrasensitive detection of proteins. Science 301(5641):1884–1886
135. Rheinländer T, Kötitz R, Weitschies W, Semmler W (2000) Magnetic fractionation of magnetic fluids. J Magn Magn Mater 219(2):219–228
136. Romanus E, Huckel M, Gross C, Prass S, Weitschies W, Brauer R, Weber P (2002) Magnetic nanoparticle relaxation measurement as a novel tool for in vivo diagnostics. J Magn Magn Mater 252(1–3):387–389
137. Kim KW, Ha HK (2003) MRI for small bowel diseases. Semin Ultrasound CT MRI 24 (5):387–402
138. Richardson JC, Bowtell RW, Mader K, Melia CD (2005) Pharmaceutical applications of magnetic resonance imaging (MRI). Adv Drug Deliv Rev 57(8):1191–1209
139. Coroiu I (1999) Relaxivities of different superparamagnetic particles for application in NMR tomography. J Magn Magn Mater 201:449–452
140. Babes L, Denizot B, Tanguy G, Le Jeune JJ, Jallet P (1999) Synthesis of iron oxide nanoparticles used as MRI contrast agents: a parametric study. J Colloid Interface Sci 212 (2):474–482
141. Kim DK, Zhang Y, Kehr J, Klason T, Bjelke B, Muhammed M (2001) Characterization and MRI study of surfactant-coated superparamagnetic nanoparticles administered into the rat brain. J Magn Magn Mater 225(1–2):256–261

142. Harisinghani MG, Barentsz J, Hahn PF, Deserno WM, Tabatabaei S, van de Kaa CH, de la Rosette J, Weissleder R (2003) Noninvasive detection of clinically occult lymph-node metastases in prostate cancer. N Engl J Med 348(25):2491–2499
143. Bjornerud A, Johansson L (2004) The utility of superparamagnetic contrast agents in MRI: theoretical consideration and applications in the cardiovascular system. NMR Biomed 17 (7):465–477
144. Jordan A, Scholz R, Wust P, Fahling H, Felix R (1999) Magnetic fluid hyperthermia (MFH): cancer treatment with AC magnetic field induced excitation of biocompatible superparamagnetic nanoparticles. J Magn Magn Mater 201:413–419
145. Andra W, d'Ambly CG, Hergt R, Hilger I, Kaiser WA (1999) Temperature distribution as function of time around a small spherical heat source of local magnetic hyperthermia. J Magn Magn Mater 194(1–3):197–203
146. Hilger I, Hergt R, Kaiser WA (2000) Effects of magnetic thermoablation in muscle tissue using iron oxide particles—an in vitro study. Invest Radiol 35(3):170–179
147. Wust P, Hildebrandt B, Sreenivasa G, Rau B, Gellermann J, Riess H, Felix R, Schlag PM (2002) Hyperthermia in combined treatment of cancer. Lancet Oncol 3(8):487–497
148. Pardoe H, Clark PR, St Pierre TG, Moroz P, Jones SK (2003) A magnetic resonance imaging based method for measurement of tissue iron concentration in liver arterially embolized with ferrimagnetic particles designed for magnetic hyperthermia treatment of tumors. Magn Reson Imaging 21(5):483–488
149. Mornet S, Vasseur S, Grasset F, Duguet E (2004) Magnetic nanoparticle design for medical diagnosis and therapy. J Mater Chem 14(14):2161–2175
150. Lübbe AS, Bergemann C, Riess H, Schriever F, Reichardt P, Possinger K et al (1996) Clinical experiences with magnetic drug targeting: a phase I study with 4′-epidoxorubicin in 14 patients with advanced solid tumors. Cancer Res 56(20):4686–4693
151. Bonadonna G, Gianni L, Santoro A, Bonfante V, Bidoli P, Casali P, Demicheli R, Valagussa P (1993) Drugs 10 years later—epirubicin. Ann Oncol 4(5):359–369
152. Lübbe AS, Bergemann C, Brock J, McClure DG (1999) Physiological aspects in magnetic drug-targeting. J Magn Magn Mater 194(1–3):149–155
153. Frey NA, Peng S, Cheng K, Sun S (2009) Magnetic nanoparticles: synthesis, functionalization, and applications in bioimaging and magnetic energy storage. Chem Soc Rev 38 (9):2532–2542

Chapter 8
Nanofluidics-Based Mass Spectrometry. Applications for Biomarker Discovery in Lysosomal Storage Diseases

Mirela Sarbu and Alina D. Zamfir

Abstract Substantial efforts are currently invested in development of micro- and nanofluidics-based systems as front end technology for electrospray ionization (ESI) mass spectrometry (MS). Since its first introduction in biological MS chip-based ESI demonstrated a high potential to discover novel biopolymer species due to the efficient ionization properties, preferential formation of multiply charged ions and the elevated reproducibility and sensitivity. In combination with high-resolution mass spectrometers or instruments able to perform multistage fragmentation, chip-electrospray confirmed its unique ability to offer structural elucidation of minor species in complex mixtures, which often represent valuable biomarkers of severe diseases. This aspect is of particular importance for the applicability of chip MS in clinical investigation where only minute amounts of biological material are available. In view of these major advantages of nanofluidics in conjunction with modern MS, this chapter reviews the strategies, which allowed a successful application of chip technology for early diagnostic of lysosomal storage diseases. The first part is dedicated to the principles of ESI MS and to advanced nanochip systems for ESI MS. The second part highlights important achievements of nanochip ESI MS in biomarker discovery and diagnostic of LSDs such as Fabry and Schindler diseases. Finally, this chapter emphasizes that advanced chip ESI MS has real perspectives to become a routine method for early diagnosis and therapy of severe pathologies such as LSDs.

M. Sarbu
Faculty of Physics, West University of Timisoara, Timisoara, Romania

Aurel Vlaicu University of Arad, Arad, Romania

A.D. Zamfir (✉)
Faculty of Physics, West University of Timisoara, Timisoara, Romania

National Institute for Research and Development in Electrochemistry and Condensed Matter, Timisoara, Romania
e-mail: alinazamfir@yahoo.com

M. Lungu et al. (eds.), *Nanoparticles' Promises and Risks*,
DOI 10.1007/978-3-319-11728-7_8

8.1 Electrospray Ionization Mass Spectrometry

Electrospray ionization (ESI) is a soft ionization technique at atmospheric pressure for which the principle of ion formation in the gas phase is very different from the electron impact ionization (EI), chemical ionization (CI), atmospheric pressure chemical ionization (APCI), or fast atom bombardment (FAB). The major dissimilarity between ESI and the classical ionization techniques listed above is that in ESI the sample is dissolved in a polar solvent solution and possibly with added electrolytes in low concentration (10^{-5}–10^{-3} mol/L) and sprayed at atmospheric pressure through a capillary held at a high potential [1–3]. Hence, electrically charged droplets are formed which, after the solvent evaporation, yield ions in the gas phase.

One of the major advantages of this type of ionization is its unique ability to generate multiply charged ions in the gas phase. In contrast to EI, CI or FAB methods, where multiple charging of molecules is a phenomenon rarely observed, in the case of electrospray, multiply charged ions are often formed as : (1) alkali metal cations: Li^+, Na^+, K^+, Cs^+, Rb^+; (2) alkaline earths Mg^+, Ca^{2+}, Sr^{2+}, Ba^{2+}; (3) singly and doubly charged ions of transitional metals and their complexes with mono- and polydentate ligands; (4) anions of inorganic and organic acids such as NO_3^-, Cl^-, $H_2PO_4^-$, $H_2SO_4^-$ and SO_4^-; (5) mono- and multiply protonated organic bases such as amines, alkaloids, peptides, proteins; (6) organic acids or organic phosphates simply and multiply deprotonated and nucleic acids; (7) singly, doubly or triply protonated or deprotonated carbohydrates.

An important consequence of multiple charging is the reduction of the ion *m*/*z* ratio, which allows the application of ESI mass spectrometry (MS) to large molecules such as peptides, proteins, oligonucleotides, organometallic compounds, polymers, anti-cancer drugs, pesticides, etc.

Among all ionization techniques, ESI is one of the most efficient as it is able to generate, at atmospheric pressure, ions directly from solution, which makes it applicable to a large class of non-volatile substrates such as most of the biomolecules, for which, prior to ESI development, mass spectrometric analysis was not feasible.

Initial experiments carried out by the physicist John Zeleny in 1917 [4] preceded the first description by Dole et al. in 1968 [5] of the electrospray principle, including *the charge residue model* (CRM) which has survived as a main explanation for the controversial ESI process. However, the well-defined breakthrough of ESI as a general ionization method came in 1984–1989 when John B. Fenn and collab. [6] presented their experiments on identification of polypeptides and proteins of 40 kDa molecular weight. Fenn showed that a molecular-weight accuracy of 0.01 % could be obtained by applying a signal-averaging method to the multiple ions formed in the ESI process. The findings were based on experiments started in 1984 in Fenn's laboratory at Yale, when electrospray and mass spectrometry were successfully combined for the first time. Fenn used his knowledge of free-jet expansion to improve Dole's method with a counterflow of gas for desolvation,

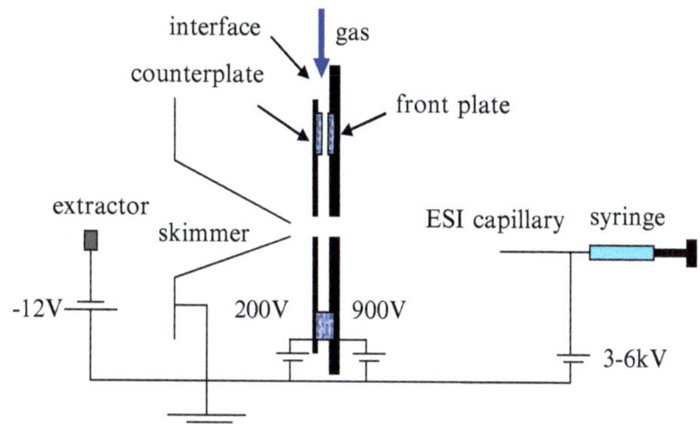

Fig. 8.1 Schematic of the ESI MS coupling

eliminating re-solvation of formed macromolecular ions. This discovery was closely followed by results from a Russian research group (Aleksandrov et al.) [7].

In ESI, basically, the liquid containing the analyte of interest is pumped through a metal capillary, which has an open end with a sharply pointed tip (Fig. 8.1). A particular case of high sensitivity represents the nano-electrospray ion sources developed by the groups of Mann [8] and Smith [9].

Since all droplets contain the same electrical charge, at the very end of the cone, they emerge into a fine spray called ESI plume. Depending on the polarity of the applied electric field, the charges may be positive or negative. The droplets are usually less than 10 μm across and contain both solvent and analyte molecules. The charged droplets move across the electric field existing between capillary and counter-electrode and, under a curtain gas flow, the solvent molecules evaporate from the droplet. According to Dole's CRM, as the droplet size decreases while the total charge on the droplet is constant, the charge surface density increases until the droplet's surface tension is exceeded by the repulsive electric forces. At this critical point, the droplet explodes into smaller, still highly charged droplets. This process, called *Rayleigh explosion*, repeats itself until the analyte molecule is stripped of all solvent molecules, and is left as a multiply charged ion.

The tip is attached to a voltage supply and its end faces a counter-electrode plate. As the voltage is increased, the liquid becomes charged and due to charge-repulsion effect, it expands out of the capillary tip forming the so-called *Taylor* cone (Fig. 8.2).

Application of the field to the liquid results in the formation of an electrical double layer. The accumulation of charge on the meniscus surface induces charged droplets spraying. By solvent evaporation and explosion of droplets gas phase ions are formed.

The number of charges retained by an analyte depends on such factors as the composition and pH of the electrosprayed solvent as well as the chemical nature of the sample [10]. For small molecules (<2,000 Da) ESI typically generates singly, doubly or triply charged ions, while for large molecules (>2,000 Da) the ESI process typically generates a series of multiply charged species and the resultant

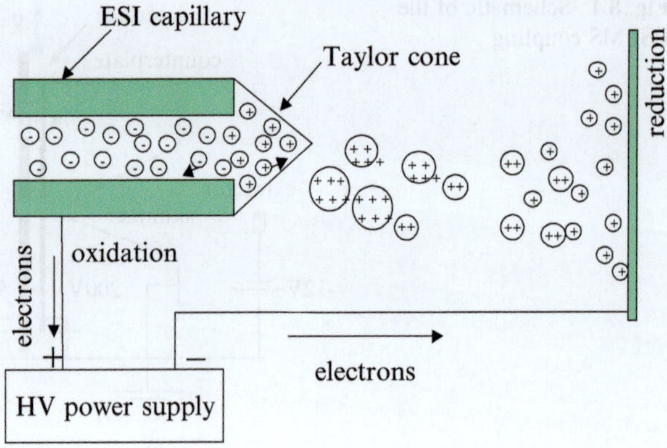

Fig. 8.2 The main physicochemical processes of electrospray ionization in positive ion mode

ESI mass spectrum contains multiple peaks corresponding to the different charge states [10–12].

This feature brings complexity to the interpretation of the ESI mass spectra but concomitantly, as a first advantage, it adds to the information and can be used to improve the accuracy of the molecular-weight determination. The method of deducing this way the molecular weight was presented in *the multiple charge theory* described by Fenn. The theory showed that different charge states could be interpreted as independent measurements of molecular weight and that an averaging method based on the solution of simultaneous equations could provide accurate molecular weight estimations for large molecules. The complex charge pattern can simply be deconvoluted and the mass of the uncharged protein is determined to dramatically higher accuracy than if the interpretation of data was based on a single ion. The second advantage of multiple charging is the formation of ions with reduced *m*/*z* ratio measurable with good resolution by almost any type of analyzer with which ESI has been interfaced: magnetic sector, single or triple quadrupole, time-of-flight (TOF), quadrupole ion trap (QIT), Fourier-transform ion cyclotron resonance (FTICR) or hybrid quadrupole time-of-flight analyzer (QTOF). All these make ESI the method of choice for large biopolymers and molecular aggregates or complexes that only have weak non-covalent interactions, such as protein–protein, enzyme–substrate or protein–ligand complexes [13–16].

The ESI process involves three main steps prior to mass analysis: the generation and charging of the ESI droplets, the droplet division by solvent evaporation and repetitive disintegrations leading ultimately to very small highly charged droplets and the production of gas-phase ions from these droplets. They are subsequently followed by the processes that modify the gas-phase ions in the atmosphere and the sub-atmospheric pressure-sampling regions of the mass spectrometer.

8.1.1 Generation of the Charged Droplets. Electrophoretic Mechanism

When applied at the tip of ESI capillary containing the analyte solution, the electric field penetrates the solution and the positive and negative ions will move in opposite directions until a charge distribution results. It counteracts the imposed field leading to field-free conditions inside the solution [10]. The ions bearing charge of the same polarity as the capillary are drifted downfield in the solution, toward the meniscus of the liquid from the tip, while ions of opposite charging are drifted away from the surface. The reciprocal repulsion between the ions and capillary surface overcomes the surface tension of the liquid, the surface begins to expand and the so-called Taylor cone is formed. If the applied field is sufficiently high, a fine jet emerges from the cone tip, which breaks up into small charged droplets [10].

8.1.2 Droplet Shrinkage by Solvent Evaporation and Coulomb Fission

The charged droplets shrink because of the solvent evaporation while the charge remains constant. The decrease of the droplet radius r_c at a constant charge q leads to an increase of the electrostatic repulsion of the charges at the droplet surface until the droplets reach the Rayleigh stability limit:

$$q_R = 8\pi\sqrt{\varepsilon_0 \sigma r_c^3} \tag{8.1}$$

Equation (8.1), the mathematical expression of Rayleigh limit, gives the conditions at which the electrostatic repulsion becomes equal to the force of tension surface σ, which keeps the droplet together. The charged droplet becomes unstable when the radius and charge satisfy Eq. (8.1). By this coulombic fission, the extra-charges are expelled until the ratio charge/volume reaches again the stability limit [10]. The outcome is the formation of very small charged droplets.

8.1.3 Theories Upon Gas-Phase Ion Formation

Two theories have been proposed to account for the formation of gas-phase ions from very small and highly charged droplets. The first one was proposed by *Dole* [5] and called *charged residue model*. It postulates the generation of extremely small droplets, which contain only one ion. Solvent evaporation from such a droplet leads to a gas-phase ion. The second theory belongs to *Iribarne* and *Thomson* and it predicts that after the radius of the droplet decrease to a given size, direct emission of an ion from the droplet becomes possible [17]. This particular model is known as the *ion evaporation theory*.

8.2 Nanochips for Electrospray Ionization Mass Spectrometry

In the development of miniaturized micro- and nanofluidic systems a significant progress in the past decade took place primarily due to numerous advantages of nanochip analysis, including the ability to analyze minute samples, the increased throughput and performance, as well as reduced costs [18–20]. In addition, several laboratory procedures such as sample preparation, purification, separation, and detection have been integrated onto a single microchip unit forming the so-called micro-total analysis systems (μTAS) to allow enhancement in sensitivity, speed, and accuracy.

In the last years, by introducing MS as a detector, the potential and the applicability of μTAS have been significantly extended. On the other hand, the option for miniaturized, integrated devices for sample infusion into MS was driven by several technical, analytical, and economical advantages [21] such as simplification of the laborious chemical and biochemical strategies required currently for MS research, minimization of sample handling and sample loss, elimination of possible cross-contamination, increased ionization efficiency, high quality of spectra, possibility for unattended high-throughput experiments, and the flexibility provided by the microchips for coupling to different MS configurations.

In the field of ESI MS, two types of chip-based devices are most popular and widely used. The first category [22, 23] is represented by the out-of-plane devices, where 100 or 400 nanospray emitters are integrated onto a single silicon substrate, from which electrospray is established perpendicular to the substrate. These systems are known under the commercial name of NanoMate robot (Fig. 8.3).

NanoMate 100 or 400 is produced by Advion BioSciences and represents the world's first fully automated nanoelectrospray system, as a robotic device that provides an automated nanoelectrospray ion source for mass spectrometers [24, 25]. In this system, by an automatic infusion, samples are infused directly into MS at low flow rates in the nanoliter range (only 50–100 nL/min). In addition to the robot itself, the main element of the system is the ESI chip. The robot holds a 96-well sample plate and a 96-pipette tip tray (Fig. 8.3). Automated sample analysis is achieved by loading a disposable, conductive pipette tip on a movable sampling probe, aspirating sample via a syringe pump, and moving the sampling probe to engage against the back of the ESI chip. ESI process is initiated by applying a head pressure and voltage to the sample in the pipette tip. Each nozzle and tip is used only once in order to eliminate carryover and contamination typical to conventional autosamplers. The ESI chip is an array of nanoESI nozzles of 2.5 or 10 μm internal

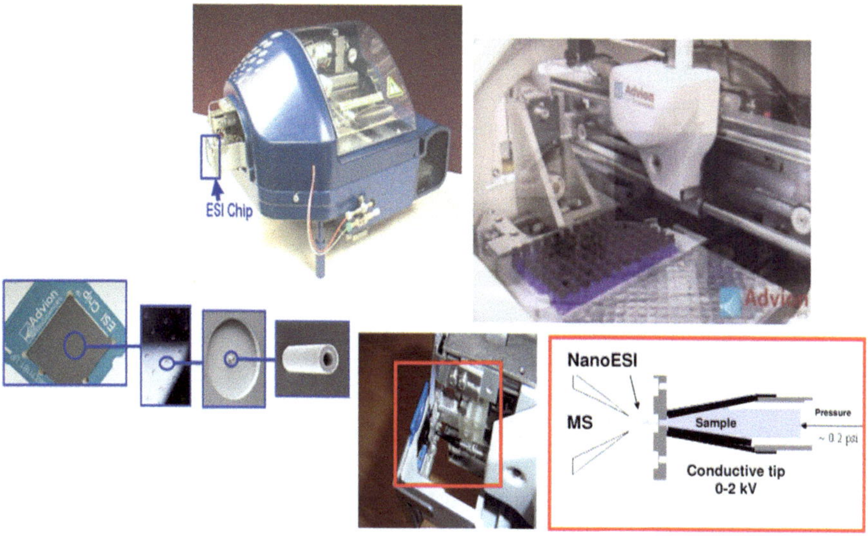

Fig. 8.3 NanoMate robot with fully automated chip-based ESI (courtesy of Advion BioSciences)

diameter etched in a planar silicon chip. The chip is fabricated from a monolithic silicon substrate using deep reactive ion etching (DRIE) and other standard microfabrication techniques. The inert coating on the surface allows a variety of acidic and organic compositions and concentrations to be used to promote ionization without any degradation of the nozzle. As visible in Fig. 8.3, a channel extends from the nozzle through the chip. In conventional electrospray devices the electric field is defined by the potential difference between the ESI tip (fluid potential) and the mass spectrometer inlet. A unique feature of the ESI chip is the incorporation of the ESI ground potential into the spray nozzle so that the electric field around the nozzle tip is formed from the potential difference between the conductive silicon substrate and the voltage applied to the fluid via the conductive pipette tip. In such a configuration, the distance that defines the electric field is about 1,000 times shorter than the one between the nozzle and mass spectrometer. Therefore, the mass spectrometer position and voltage, though crucial for efficient ion transfer into analyzer, do not play any role in the formation of chip electrospray, thus essentially decoupling the ESI process from the inlet of the mass spectrometer [26].

These devices are amenable to high-throughput sample delivery to ESI-MS by automated infusion and have the potential to completely replace flow-injection analysis assays. Moreover, the technical quality of the nanosprayers obtained by advanced silicon nanotechnology is so high, and the experiments so reproducible that such devices have been found in some instances to give more robust and quantitative analyses than LC- or CE-MS and to be able to suppress the need for separation prior to MS analysis [27]. Due to efficient ionization properties, silicon-based chip MS preferentially forms multiply charged ions, and the in-source

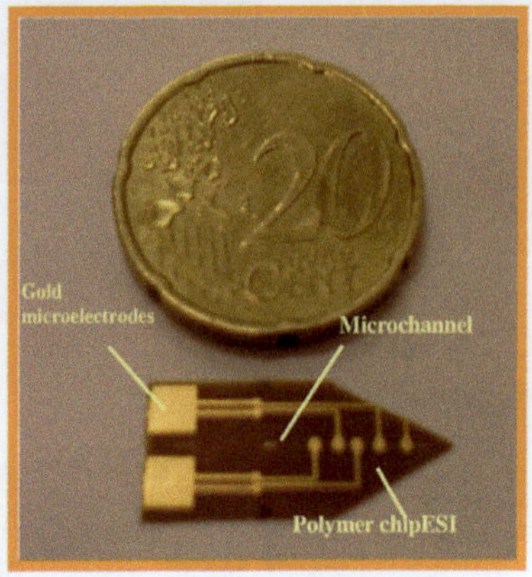

Fig. 8.4 Planar polymer microchip for ESI produced by DiagnoSwiss (courtesy of DiagnoSwiss)

fragmentation of labile groups attached to the main structural backbone is minimized, enhancing the sensitivity of the analysis.

Another popular category of chips for ESI MS consists of planar or thin microchips, made from glass [28] or polymer [29] material, embedding a microchannel at the end of which electrospray is generated in-plane, on the edge of the microchip.

During the last years, the progress in polymer-based microsprayer systems was promoted by development of simpler methods for accurate plastic replications and ease to create lower-cost disposable chips. For instance, DiagnoSwiss Lausanne has developed a disposable polymer microchip with integrated microchannels and electrodes, which was coupled to both a quadrupole time-of-flight and a Fourier transform ion cyclotron resonances mass spectrometer [30]. The chip presented in Fig. 8.4 was microfabricated by semiconductor techniques including photolithography [30].

For the microchip fabrication, the starting material was a polyimide foil of 75 mm thickness, which is coated on both sides with 5 mm copper. A photoresist was patterned on the copper-coated polyimide foil through a printed slide acting as a mask. Photoresist was then developed and chemical etching was used to remove the deprotected copper where microchannels are to be patterned. Polyimide was plasma-etched to the desired depth. As both sides of the substrate were exposed to the plasma, through-holes were fabricated to act as sample reservoirs and/or provide access to the microchannel. The final microchannels were 120 mm wide, 45 mm deep (nearly "half moon" cross section), with 100 mm gold-coated microelectrodes placed at the bottom of the microchannel (Fig. 8.5).

A 35 mm polyethylene/polyethylene terephthalate was laminated to close the channels. As described before [30] one end of each channel was manually cut in a tip shape, so that the outlet of the microchannel was located on the edge of the chip. For sample dispensing, either a reservoir was pasted over the inlet of the microchannel or the chip was sandwiched in a home-made chip holder with an integrated reservoir (Fig. 8.6).

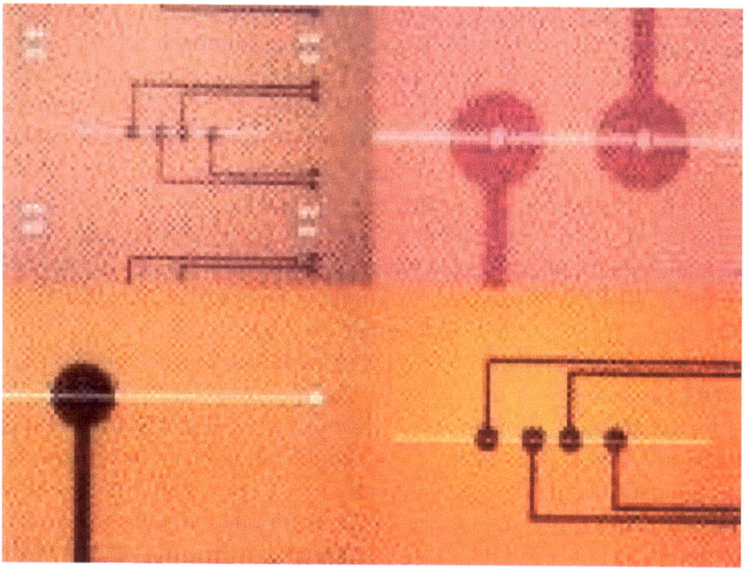

Fig. 8.5 Plasma etched disposable polyimide microchip for electrospray with integrated 100 μm gold electrodes (courtesy of DiagnoSwiss)

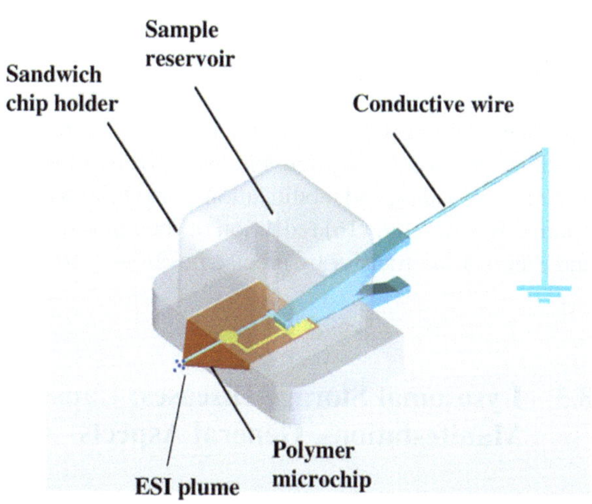

Fig. 8.6 Schematic of the thin planar polymer chip for ESI incorporated into the sandwich holder for coupling with MS. Reprinted with permission from [30]

For MS coupling one end of each channel was manually cut in a tip shape, which was visually inspected with a stereomicroscope. This way, the outlet of the microchannel is located on the edge of the chip, providing an in-plane electrospray.

Though compatible to automation for high-throughput analysis, these designs are best suited to act like micrototal analysis systems by integration of other analytical functions prior to sample delivery to the MS system, such as sample cleanup, analyte separation by CE, and chemical tagging [30]. Moreover, in comparison with glass nanospray capillaries, these thin microsprayers were found to provide superior stability of the spray with time, improved S/N at various flow rates. Several approaches for interfacing MS to silicon, glass, or polymer chips, providing flow rates which include them in categories from nano- to microsprayers, were reported. Different configurations such as single and triple quadrupole, TOF, IT, and ultra-high-resolution FTICR mass spectrometers were adapted to chip-based ESI and contributed significant benefits for various studies. However, as nicely illustrated already by a number of reviews [31–33] the general areas of implementation and applications of chip technologies either silicon-based, glass, or polymer microchips have by far been primarily proteomics, genomics, cell analysis, drug discovery, glycomics and metabolomics.

The potential of the modern chip-based ESI systems considerably broadened the area of MS applicability in life sciences. The option for miniaturized, integrated devices for sample infusion into MS is driven by several technical, analytical and economical advantages [34] such as: (1) simplification of the laborious chemical and biochemical strategies required currently for MS research; (2) high-throughput nanoanalysis/identification of biomolecules; (3) elimination of the time-consuming optimization procedures; (4) increase of the sensitivity by drastically reduction of the sample and reagent consumption, sample handling and potential sample loss; (5) high reproducibility of the experiments; (6) potential to discover novel biologically relevant structures due to increased ionization efficiency; (7) high signal-to-noise ratio; (8) reduced *in-source* fragmentation; (9) flexibility and broad area of applicability; (10) low cost of analysis and chip production; (11) possibility for unattended high-throughput experiments reducing the man power and man intervention; (12) possibility to perform several stages of sample preparation in a single integrated unit followed by direct MS structural analysis; (13) elimination of possible cross-contamination and carry-overs; (14) flexibility for different configurations, upgrading and modifications; (15) minimal infrastructure requirements for optimal functioning; (16) reduction of the ion source size facilitating manipulation and efficient ion transfer by precise positioning toward the MS sampling orifice.

8.3 Lysosomal Storage Diseases: Causes and Clinical Manifestations. General Aspects

The human body is made up of billions of cells, which contain hundreds of small organelles known as lysosomes. The term lysosome was introduced in 1955 by De Duve [35] to describe what was considered to be a hypothetical particle, unobserved by then with the microscope. Later studies in the field clearly demonstrated that the lysosome represents the storage sites not only for acid phosphatases, but also for various types of enzymes belonging to the group of lytic enzymes [36–38].

Lysosomes function as the digestive system of the cell; these organelles are capable of breaking down large complex molecules — proteins, nucleic acids, carbohydrates and lipids — into smaller, simpler components that can be reused by the cell for a while, then degraded and recycled. The recycling process carried out by lysosomes thus enables the body's cells to function normally. When degradation and recycling processes fail, there are a number of problems due to accumulation and storage within lysosomes of the materials ready for breakdown. When the accumulated amount is very high, it inhibits the catabolic enzymes and permeases that are not genetically deficient, leading to secondary substrate accumulation [39, 40]. Therefore, the diseases caused by such disruption of normal functioning are called "lysosomal storage diseases." Although the lysosome have been discovered only 5–6 decades ago, the first description of a lysosomal storage disease was in 1881 for Tay-Sachs disease, followed in 1882 by the description of Gaucher disease and in 1898 of the Fabry disease, though at that moment it was not clear that the various diseases share the common feature of lysosomal storage. This scientific breakthrough represented the starting point for investigations and discoveries in terms of the intracellular biology of these enzymes and their substrate and also in understanding the pathophysiological basis of the lysosomal storage diseases (LSDs).

LSDs comprise a heterogeneous group of more than 50 inherited genetic diseases characterized by mutations in gene that encode mainly lysosomal hydrolases [41], but also some lysosome membrane proteins or other proteins involved in the processing, trafficking and targeting [42]. As a consequence, significant decrease (<10 % normal levels) or a complete absence in the activity of a soluble lysosomal enzyme is noticed. The common feature for all LSDs is the initial inappropriate accumulation of specific macromolecular substrates inside organelles of the endosomal–autophagic–lysosomal system, which disrupts the normal functioning of the cell, leading to cell dysfunction and clinical abnormalities specific for LSDs and ultimately, premature death.

The most common lysosomal storage disease classification is according to the nature of the major compound accumulated. Large classes of LSD correspond to mucopolysaccharidoses (MPS), sphingolipidoses, mucolipidoses, glycoproteinoses, oligosaccharidoses, and glycogen storage diseases. Although the absence and/or the deficiency of a lysosomal enzyme activity represent the common basic pathogenesis for this class of congenital diseases, it is not possible to generalize LSD symptoms. Even within a single disease, the clinical manifestations vary depending on the amount of the accumulated substrate; as the latter is higher, the number and cell types affected by storage are greater, so that nearly every bodily system is affected ultimately.

Patients with LSDs appear normal at birth and become symptomatic over the first 3–6 months or even years of life. Several disorders, including Pompe, Tay-Sachs, Gaucher, may exist in three forms, infantile, juvenile and adult. The speed and severity of the evolving symptoms depends on the basic disease and the value of the residual enzymatic activity. As a general trait, infantile form characterized by earlier onset of symptoms arising from brain pathology represents the most severe form of the disease (i.e. neonatal Gaucher disease infants die in utero or shortly after birth), whereas a later onset mainly of peripheral symptoms is characteristic for milder adult forms (i.e. heterozygous female Fabry patients who may remain asymptomatic throughout life). Juvenile forms are intermediate between infantile and adult forms.

Many of the conditions show significant similarities. Progressive central nervous system impairment, with or without systemic involvement, represents by far one of the most devastating aspects of LSDs and also one of the most difficult disease to treat. The incidence of central nervous system (CNS) involvement is about 60–70 % in all LSDs [43]. CNS disease can result in developmental delay, mild-to-severe mental retardation, seizures and profound neurodegeneration. Seizures were found in oligosaccharidoses and sphingolipidoses, while in mucopolisachharidoses were not reported. Severe seizures are frequently associated with a rapid mental decline, which in 1–2 years leads to the vegetative state. Another feature of LSDs is the visceromegaly. Splenomegaly, especially hepatosplenomegaly, is the hallmark of lysosomal storage diseases. Heart diseases are also commonly present in LSDs. Reddish purple cutaneous vascular lesions, known as angiokeratomas, are also characteristics for some LSDs, most often encountered in Fabry disease. They are primarily distributed on the buttocks, groin, umbilicus and upper thighs and become larger and more numerous with age.

A number of other clinical manifestations are specific to certain diseases, such as: optic atrophy seen in Krabbe disease, retinopathy characteristic for mucolipidoses type II (i-cell disease), supranuclear ophthalmoplegia (pain in eye movement) common in type II and III of Niemann-Pick disease, ichthyosis have been reported in multiple sulfatase deficiency, while proteinuria and chronic renal failure is distinctive for Fabry disease. Aggressive behavior is prominent in mucopolisaccharidoses type II (Hunter disease), type III (Sanfilippo disease) and alpha-fucosidosis.

Most LSDs are autosomal recessive disease, which means that their transmission and manifestation requires the presence of two copies of the same altered gene, one copy from each parent; in the case of heterozygotes (carrier-individuals with a single malfunctioning gene copy), the enzyme activity generated by their other (normal) allele is often sufficient, so they do not exhibit evidence of tissue storage. Exceptions are Fabry, Hunter and Danon diseases which are X-linked heterosomal recessive disorders. While for carrier females of Hunter's syndrome, clinical problems related to the presence of an altered protein are not evident, this is not the case of most of the females who are carriers of Fabry's or Danon's diseases. The last ones may experience disease-related complications, which in a few cases can be as severe as those found in classically affected males.

Although lysosomal storage diseases are considered as a group of rare diseases, their combined prevalence is between 1:5,000 and 1:8,000 [44, 45]. This relatively high incidence caused investigations regarding the feasibility of several screening methods for these diseases [46, 47]. Necessity of screening methods emerge from: (1) the high frequency of certain diseases among a particular ethnic group (i.e. Gaucher, Tay-Sachs and Niemann-Pick diseases are common among

Ashkenazi Jews, gangliosidosis are predominantly on Japanese, while Pompe is characteristic for Chinese and African), (2) the opportunity to interfere before the development of significant neurological sequels on cases diagnosed early and potentially treatable (i.e. bone marrow transplant for the juvenile form of Krabbe disease); (3) the possibility of prevention in subsequent pregnancies in families at risk.

Early diagnosis before serious physical and/or mental deterioration is becoming more critical as efficient therapy becomes a reality for some disorders. Due to the wide clinical spectrum of symptoms, the diagnosis of LSDs merely by the clinical manifestation is difficult; sophisticated laboratory biochemical and molecular genetic testing being required for establishing a definitive diagnosis of lysosomal storage diseases. In the first approach, various clinical samples can be used for measuring the accumulated primary substrate or the lysosomal enzyme activities, such as urine, blood, amniotic fluid, skin fibroblasts and tissue biopsies. The most commonly screening tests emphasize the presence of certain oligosaccharides, glycosphingolipids and glycosaminoglycans in urine and/or blood. This is possible due to the development of tandem mass spectrometry technique for the identification and quantification of lysosomal substrates and metabolites [47, 48] (see Sect. 8.4 for diagnosis of LSD by mass spectrometry). Extremely accurate prenatal diagnosis of LSDs is also possible by performing enzymatic assays on chorionic villus samples, cultured trophoblasts, and cultured amniotic fluid cells. Moreover, in cases of newborn infants who have no family history of LSDs, early diagnosis tests are based on measuring the activity of certain proteins and/or lysosomal enzymes, including lysosome-associated membrane proteins (LAMP-1 and -2) and saposins in blood samples obtained from heal sticks [49].

Although so far are no complete cures for storage disorders, scientists worldwide are working to develop ways to make the lysosome function normally again. Presently, patients often undergo a variety of therapies and palliative cares, as dialysis, surgery or physical therapy that can helpful in managing the symptoms, have a positive impact on the patient's quality of life, by they do not change the biochemical cause of the disease and usually do not prevent disease progression. However, over the past decades there has been a remarkable development in the number of disease-specific therapies that can address to principal enzyme and/or storage problem. These therapies include: (1) *enzyme replacement therapy (ERT)*, based on intravenous administration of a fully functional wild-type lysosomal enzyme, and become standard of care for Gaucher, Fabry, Pompe, mucopolysaccharidosis (MPS) type I and VI diseases; (2) *enzyme enhancement therapy (EET)*, where low molecular-weight pharmacological chaperons are used to prevent proteins misfolding or mistracking by stabilizing the conformation and restoring the function of a protein. The major advantage of this therapy against ERT rises from the possibility of treating LSDs that affect the brain since, unlike therapeutic enzymes, low molecular-weight chaperones might cross the blood–brain barrier; (3) *hematopoietic stem cell transplantation (HSCT)* involves the production and secretion of deficient enzyme by the engrafted donor leukocytes in the host tissue, which later is taken up by residual enzyme-deficient host cells. This therapy is used mainly in MPS I (Hurler) and MPS VII diseases, while for MPS II and MPS VI is used in conjunction with ERT [50]; (4) *substrate reduction therapy (SRT)* involves the use of certain selective inhibitors, as L-cycloserine, to reduce the substrate synthesis to a level approximately equal to the rate of breakdown. Since this type of therapy presents significant side effects, its use is limited, however is successfully used in the Gaucher disease [51]; (5) *gene therapy* which requires in vivo injection of a gene transfer vector directly into a tissue or into the circulation, or *ex vivo* genetically modification of cells followed by their transplant into an affected patient in order to create a repository of enzyme that can be secreted into the circulation [52, 53]. Although up to now have been performed preclinical studies on animal models of lysosomal storage diseases, and just a few clinical trials on human with lysosomal storage diseases, gene therapy is a promising treatment modality for the LSDs.

Taking into account that only 5–6 decades have passed since the discovery of the lysosome and its correlation with certain diseases already described, it is obvious the effort on investigating the lysosomal storage diseases and the advanced level reached on their treatment.

8.4 Applications of Mass Spectrometry to Diagnosis of Lysosomal Storage Diseases

Considering the advanced level achieved in terms of treatment methods, presymptomatic diagnosis and early intervention are crucial. As briefly mentioned above, in the screening of LSDs, several assays are used, as the fluorometric assays with 4-methylumbelliferone (4-MU) substrates, tandem MS (MS/MS) or liquid chromatography MS/MS (LC-MS/MS), multiplex enzyme assays, digital microfluidic fluorometric assays, and immune-quantification assays. Unlike the fluorometric assay, high-throughput mass spectrometry methods based on either ESI or matrix-assisted laser desorption/ionization (MALDI), originally developed to screen newborns for lysosomal storage disorders, are most specific, powerful, reliable and efficient tools. On one hand, these methods can be used from different perspectives: for quantification, by measuring in urine or blood samples the substrate concentration, or the enzyme activity, as well as for investigation of protein expression on cultured cells; on the other hand, the chosen or synthesized substrates for these assays are structurally closer to the natural substrates, as the artificial 4-MU substrates. The procedure for the measurement of enzyme activities involves a simple reasoning: based on the fact that enzyme products are specific to each enzyme reaction, they are quantified against internal standards with known concentrations. Since the method was optimized and refined into a standardized protocol, in 2008 Genzyme Corporation began the production in large quantity and the distribution of the analyte specific regents (ASR), namely the substrates, internal standards and products. Due to the complexity of the initial procedures, based on achieving separate enzymatic reactions, followed by their combination and purification by liquid–liquid extraction (LLE) and solid phase extraction (SPE) to remove the salts and detergents prior to mass spectrometry analysis, several technical modifications were carried out focusing on simplifying the sample preparation processes. In the last 6 years, to simplify the sample preparation procedures and to reduce the running costs, several technical modifications, i.e. the conjunction of mass spectrometry with separation techniques such as capillary electrophoresis (CE) and high-performance liquid chromatography (HPLC), were carried out. An attempt in this direction belongs to Marca et al. [54] who proposed a robust method suitable for large-scale studies (screening) on dried blood spot samples, based on five reaction mixtures from separate reactions, followed by their combination, centrifugation and analysis on an LC MS/MS. The simplicity of their study emerges from the only 4 min of analysis time, as well as from the supernatant analysis without further manual purification after enzyme reaction. Two years later, in 2010 a similar approach [55], aimed to eliminate the off-line LLE and SPE steps, based on high sensitivity, specificity and throughput, involved the combination of ultra-high-performance liquid chromatography (UHPLC) and multiplexed sample introduction. With the same objective of simplifying the analysis procedures, multiplex approaches have been developed and piloted in newborn screening (NBS) studies. Simplified newborn screening protocol based on high-throughput hexaplex assay for Fabry, Gaucher, Niemann-Pick A/B, Pompe, Krabbe and MPS I diseases have been reported three years ago [56]. The protocol based on a dual-channel TLX-2 system (two TurboFlow and two analytical columns) coupled with a TSQ Quantum Ultra AM mass spectrometer eliminated the use of the organic compounds from the screening of LSDs by MS/MS and reduced the need for large amounts of consumables.

There were still several multiplex studies based on LLE and SPE purification steps, such as the screening of the three enzymes involved in Pompe, Fabry and MPS-I diseases [57], and the study based on 4 + 1 multiplex MS/MS assay [58]. In 2013, it was reported a most comprehensive and robust multiplex approach using up to nine lysosomal enzymes. Besides the Gaucher, Pompe, Krabbe, Fabry and Niemann-Pick A/B enzymes, they added to the multiplex assay the MPS-I, MPS-II, MPS-IVA and MPS-VI enzymes [59].

Since mass spectrometry methods have been developed to screen newborns for LDSs, and now they are currently feasible to diagnose Pompe, Fabry, Gaucher, Krabbe, Niemann-Pick A/B, as well as mucopolysaccharidosis I diseases, this methodology could also discriminate between affected patients and controls of various ages. In 2013 a study employed tandem mass spectrometry (MS/MS) assays to screen for Pompe, Fabry, Gaucher, Krabbe, and Niemann-Pick A/B diseases, as well as MPS I, in a total of 218 infants, children, and adults, from which 205 were control patients and 13 affected patients [60]. A distinction between the enzyme activity in non-affected infants, children, and adults of any age and the enzyme activity in patients suffering from Pompe, Fabry, Gaucher, and Krabbe diseases could be observed. Also, as compared to enzyme activities in newborns, in individuals older than 18 years those were significantly lower (about 50 %), the GLA enzyme (involved in Fabry disease) being the most affected by age, only about 38 % of the median. The method based on API 3200 triple quadrupole instrument mass spectrometer and multiple-reaction monitoring mode highlighted that affected patients presented enzymatic activities corresponding to less than 20 % of the age-matched controls.

The implementation of tandem mass spectrometry, based on substrate concentrations and enzyme activity measurements, into newborn screening programs resulted in an exponential increase of the number of detectable metabolic disorders. The synthesis of new reagents and substrates more closely related to homologous with the natural enzyme precursors, improved the assay performance by enhancing the mass spectrometry sensitivity and thus, decreasing the detection limit. Moreover, multiplex MS approaches have reduced the costs and time of analysis due to the possibility of MS/MS method to detect multiple enzyme products simultaneously. Compared with standard methodologies, diagnosis by tandem mass spectrometry in dried blood spots exhibits considerable technical advantages. Nowadays, it is currently feasible to diagnose Pompe, Fabry, Gaucher, Krabbe, and Niemann-Pick A/B diseases, as well as mucopolysaccharidosis I, by MS/MS.

8.5 Chip-Based Mass Spectrometry for Rapid Diagnosis of Fabry Disease

In the last 5–6 years, different MS-based strategies for monitoring the α-galactosidase A (GLA) enzyme in dried blood spots (DBS) [58, 61–63], as well as for measurement of the storage products in plasma [64] or urine have been developed and implemented. Despite the facts that were capable to provide early detection of LSDs in general and of Fabry disease in particular, false-negative results [65] have been reported in a number of previous studies based on DBS assays. In order to avoid the generation of false-negative or false-positive response, it is necessary to ensure through the implemented methods a high level of reproducibility and sensibility necessary for accurate comparative screening patient vs. healthy control, as well as to avoid the carry-over from sample to sample which might occur during successive ESI infusions. In 2013, a novel platform for rapid and reliable diagnostic of Fabry disease, that eliminates all the drawbacks of the DBS-MS method mentioned above, have been successfully implemented [63] and it is based on enzyme assay and fully automated chip-nanoESI MS, collision-induced dissociation (CID) and electron transfer dissociation (ETD) MS/MS for a total of 13 DBSs, 11 from healthy donors and 2 from Fabry patients.

The enzymatic assay workflow comprised five stages: (1) obtaining the DBSs by uniform spotting 75 μL of whole blood onto Whatman 903® specimen collection paper, drying at room temperature and cutting with a 3 mm diameter puncher from the middle of the circle, then storing them at 4 °C; (2) enzymatic extraction from

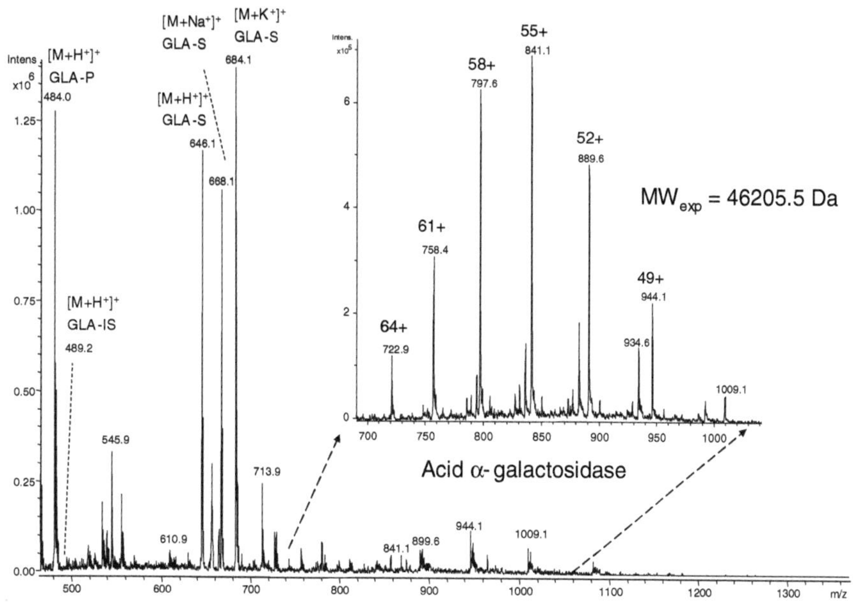

Fig. 8.7 Fully automated chip-nanoESI on a NanoMate robot coupled to HCT MS of the assay products after enzymatic reaction in a DBS sample from a healthy donor. *Inset*: zoom out of *m*/*z* (720–1,010) range corresponding to α-galactosidase protein ($MW_{exp} = 46{,}205.5$ Da). Positive ion mode detection; ESI MS parameters: ChipESI, 1.4 kV; Cap Exit, 50 V. Back nitrogen pressure 0.30 psi. Nitrogen nebulizer on MS at 50 psi. Reprinted with permission from [63]

DBS using an aqueous buffer solution; (3) enzymatic reaction between the DBS α-galactosidase extracts and the substrate/internal standard (GLA-S/GLA-IS) cocktail — molar ratio 500:1 — resulting in cleavage off the terminal galactose from GLA-S generating the product (GLA-P). GLA-IS, the deuterated counterpart of GLA-P, served for positive control; (4) purification steps of the final enzymatic mixture to remove buffers, salts and detergents using silica gel; (5) fully automated chip-nanoESI MS and tandem MS analysis of the reaction products on a high capacity ion trap (HCT) ultra PTM mass spectrometer coupled via an in-laboratory mounting system to a NanoMate robot incorporating ESI 400 chip technology.

The first MS approach of the method demanded MS screening of the α-galactosidase reaction products, both in healthy control (Fig. 8.7) and in patient (Fig. 8.8). The presence of GLA-P, GLA-IS and GLA-S ions is illustrated in the spectrum from Fig. 8.7, which results in a GLA-P/GLA-IS ratio of 26. The presence of GLA-P in the healthy subjects discloses an active enzyme capable to cleave GLA-S, a fact confirmed also by the presence in the spectrum of a series of signals having an envelope shape, which is characteristic for a protein. Using the same condition for ionization and detection, the spectrum of DBS reaction products from Fabry patient presented a large discrepancy compared with the healthy one.

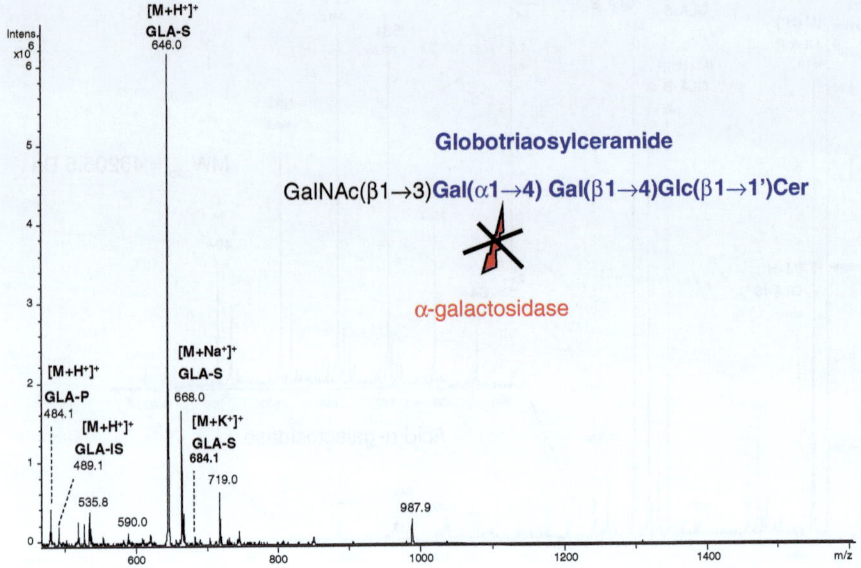

Fig. 8.8 Fully automated chip-nanoESI on a NanoMate robot coupled to HCT MS of the assay products after enzymatic reaction in a DBS sample from a patient diagnosed with Fabry disease; in vivo, the major non-degraded substrate is globotriaosylceramide. Experiment conditions as in Fig. 8.7. Reprinted with permission from [63]

The low intensity signal of GLA-P in Fig. 8.8 illustrates a diminished activity of the enzyme, while the lack of protein signals indicates the presence of the enzyme in a quantity which is below even the detection limit of the chip-nanoESI HCT MS. There were differences also regarding the GLA-P/GLA-IS ratio, which in Fig. 8.8 was only 2. This 13-fold reduction in GLA-P/GLA-IS ratio calculated according to absolute intensity of the signals from MS screening in the comparative control vs. patient highlighted that the platform can reliably monitor the activity/expression of α-galactosidase in Fabry patients by chip-based nanoESI MS. Moreover, confirmation of GLA-P as well as the GLA-IS structures was further carried out by extensive structural characterization using two complementary fragmentation techniques, CID and ETD, the latter being for the first employed in rapid diagnosis of LSDs.

The sensitivity and reproducibility of the NanoMate HCT MS platform enable discrimination between different Fabry cases, an aspect with major clinical importance when choosing an appropriate therapy. For instance, this rapid detection of rare diseases tool exhibit the capability to distinguish the situations in which the GLA enzyme is completely absent from those in which the enzyme is present at the trace level, as well as between the patients unable to produce the GLA enzyme from those that produce an inactive GLA enzyme.

8.6 Chip-Based Mass Spectrometry for Rapid Diagnosis of Schindler Disease

Schindler disease is a recently recognized lysosomal storage disorder caused by the deficient activity of α-*N*-acetylgalactosaminidase (NAGA) enzyme. From the three different phenotypes identified to date, the most severe form is the type I with an infantile-onset neuroaxonal dystrophy. Schindler disease type II represents a milder form with an adult-onset characterized by *angiokeratoma corporis diffusum* and mild intellectual impairment, while the type III represents an intermediate form with manifestations ranging from psychomotor retardation in infancy to a milder autism and marked behavioral difficulties. The NAGA deficiency causes a 100 times higher concentration of *O*-glycans in urine than in healthy controls. For this reason, screening, structural characterization and complete identification of *O*-glycosylated amino acids and peptides extracted from patients' urine are of major diagnostic importance.

Mass spectrometry based on nanofluidics have revealed in the last decade the ability to discover in Schindler disease patient urine *O*-glycosylated amino acids and *O*-glycopeptides as disease markers useful for a rapid diagnosis [29, 30, 66–68]. Besides, MS experiments based both on out-of-plane nanochip devices and polymer thin microchips for screening and sequencing by CID of glycoconjugate pools extracted and purified from the urine of Schindler disease patients and normal controls were characterized by an unsurpassed sensitivity, throughput, reproducibility, quality and reliability of the data [29, 30, 66–68].

For mapping, sequencing and structural elucidation of a complex mixture of *O*-glycosylated amino acids and peptides extracted and purified from the urine of a healthy subject (denoted *Ty*), the thin chip polymer-based microsprayer coupled to a hybrid QTOF MS has been used [29]. An aliquot of 10 μl of *O*-GalNAc glycosylated amino acids and peptides mixture having 5 pmol μl^{-1} sample concentration was dispensed into the reservoir pasted over the inlet of the chip microchannel as described in Sect. 8.2.

The negative ion mode electrospray process was initiated at 2.8 kV ESI voltage and 100 V potential of the sampling cone. The flow rate of about 200 nl min^{-1} together with the source parameters generated a constant and stable spray. The spectrum combined over 20 scans (40 s), equivalent to a sample consumption of 0.66 pmol, exhibited a high signal/noise (S/N) ratio and yielded data on 26 different saccharide components which were expressing the *O*-GalNAc-Ser/Thr core-motif extended by either sialylation or fucosylation. The mixture was found to be dominated by serine (Ser-) and threonine (Thr-) linked disialosaccharides with chain lengths ranging from tetra- to octasaccharide.

In order to test the limit of microchip sensitivity for glycoconjugate detection, *Ty* solution was diluted in pure methanol yielding an aliquot at 1.25 pmol μl^{-1} concentration and further analyzed (Fig. 8.9).

The signal was acquired for over 20 scans, which resulted in 0.16 pmol sample consumption. Even under these limited concentration conditions, the potential of

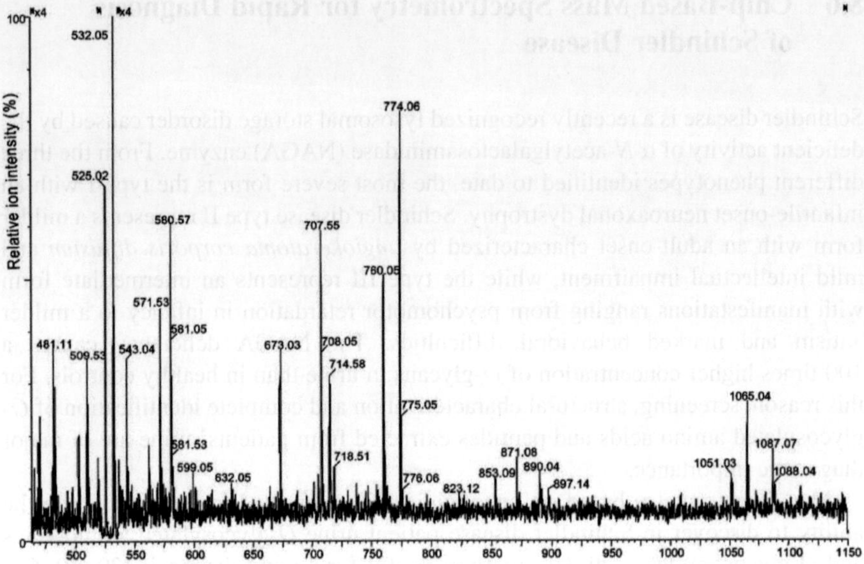

Fig. 8.9 Microchip ESI QTOF MS of the *Ty* mixture of *O*-glycosylated amino acids and peptides from human urine. ESI voltage 2.8 kV; sampling cone potential 100 V; signal acquisition 20 scans. Solvent: methanol; average sample concentration: 1.25 pmol μl^{-1}. Reprinted with permission from [29]

the microchip ESI QTOF MS methodology afforded the identification of 13 different components in the mixture with a fair S/N ratio. The potential raised from the high sensitivity and the advantage regarding the minimization of the in-source decay of the labile attachments such as *N*-acetyl neuraminic acid (Neu5Ac) of the microchip ESI-MS, make suitable the use of this microchip platform in clinical applications for comparative compositional analysis of homologous biological glycoconjugate mixtures from either tissues/body fluids or natural products.

Another study based on the polymer microchip coupled this time with a FTICR 9.4T mass spectrometer has been designed and implemented for compositional mapping of highly heterogeneous mixtures of *O*-glycosylated sialylated peptides from the urine of a healthy individual (denoted *Ty2*) and an age-matched patient (denoted *Gy2*) diagnosed with Schindler disease type I [30]. In the reported case, special considerations of the thin chipESI FTICR MS parameters with respect to the direct spray configuration of the FTICR instrument were necessary. The negative ion mode electrospray process could be initiated by applying 1,500 V to the transfer capillary, while the fine positioning of the microsprayer toward the MS inlet turned out to be crucial for efficient transfer of the ionic species into the MS and long-term stability of the electrospray.

For the *Ty2* mixture of *O*-GalNAc glycosylated sialylated peptides was used a concentration of approximately 7 pmol μl^{-1} in methanol, and the chosen conditions for ionization generated a constant and stable spray accompanied by a high ion intensity. The higher abundance of the disialylated saccharides toward the monosialylated species demonstrated the feasibility of the optimized chipESI FTICR MS approach to exhibit only a very low degree of in-source fragmentation related to the labile sialic acid moiety, and therefore to satisfy the fundamental criteria for detection of intact sialylated glycoconjugates.

For exploring the limits of sensitivity at which the level of mixture complexity could still be reliably assessed, a broad study on four successive dilutions from 20 up to 3 pmol μl^{-1} (calculated using the average relative molecular mass of the components, $M_r = 2{,}500$) was conducted by chipESI FTICR MS method optimized in negative ion mode for the *Gy2* mixture of glycopeptides. In comparison with the chipESI FTICR spectrum of the normal urine glycopeptides, in the spectrum of the patient urine a higher content of pentasaccharides was observed, while hexasaccharides linked to Ser and Thr bearing two sialic acid moieties were detected at higher abundance. Moreover, a nonasaccharide bearing three sialic acid moieties assigned to the sodiated dehydrated $Neu5Ac_3Hex_2HexNAc_4$-Ser was for the first time detected with a mass accuracy of 4.6 ppm.

The potential of the chipESI FTICR MS method to exhibit not only a high sensitivity, high mass accuracy and resolution but also superior reproducibility of the experiments was also demonstrated in this study; no significant difference with respect to the ionic species observed between the spectra resulted from 12 and 6 pmol μl^{-1} sample concentration (Figs. 8.10a, b) was observed.

Unlike conventional capillary-based nanospray characterized by poorer reproducibility of the glass-capillaries shape and signal interruption due to blockages of the tip, polymer chip produces a spray of high stability over a long time frame of signal acquisition, even for complex biological analytes. This aspect is more clearly illustrated by the spectrum in Fig. 8.10b, where 100 scans were acquired for a sample concentration of only 6 pmol μl^{-1}.

The coupling of fully automated chip-based nanoESI system (NaoMate robot) to high-performance mass spectrometry was introduced in clinical analyses and in particular for LSDs, almost 10 years ago [66–68].

In 2004, the newly conceived NanoMate/FTICR coupling was tested for high-performance mapping and sequencing by sustained off-resonance collision-induced tandem mass spectrometry (SORI-CID MS^2) of *O*-glycosylated sialylated peptides and amino acids from the urine of a patient suffering from Schindler disease and from the urine of a healthy age-matched individual for comparative assay [66]. The mixtures were dissolved in methanol to different concentrations and transferred to the glass plate of the NanoMate™ 100 robot. The initiation of the electrospray required a starting back pressure of about 6,200 Pa, which, in order to enhance the stability of the spray, was slightly decreased to lower values during acquisition [66]. For the initiation of the chip spray and the efficient transferring of the ionic species into the mass spectrometer, a fine adjustment of the NanoMate chip position with respect to the FTICR ion transfer capillary has been reported to be a critical step.

An extensive study on the limit of sensitivity of the method was carried out by stepwise dilutions to 5 pmol μl^{-1} (Fig. 8.11), 2.5 pmol μl^{-1}, 1 pmol μl^{-1} and 0.5 pmol μl^{-1}, respectively; the highest sensitivity obtained in this set of experiments

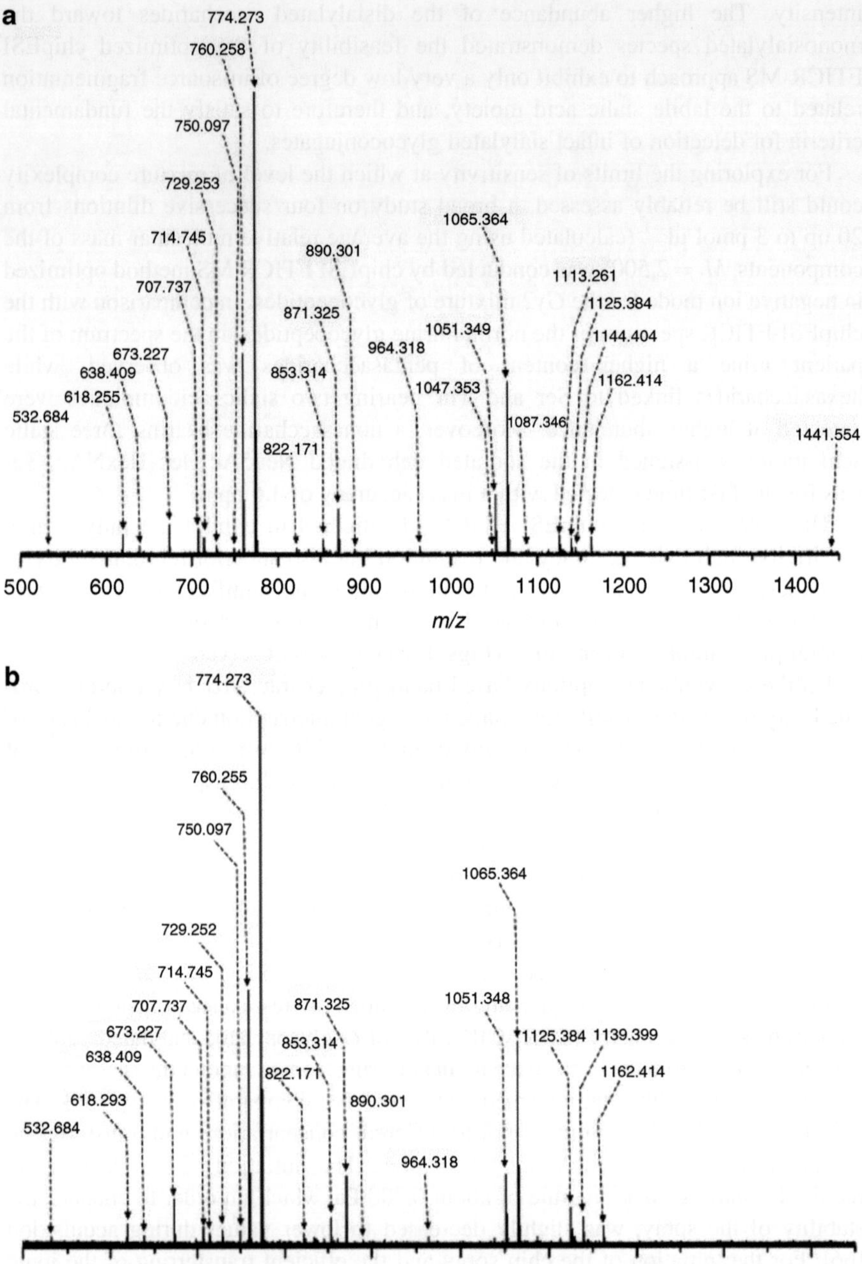

Fig. 8.10 (−)ChipESI FTICR MS of the *Gy2* mixture of *O*-glycosylated peptides from the urine of a patient diagnosed with Schindler disease. (**a**) Sample concentration: 12 pmol μl^{-1} in methanol. Number of scans: 50; (**b**) sample concentration: 6 pmol μl^{-1} in methanol. Number of scans: 100. Reprinted with permission from [30]

m/z Exp.	Theor.	Charge state	Proposed structure
525.1738	525.1755	2−	$Neu5Ac_2HexHexNAc$-Ser
532.1818	532.1834	2−	$Neu5Ac_2HexHexNAc$-Thr
571.7029	571.7045	2−	$Neu5Ac_2HexHexNAc$-Thr-Pro −H_2O
580.7084	580.7097	2−	$Neu5Ac_2HexHexNAc$-Thr-Pro
608.2083	608.2020	3−	$Neu5Ac_2Hex_2HexNAc_4$-Ser −H_2O (Na)
612.8803	612.8739	3−	$Neu5Ac_2Hex_2HexNAc_4$-Thr −H_2O (Na)
707.7408	707.7416	2−	$Neu5Ac_2Hex_2HexNAc_2$-Ser
714.7485	714.7495	2−	$Neu5Ac_2Hex_2HexNAc_2$-Thr
760.2624	760.2629	1−	Neu5AcHexHexNAc-Ser
763.2740	763.2758	2−	$Neu5Ac_2Hex_2HexNAc_2$-Thr-Pro
774.2792	774.2786	1−	Neu5AcHexHexNAc-Thr
826.9557	826.9446	3−	$Neu5Ac_3Hex_3HexNAc_5$-Ser −H_2O (Na)
831.6220	831.6165	3−	$Neu5Ac_3Hex_3HexNAc_5$-Thr −H_2O (Na)
834.2789	834.2719	3−	$Neu5Ac_3Hex_3HexNAc_5$-Ser −H_2O (Na)
890.3063	890.3077	2−	$Neu5Ac_2Hex_3HexNAc_3$-Ser
912.8175	912.8067	2−	$Neu5Ac_2Hex_2HexNAc_4$-Ser −H_2O (Na)
923.8091	923.7977	2−	$Neu5Ac_2Hex_2HexNAc_4$-Ser −H_2O (2Na)
930.8161	930.8055	2−	$Neu5Ac_2Hex_2HexNAc_4$-Thr −H_2O (2Na)
1051.3576	1051.3584	1−	$Neu5Ac_2HexHexNAc$-Ser
1065.3747	1065.3740	1−	$Neu5Ac_2HexHexNAc$-Thr

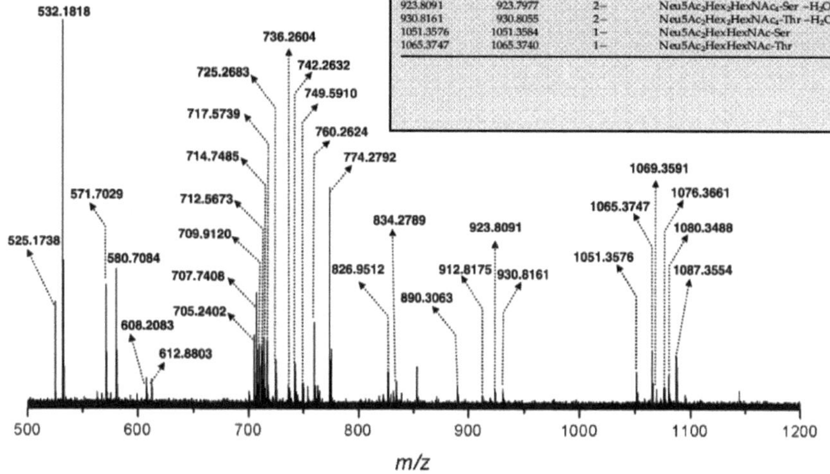

Fig. 8.11 Automated (-) chipESI FTICR MS of a fraction from the urine of a patient suffering from Schindler disease. Sample concentration: 5 pmol μl^{-1}, in methanol. Number of scans: 150. Average sample consumption for MS experiment: 2.5 pmol. *Inset table*: Ions detected and identified with a mass accuracy below 12 ppm in the mixture of *O*-glycosylated peptides at 5 pmol μl^{-1}. Reprinted with permission from [66]

was for 0.5 pmol μl^{-1}. Hence as compared with previously reported data on a similar glycopeptide mixture obtained by regular capillary-based (-) nanoESI FTICR MS [69], the NanoMate system was found to exhibit a superior sensitivity, a higher ionization efficiency of even less abundant glycopeptide species and a preferential formation of multiply charged ions. A considerable number of triply charged ions could be formed and detected with a high S/N ratio, whereas, by capillary-based (-) nanoESI FTICR MS, no triply charged glycopeptide ions were formed/detected. Additionally, four new components, not identified by any other method before were detected by this novel approach and identified under a high mass accuracy.

For the identification of potential diagnostic marker components, the *Ty* mixture of glycopeptides extracted from the urine of a healthy control person was further investigated by (-) nanoESI chip FTICR MS under the same ionization/detection conditions as those from patient urine analysis.

If in patient urine glycoconjugate species expressing saccharide chain lengths up to undecasaccharide have been identified, in healthy control urine, besides their

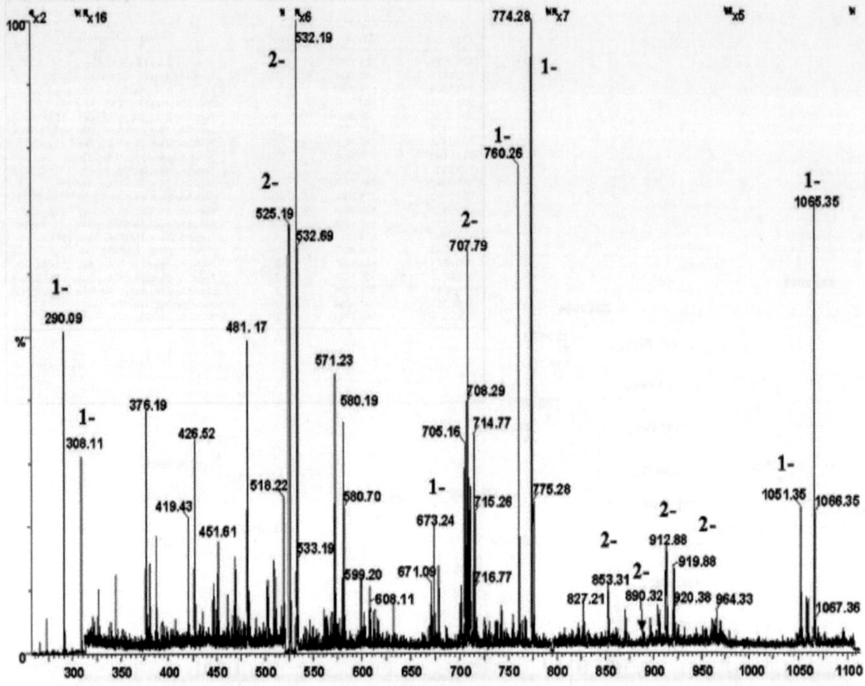

Fig. 8.12 Automated chip-based (-) nanoESI QTOF MS of *BPy O*-glycopeptide mixture. Sample concentration: 3 pmol μl^{-1}, in methanol. Signal acquisition: 1.3 min. Reprinted with permission from [67]

reduced number, the dominating species were those with shorter chains (up to hexasaccharide units) and lower degree of sialylation.

In the first study [67], complex carbohydrate mixtures from the urine of a patient suffering from hereditary *N*-acetylgalactosamine deficiency (Schindler's disease) was submitted to glycoscreening by fully automated chip-based (-) nanoESI QTOF MS. Using a sample concentration of about 3 pmol μl^{-1} for *BPy* fraction of *O*-glycopeptide mixture extracted from the urine of a patient suffering from Schindler disease, the mass spectrum displayed a high level of heterogeneity concerning the glycan chain lengths, originating from tri- up to decasaccharide units *O*-linked either to Ser, Thr, or the threonine-proline (Thr-Pro) dipeptide (Fig. 8.12). Remarkably, in such a configuration, a high signal/noise ratio could be obtained within only 30 s of acquisition. CID at variable collision energies (VE-CID) was further used for the fragmentation of the disialooctasaccharide *O*-linked to Ser at *m/z* 890.32 (Fig. 8.13).

MS experiments carried out under the same conditions of sample solution and ionization/detection indicated a run-to-run and nozzle-to-nozzle reproducibility of nearly 100 %. At the same time, this approach significantly improved the speed in sample delivery and data collection, simplified the sample handling steps and minimized the analyte loss.

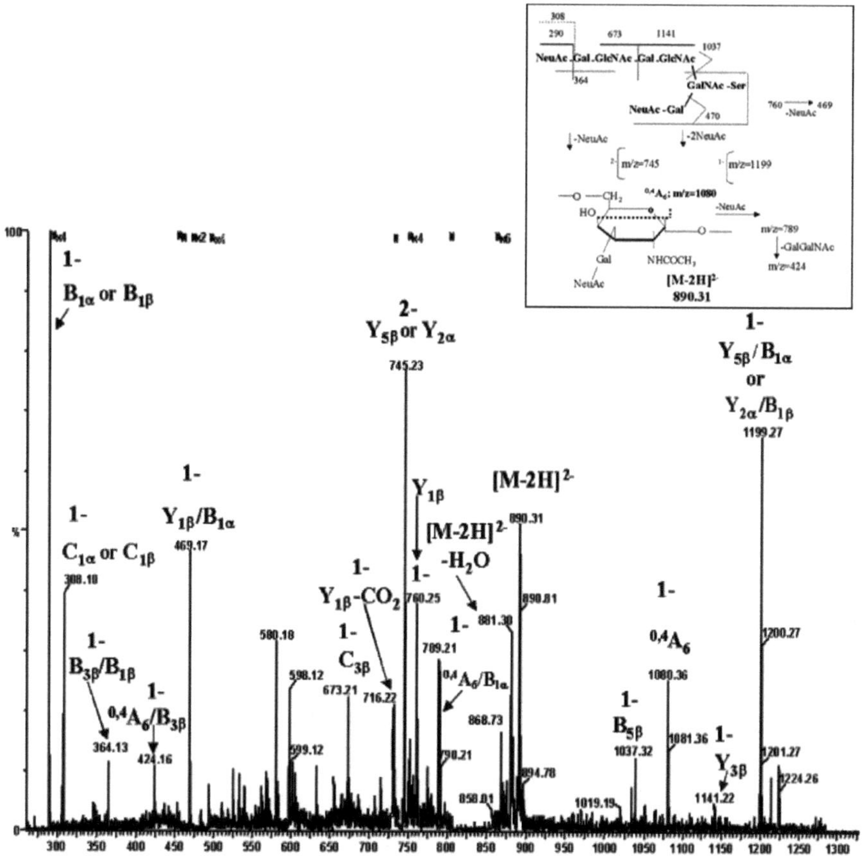

Fig. 8.13 Fragmentation spectrum obtained by automated chip-based (-) nanoESI QTOF MS/MS derived by using $NeuAc_2Gal_3GlcNAc_2GalNAcSer$ detected as a doubly charged ion at *m*/*z* 890.32 across the 25–40 eV collision energy range. *Inset*: Fragmentation scheme of $NeuAc_2Gal_3Glc$-$NAc_2GalNAc$-Ser. Reprinted with permission from [67]

The combination of fully automated chip-based nanoESI NanoMate™ 100 robot with the quadrupole time-of-flight MS for the glycopeptides screening in the urine of a Schindler disease patient was reported in two studies [67, 68], in one of them the platform being also off-line coupled with capillary electrophoresis separation technique [68].

NanoMate QTOF MS methodology was found advantageous for discovery of novel carbohydrate species in complex biological samples due to the increased sensitivity, reproducibility, ionization efficiency and by the ability to generate a sustained and constant electrospray.

An efficient and versatile method for separation of complex mixtures obtained from biological sources is represented by capillary electrophoresis (CE). The compatibility of the same platform described above has been tested in conjunction with off-line capillary electrophoresis for compositional mapping and structural investigation of a fraction denoted *BQ5*, obtained from the urine of a Schindler disease type I patient [68]. Off-line CE MS, which involves collection of CE separated fractions and their subsequent MS analysis, might be considered as a convenient approach due to its flexibility toward system optimization since: (1) the experimental conditions for the CE and MS instruments can be set separately; (2) different types and configurations of MS ion sources can be used. However, off-line CE MS exhibits a major drawback, namely the lack of sensitivity since the separated components are collected and inevitably diluted in the CE buffer [68].

By combining for the first time the off-line CE, NanoMate robot and QTOF MS in a 3-stage platform this disadvantage was eliminated due to the elevated sensitivity of the NanoMate QTOF MS.

All the achievements presented here highlight the emerging role of chip-based mass spectrometry as a viable routine tool for early diagnosis of lysosomal storage diseases. In combination with high-performance mass analyzers, these devices characterized by reproducibility, speed, sensitivity and low cost of analysis will certainly become in the near future widespread used in the management of these rare, however, devastating diseases.

Acknowledgements The financial support provided to M.S. through the strategic grant POSDRU/159/1.5/S/137750, Project "Doctoral and Postdoctoral programs support for increased competitiveness in Exact Sciences research" co-financed by the European Social Fund within the Sectorial Operational Programme Human Resources Development 2007–2013 is gratefully acknowledged. A.D.Z. thanks to the Romanian National Authority for Scientific Research, ANCS/UEFISCDI, for the financial support through the research projects PN-II-ID-PCE-2011-3-0047, PN-II-RU-TE-2011-2-0008, PN-II-PT-PCCA-2013-4-0191 contract No. 118/2014 and to the European Commission, project FP7 Marie Curie-PIRSES-G A-2010-269256.

References

1. So PK, Hu B, Yao ZP (2013) Mass spectrometry: towards in vivo analysis of biological systems. Mol Biosyst 9(5):915–929
2. Banerjee S, Mazumdar S (2012) Electrospray ionization mass spectrometry: a technique to access the information beyond the molecular weight of the analyte. Int J Anal Chem 282574:2012
3. Dunn WB (2011) Mass spectrometry in systems biology an introduction. Methods Enzymol 500:15–35
4. Zeleny J (1917) Instability of electrified liquid surfaces. Phys Rev 10:1–6
5. Dole M, Mack LL, Hines RL et al (1968) Molecular beam of macroions. J Chem Phys 49:2240–2249
6. Fenn JB, Mann M, Meng CK et al (1989) Electrospray ionization for mass spectrometry of large biomolecules. Science 246:64–71
7. Aleksandrov ML, Gall LN, Krasnov VN et al (1984) On the working characteristics of an ion source with eletrohydrodynamic introduction of liquids into the mass spectrometer. Bioorg Khim 10:710–712
8. Wilm M, Mann M (1996) Analytical properties of the nanoelectrospray ion source. Anal Chem 68(1):1–8

9. Feng B, Smith RD (2000) A simple nanoelectrospray arrangement with controllable flowrate for mass analysis of submicroliter protein samples. J Am Soc Mass Spectrom 11(1):94–99
10. Wilm M (2011) Principles of electrospray ionization. Mol Cell Proteomics 10(7): M111.009407
11. Crotti S, Seraglia R, Traldi P (2011) Some thoughts on electrospray ionization mechanisms. Eur J Mass Spectrom (Chichester, Eng) 17(2):85–99
12. Liuni P, Wilson DJ (2011) Understanding and optimizing electrospray ionization techniques for proteomic analysis. Expert Rev Proteomics 8(2):197–209
13. Jaquillard L, Saab F, Schoentgen F, Cadene M (2012) Improved accuracy of low affinity protein-ligand equilibrium dissociation constants directly determined by electrospray ionization mass spectrometry. J Am Soc Mass Spectrom 23(5):908–922
14. Hofstadler SA, Sannes-Lowery KA (2006) Applications of ESI-MS in drug discovery: interrogation of noncovalent complexes. Nat Rev Drug Discov 5(7):585–595
15. Beck JL, Colgrave ML, Ralph SF, Sheil MM (2001) Electrospray ionization mass spectrometry of oligonucleotide complexes with drugs, metals, and proteins. Mass Spectrom Rev 20 (2):61–87
16. Potier N, Rogniaux H, Chevreux G, Van Dorsselaer A (2005) Ligand-metal ion binding to proteins: investigation by ESI mass spectrometry. Methods Enzymol 402:361–389
17. Nguyen S, Fenn JB (2007) Gas-phase ions of solute species from charged droplets of solutions. Proc Natl Acad Sci U S A 104(4):1111–1117
18. Lorenz UJ, Zewail AH (2014) Nanofluidics. Observing liquid flow in nanotubes by 4D electron microscopy. Science 344(6191):1496–1500
19. Segerink LI, Eijkel JC (2014) Nanofluidics in point of care applications. Lab Chip 14:3201–3205
20. Sandetskaya N, Allelein S, Kuhlmeier D (2013) Application of nanotechnology in miniaturized systems and its use for advanced analytics and diagnostics - an updated review. Recent Pat Food Nutr Agric 5(3):220–238
21. Gao D, Liu H, Jiang Y, Lin JM (2013) Recent advances in microfluidics combined with mass spectrometry: technologies and applications. Lab Chip 13(17):3309–3322
22. Lorenz M, Ovchinnikova OS, Van Berkel GJ (2014) Fully automated laser ablation liquid capture surface analysis using nanoelectrospray ionization mass spectrometry. Rapid Commun Mass Spectrom 28(11):1312–1320
23. Zamfir AD (2014) Neurological analyses: focus on gangliosides and mass spectrometry. Adv Exp Med Biol 806:153–204
24. Zhang S, Van Pelt CK, Wilson DB (2003) Quantitative determination of noncovalent binding interactions using automated nanoelectrospray mass spectrometry. Anal Chem 75 (13):3010–3018
25. Van Pelt CK, Zhang S, Fung E et al (2003) A fully automated nanoelectrospray tandem mass spectrometric method for analysis of Caco-2 samples. Rapid Commun Mass Spectrom 17 (14):1573–1578
26. Zhang S, Van Pelt CK, Henion JD (2003) Automated chip-based nanoelectrospray-mass spectrometry for rapid identification of proteins separated by two-dimensional gel electrophoresis. Electrophoresis 24(21):3620–3632
27. Sarbu M, Ghiulai RM, Zamfir AD (2014) Recent developments and applications of electron transfer dissociation mass spectrometry in proteomics. Amino Acids 46(7):1625–1634
28. Sainiemi L, Sikanen T, Kostiainen R (2012) Integration of fully microfabricated, three-dimensionally sharp electrospray ionization tips with microfluidic glass chips. Anal Chem 84(21):8973–8979
29. Zamfir AD, Lion N, Vukelic Z et al (2005) Thin chip microsprayer system coupled to quadrupole time-of-flight mass spectrometer for glycoconjugate analysis. Lab Chip 5 (3):298–307

30. Bindila L, Froesch M, Lion N et al (2004) A thin chip microsprayer system coupled to Fourier transform ion cyclotron resonance mass spectrometry for glycopeptide screening. Rapid Commun Mass Spectrom 18(23):2913–2920
31. Lion N, Reymond F, Girault HH, Rossier JS (2004) Why the move to microfluidics for protein analysis? Curr Opin Biotechnol 15(1):31–37
32. Feng X, Liu BF, Li J, Liu X (2014) Advances in coupling microfluidic chips to mass spectrometry. Mass Spectrom Rev. doi:10.1002/mas.21417 [Epub ahead of print]
33. Cortes DF, Kabulski JL, Lazar AC, Lazar IM (2011) Recent advances in the MS analysis of glycoproteins: capillary and microfluidic workflows. Electrophoresis 32(1):14–29
34. Zamfir AD, Bindila L, Lion N et al (2005) Chip electrospray mass spectrometry for carbohydrate analysis. Electrophoresis 26(19):3650–3673
35. de Duve C (1959) Lysosomes, a new group of cytoplasmic particles. Subcellular particles. Ronald Press, New York
36. de Duve C (1975) Exploring cells with a centrifuge. Science 189(4198):186–194
37. de Duve C (1964) From cytases to lysosomes. Fed Proc 23:1045–1049
38. de Duve C (1983) Lysosomes revisited. Eur J Biochem 137:391–397
39. Walkley SU, Vanier MT (2009) Secondary lipid accumulation in lysosomal disease. Biochim Biophys Acta 1793(4):726–736
40. Lamanna WC, Lawrence R, Sarrazin S, Esko JD (2011) Secondary storage of dermatan sulfate in Sanfilippo disease. J Biol Chem 286(9):6955–6962
41. Winchester B (2004) Lysosomal disorders of the brain. Oxford University Press, Oxford
42. Saftig P, Klumperman J (2009) Lysosome biogenesis and lysosomal membrane proteins: trafficking meets function. Nat Rev Mol Cell Biol 10(9):623–635
43. Platt FM, Boland B, van der Spoel AC (2012) Lysosomal storage disorders: the cellular impact of lysosomal dysfunction. J Cell Biol 199(5):723–734
44. Meikle PJ, Hopwood JJ, Clague AE, Carey WF (1999) Prevalence of lysosomal storage disorders. JAMA 281(3):249–254
45. Pinto R, Caseiro C, Lemos M et al (2004) Prevalence of lysosomal storage diseases in Portugal. Eur J Hum Genet 12(2):87–92
46. Meikle, P. J, Fuller, M, Hopwood, J. J. Mass spectrometry in the study of lysosomal storage disorders. Cell. Mol. Biol. (Noisy-le-grand), 49(5), 769-77, 2003.
47. Meikle PJ, Ranieri E, Meikle PJ, Simonsen H et al (2004) Newborn screening for lysosomal storage disorders: clinical evaluation of a two-tier strategy. Pediatrics 114(4):909–916
48. Nielsen TC, Rozek T, Hopwood JJ, Fuller M (2010) Determination of urinary oligosaccharides by high-performance liquid chromatography/electrospray ionization–tandem mass spectrometry: application to Hunter syndrome. Anal Biochem 402(2):113–120
49. Wenger DA, Coppola S, Liu SL (2002) Lysosomal storage disorders: diagnostic dilemmas and prospects for therapy. Genet Med 4(6):412–419
50. Wynn R (2011) Stem cell transplantation in inherited metabolic disorders. American Society of Hematology, ASH Education Book, Washington, DC
51. Cox TM (2005) Substrate reduction therapy for lysosomal storage diseases. Acta Paediatr Suppl 94(447):69–75
52. Sands MS, Davidson BL (2006) Gene therapy for lysosomal Storage diseases. Mol Ther 13 (5):839–849
53. Seregin SS, Amalfitano A (2011) Gene therapy for lysosomal storage diseases: progress, challenges and future prospects. Curr Pharm Des 17(24):2558–2574
54. la Marca G, Casetta B, Malvagia S et al (2009) New strategy for the screening of lysosomal storage disorders: the use of the online trapping-and-cleanup liquid chromatography/mass spectrometry. Anal Chem 81(15):6113–6121
55. Kasper DC, Herman J, De Jesus VR et al (2010) The application of multiplexed, multi-dimensional ultra-high-performance liquid chromatography/tandem mass spectrometry to the high-throughput screening of lysosomal storage disorders in newborn dried bloodspots. Rapid Commun Mass Spectrom 24(7):986–994

56. Metz TF, Mechtler TP, Orsini JJ et al (2011) Simplified newborn screening protocol for lysosomal storage disorders. Clin Chem 57(9):1286–1294
57. Duffey TA, Bellamy G, Elliott S et al (2010) A tandem mass spectrometry triplex assay for the detection of Fabry, Pompe, and mucopolysaccharidosis-I (Hurler). Clin Chem 56 (12):1854–1861
58. Orsini JJ, Martin MM, Showers AL (2012) Lysosomal storage disorder 4 + 1 multiplex assay for newborn screening using tandem mass spectrometry: application to a small-scale population study for five lysosomal storage disorders. Clin Chim Acta 413(15–16):1270–1273
59. Spacil Z, Tatipaka H, Barcenas M et al (2013) High-throughput assay of 9 lysosomal enzymes for newborn screening. Clin Chem 59(3):502–511
60. Brand GD, de Matos HC, da Cruz GC et al (2013) Diagnosing lysosomal storage diseases in a Brazilian non-newborn population by tandem mass spectrometry. Clinics (Sao Paulo) 68 (11):1469–1473
61. Dajnoki A, Fekete G, Keutzer J et al (2010) Newborn screening for Fabry disease by measuring GLA activity using tandem mass spectrometry. Clin Chim Acta 411 (19–20):1428–1431
62. Spáčil Z, Elliott S, Reeber SL et al (2011) Comparative triplex tandem mass spectrometry assays of lysosomal enzyme activities in dried blood spots using fast liquid chromatography: application to newborn screening of Pompe, Fabry, and Hurler diseases. Anal Chem 83 (12):4822–4828
63. Flangea C, Mosoarca C, Cozma C et al (2013) Testing the feasibility of fully automated chip-based nanoelectrospray ionization mass spectrometry as a novel tool for rapid diagnosis of Fabry disease. Electrophoresis 34(11):1572–1580
64. Zarate YA, Hopkin RJ (2008) Fabry's disease. Lancet 372(9647):1427–1435
65. Marsden D, Levy H (2010) Newborn screening of lysosomal storage disorders. Clin Chem 56 (7):1071–1079
66. Froesch M, Bindila LM, Baykut G et al (2004) Coupling of fully automated chip electrospray to Fourier transform ion cyclotron resonance mass spectrometry for high-performance glycoscreening and sequencing. Rapid Commun Mass Spectrom 18(24):3084–3092
67. Zamfir A, Vakhrushev S, Sterling A et al (2004) Fully automated chip-based mass spectrometry for complex carbohydrate system analysis. Anal Chem 76(7):2046–2054
68. Bindila L, Almeida R, Sterling A et al (2004) Off-line capillary electrophoresis/fully automated nanoelectrospray chip quadrupole time-of-flight mass spectrometry and tandem mass spectrometry for glycoconjugate analysis. J Mass Spectrom 39:1190–1201
69. Froesch M, Bindila L, Zamfir A, Peter-Katalinić J (2003) Sialylation analysis of O-glycosylated sialylated peptides from urine of patients suffering from Schindler's disease by Fourier transform ion cyclotron resonance mass spectrometry and sustained off-resonance irradiation collision-induced dissociation. Rapid Commun Mass Spectrom 17(24):2822–2832

Chapter 9
Multidynamic Liposomes in Nanomedicine: Technology, Biology, Applications, and Disease Targeting

Shyamasree Ghosh and Waliza Ansar

Abstract Biocompatible and biodegradable liposome nanomedicine are finding large-scale application in delivery of drugs, RNA, antisense oligonucleotides, DNA, plasmids carrying gene, antioxidants to targeted sites for diagnosis, and treatment at cellular and molecular level of disease both in human and livestock.

Conventional drug delivery system suffers from the disadvantage of narrow therapeutic index, poor serum solubility, and nonspecific target effects. Circumventing this problem, liposome drug nanoparticles are designed with the advantage of enhanced permeability, retention effect, selective targeted delivery, stimuli-induced release that can evade the host defense system, cross the blood–brain barrier, thus conferring, synergetic effects, small interfering RNA (siRNA) co-delivery, multimodality therapies, and oral delivery of therapeutics at disease sites. Liposomes with diverse range of modifications are being designed to generate improvised medicine delivery systems. Immuno-liposomes, linked with antibodies are promising tools for targeted drug delivery to specific cancer cells. In the recent times several anti-cancer agents in nanodelivery systems are under preclinical and clinical studies. Lipid-based liposomal delivery systems of self-amplifying RNA vaccines are showing promising effects.

Globally research is being carried out in generating liposomes with higher compatibility within the biological system in targeting a wide array of diseases including cancer, overcoming multidrug resistance, inflammatory diseases, thrombosis, psoriasis, muscular dystrophy, spinal cord injury, neurologic diseases, diabetes, parasite induced diseases, and cardiovascular diseases.

Authors have equal contributions

S. Ghosh (✉)
School of Biological Sciences, National Institute of Science, Education and Research (NISER), Institute of Physics Campus, Sachivalaya Marg, PO: Sainik School, Bhubaneswar 751 005, India
e-mail: sree.s@niser.ac.in; shyamasree_b@yahoo.com

W. Ansar
Post Graduate Department, Asutosh College, 92, Syamaprosad Mookerjee Road, Kolkata 700026, India

M. Lungu et al. (eds.), *Nanoparticles' Promises and Risks*,
DOI 10.1007/978-3-319-11728-7_9

In this chapter, we have summarized the recent developments in the field of liposomal nanomedicine under the following topics: (i) diverse formulation of liposome nanomedicine, (ii) agents delivered by nanoliposomes, (iii) nanoliposomes and their application as vaccines, (iv) liposomal nanomedicine in targeting diseases, (v) liposomal nanomedicine and stem cell therapy, (vi) their application in preclinical and clinical studies, risks, and (vii) safety issues.

Abbreviations

5-FU	5-Fluorouracil
aHSC	Allogeneic hematopoietic stem cell
aHSCRs	Allogeneic hematopoietic stem cell recipients
aHSCT	Allogeneic hematopoietic stem cell transplantation
AL	Anionic liposomes
AmB	Amphotericin B
ASOs	Antisense oligonucleotides
Au	Gold
BC	Breast cancer
BCSCs	Breast cancer stem cells
CAS	Caspofungin
CDDP	*cis*-Diamminedichloroplatinum(II). CDDP, a chemotherapy drug, is another name for Cisplatin or cisplatinum
cGMP	Cyclic guanosine monophosphate
CHEMS	Cholesteryl hemisuccinate
CHOL	Cholesterol
CLP	Cationic liposomes
CS/DNA	Chitosan/DNA
CSF	Colony-stimulating factor
CSCs	Cancer stem cells
CTL	Cytotoxic T cell
CXC	A subfamily of chemokine
DCP	Dicetyl phosphate
DexP	Dexamethasone palmitate
DMRIE/DOPE	1,2-Dimyristyloxy-propyl-3-dimethyl-hydroxy ethyl ammonium bromide/dioleoylphosphatidylethanolamine
DMSO	Di methyl sulfoxide
DOPC	Dioleoyl phosphatidyl Choline
DOPE	Dioleoyl phosphatidyl ethanolamine
DOTAP	1,2-Dioleoyl-3-trimethylammonium-propane
DOX	Doxorubicin
DPPC	Di-palmitoyl phosphatidyl choline
DTPA	Diethylene triamine pentaacetic acid
ELISA	Enzyme-linked immunosorbent assay

ENV	*Env* is a viral protein that serves to form the viral envelope. The expression of the *env* gene enables retroviruses to target and attach to specific cell types and to infiltrate the target cell membrane
EPR	Enhanced permeation and retention
FDA	Food and drug administration
GD	A type of ganglioside
GFRs	Growth factor receptors
GCs	Glucocorticoids
GC	Glucosyl ceramide
G-CSF	Granulocyte colony-stimulating factor
GMP	Guanosine monophosphate
GPCRs	G-protein coupled receptors
HGF	Hepatocyte growth factor
HLA	Human leukocyte antigen
HRP	Horseradish peroxidase
HSC	Hematopoietic stem cell
HSCT	Hematopoietic stem cell transplantation
HSPCs	Hematopoietic stem/progenitor cells
IFN	Interferon
IL	Interleukin
LAmB	Liposomal Amphotericin B
LN	Liposomal nanoparticles
LPD	Liposome protamine/DNA lipoplex
MAS	Macrophage activation syndrome
MCFAs	Medium-chain fatty acids
MDR	Multi drug resistant
MRI	Magnetic imaging resonance
MRP	Multi drug resistant associated protein
MLV	Multilamellar vesicle
NLP	Neutral liposomes
NGR	Asparaginen-Glycine-Argine peptide motif
PD	Parkinson's disease
PBS	Phosphate buffered saline
PC	Phosphatidyl choline
pDNA	Plasmid DNA
PE	Phosphatidyl ethanolamine
PEG	Polyethylene glycol
PEGylation	Process of covalent attachment of PEG polymer chains to another molecule normally a drug or therapeutic protein
PLD	PEGylated liposomal doxorubicin
PEI	Polyethylenimine
PFV	Preformed vesicle, a type of liposome
PL	Poly-lysine
PLA	Polylactic acid

PS	Phosphatidylserine
RBC	Red blood cells
rMSCs	Rat mesenchymal stem cells
rBMSCs	Rat bone marrow stem cells
RE	Reticulo-endothelial
RGD	Arginylglycylaspartic acid, a tripeptide composed of L-arginine, glycine, and L-aspartic acid
SC	Stem cells
siRNA	Small interfering RNA
SLA	Soluble Leishmania antigens
SPECT/CT	Micro-single photon emission computed tomography/computed tomography
TLR	Toll-like receptors
TNF	Tumor necrosis factor
TSL	Thermosensitive liposomes
VEGF	Vascular endothelial growth factor
VEGFR	Vascular endothelial growth factor receptor

9.1 Introduction

The application of nanotechnology to medicine, i.e., nanomedicine and its impact on public health has been evidenced globally. Nanosized biological tools find tremendous application in the diagnosis, prevention, and treatment of diseases. The dynamic several distinct application areas include drug delivery, in vitro diagnostics, drug based therapies, in vivo imaging, and active implants. Over the past few decades, encouraging success has been made in the field of nanomedicine, resulting in a number of more effective and less toxic products and imaging systems enabling therapeutic and diagnostic interventions in hard and soft tissue implants, bone substitute materials, dental restoratives, antibiotic, antiallergic, anticancerous therapeutic materials. The main goal of nanotherapeutics is to improve pharmacokinetics, drug bioavailability, efficacy, and safety to promote the treatment of diseases by successful application of nanotherapeutic products like liposomes, their regulatory pathway, for timely clinical translations and the realization of health care benefits for human and life stock [1–7].

Liposomes, composed of phospholipid bilayers encompassing an aqueous core have been reported of their remarkable ability to affect growth and proliferation, and carry drugs and both hydrophobic and hydrophilic compounds. This property of liposomes finds wide-scale application in the therapeutic and clinical research and as drug delivery devices. The name liposome is derived from two Greek words: “Lipos” meaning fat and “Soma” meaning body. The first liposomes were described by British hematologist Dr Alec D Bangham in 1961 which was published in 1964, at the Babraham Institute, in Cambridge. In the last two decades considerable research globally have led to an array of techniques for encapsulating therapeutic agents including drugs, plasmid DNA (pDNA), antisense oligonucleotides (ASOs), and small interfering RNA (siRNA) with liposomes of 100 nm diameter generating high drug to liposomes nanoparticles ratio thus allowing gene delivery, gene correction, drug delivery with more circulation time, greater retention, and accumulation of therapeutic agents in disease niche. Different drug molecules, including antimicrobial and anticancer agents, peptide hormones, chelating agents, vaccines, enzymes, proteins, and genetic materials, have been incorporated into the aqueous or lipid phases of liposomes, with varied sizes, compositions, and other characteristics, to provide selective delivery to the target site for in vivo therapeutics. Biocompatible, biodegradable, nonimmunogenic, liposomes or drug encapsulated nanoparticle liposomes, coated with bioactive moieties has revolutionized the drug delivery system thereby reducing drug side effects and increasing efficiency and therapeutic property. Researches on diverse modifications of liposomes for targeted drug delivery are in process. In this chapter, we summarized the recent developments in the design and modification of liposomes and their potential application in delivery of drugs and diverse types of therapeutic agents applied in targeting a vast array of diseases.

9.1.1 Liposome Nanomedicine: Formulation

Liposomes are the oldest nanotherapeutic bioagent that have been used clinically. The designing of liposomal products incorporates consideration of the critical parameters that determine the in vivo liposomal performance for parenteral administration including particle size, surface charge, membrane composition, and fluidity. The first liposomal formulation of the cytotoxic anticancer drug doxorubicin (DOX) were launched in the early 1990s [6, 7]. Liposome encapsulation of DOX increases its half-life and distribution in tumor tissues, with reduced acute and cumulative cardiotoxicity. Doxil, the PEGylated liposomal formulation of DOX is being applied in the treatment of refractory Kaposi's sarcoma, ovarian and recurrent breast cancers (BCs) [5–7]. Liposome encapsulated therapeutic AmBisome from Gilead Sciences Ltd has been clinically approved over amphotericin B to overcome the associated side effects like infusion-related reactions and nephrotoxicity on chronic usage of amphotericin B. The increased uptake of amphotericin B-loaded liposomes by the reticulo endothelial (RE) system is noted for the small size and negative charge of the liposomes [5].

Several techniques, including the dense gas and supercritical fluid techniques, ether/ethanol injection, Bangham method, detergent depletion, reverse-phase evaporation, and emulsion methods, have been reported for preparing liposomes (Fig. 9.1a, b) with high-entrapment efficiency with increased therapeutic index, narrow particle size distribution, long-term stability via encapsulation, site avoidance effect with reduced elimination, selective passive targeting to tumor tissues (liposomal DOX), and target-specific active drug loading property. Due to the differences in preparation methods and lipid composition, liposomes were

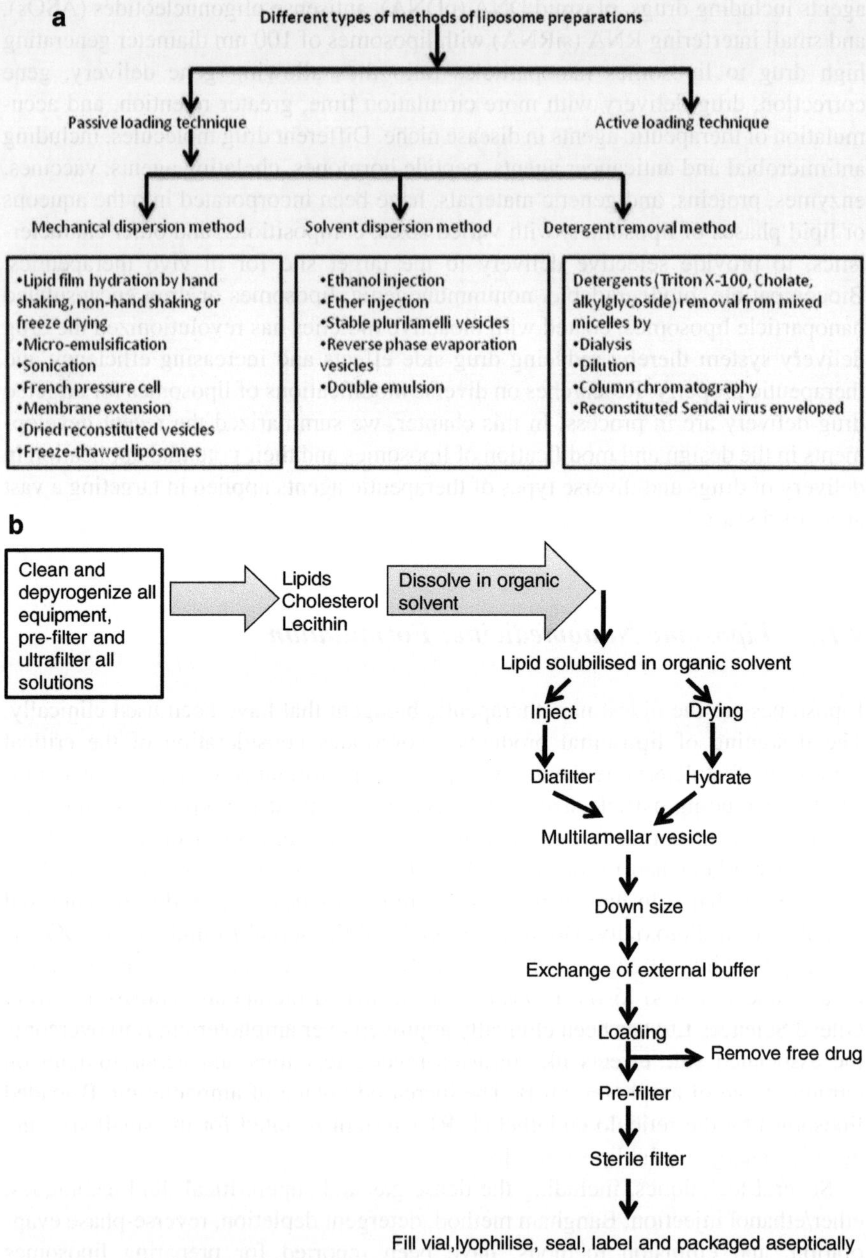

Fig. 9.1 (**a**) The different methods employed for preparation of liposomes are tabulated schematically [5–9]. (**b**) Steps of liposome preparations represented in a flow diagram, starting with the mixture of lipids and terminating in packaging of liposomes [5–9]

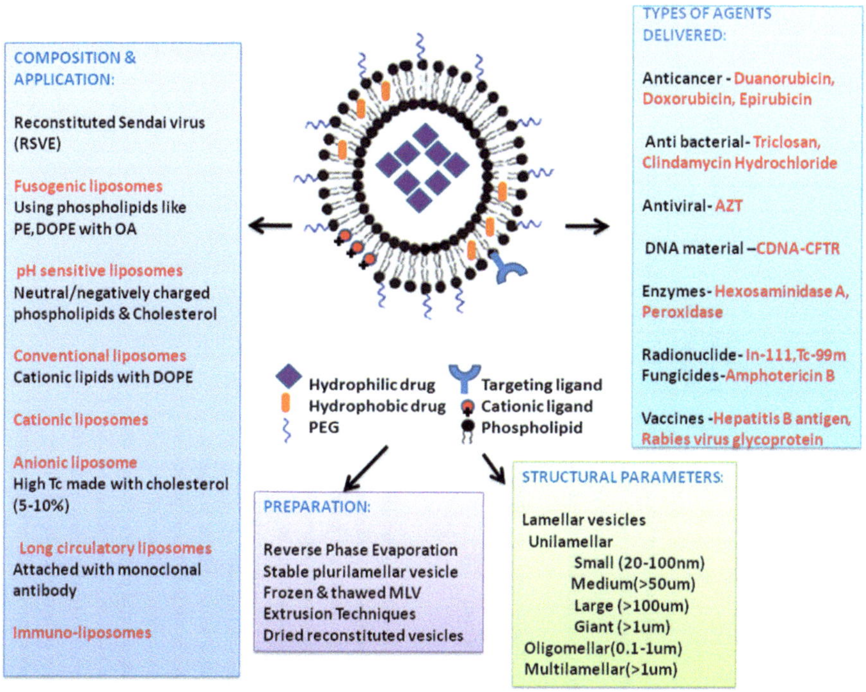

Fig. 9.2 Composition, application, preparation, structural parameters, and agents delivered by nanoliposome

classified according to their lamellarity (uni- and multi-lamellar vesicles, MLV), intermediate, or large; and surface charge (anionic, cationic, or neutral) (Fig. 9.2). Liposomes without any aqueous material are called micelles; however, reverse micelles can be made to encompass an aqueous environment. Different types of liposomes, including cationic, temperature sensitive liposomes, pH sensitive and virosomes, have been designed by modulating the formulation techniques and lipid composition. Liposomes are prepared from diverse sources including mixed lipid chains (like egg phosphatidyl ethanolamine, PE and phosphatidyl choline, PC), naturally derived phospholipids and of pure surfactant components like DOPE (dioleoyl phosphatidyl ethanolamine), polymers like poly (lactic-*co*-glycolic acid) dilinoleoylphosphatidylcholine, synthetic zwitterionic materials such as poly (carboxybetaine) and poly(sulfobetaine), Cholesterol succinyl chitosan like polysaccharide, Carboplatin-phospholipid, and many others as noted in Tables 9.1 and 9.2 [5–7]. Cationic lipid-based liposomes can easily complex with negatively charged nucleic acids via electrostatic interactions, resulting in nanocarriers with biocompatibility, low toxicity, and easy large-scale production required for in vivo clinical trials. Liposomes in the nanosize range have been recognized as a versatile drug delivery system of both hydrophilic and lipophilic molecules.

Table 9.1 Different liposomal formulations as carriers of nanomedicine

Liposome type	Formulation	Agent transported	Mechanism	Disease/cell targeted/agent	Immunological response	References
Polymer core lipid shell	Polymer PLGA–lecithin–PEG core–shell NPs	(1) Docetaxel Doxil/Caelyx and Genexol-PM (2) Clotrimazole (3) Doxorubicin (4) Paclitaxel (5) Fluorouracil	Lipid monolayer acts as a molecular fence and contributes to keep the drug molecules in the hydrophobic core, and water out of the core which would hydrolyze the PLGA polymer and increase erosion and drug release The mechanism of drug release from a polymeric core is typically bulk or surface erosion, or in some cases a combination of both	(1) Hela and HEpG2 (2) Fungal infections in rats (3–5) Metastatic breast cancer, head and neck cancer, gastric cancer, hormone-refractory prostate cancer, and non-small-cell lung cancer	Controlled drug release and physical stability in PBS and plasma. NPs well tolerated by human cell line models, HeLa and HepG2	[103–107]
Hollow core shell type lipid polymer lipid NPs	Polymer: poly(lactic-*co*-glycolic acid) dilinoleoylphosphatidylcholine 1,2-ditetradecanoyl-*sn*-glycero-3-phosphoethanolamine-*N*-diethylenetriaminepentaacetic acid poly (2-methyloxazoline)-block-poly (dimethylsiloxan)-block-poly (2-methyloxazoline) Lipids: Stearic acid Lecithin decanoic acid triolein	Doxorubicin and combretastatin paclitaxel indium 111 and yttrium 90 Fluoroquinolone antibiotics, cisplatin, and gemcitabine	PEGs are being employed in long-circulating NPs formulation, thereby reducing plasma protein adsorption, macrophage uptake, and particle aggregation, while increasing circulation time. Antibodies/other targeting ligands are attached to the surface of nanoliposomes by using various covalent and noncovalent coupling techniques First, the phospholipid forms a bilayer in aqueous solution and attaches to the polystyrene particle surface by adsorption to form homodispersed and stable phospholipid vesicle-covered particles. Second, after bilayer attachment, hydrophobic attractions between the polystyrene surface and hydrocarbon chain of the phospholipid bilayer collapse the bilayer structure and leave a monolayer covering the polymer particle	Adenocarcinoma, breast cancer, resistant cancers.	Complement system activation, plasma/serum protein binding, and coagulation cascade activation	[108–110]

Liposome type	Formulation	Agent transported	Mechanism	Disease/cell targeted/agent	Immunological response	References
Erythrocyte membrane like polymeric NPs	Sub-100-nm polymeric particles are coated with RBC membrane derived vesicles to mimic complex surface chemistry of erythrocyte membrane. Synthetic zwitterionic materials such as poly (carboxybetaine) and poly(sulfobetaine) have been proposed as alternatives to PEG because of their strong hydration that is highly resistant to nonspecific protein adsorption	Doxorubicin	Provides a bilayered medium for transmembrane protein anchorage and avoids chemical modifications that could compromise the integrity and functionalities of target proteins. The drug release is gradual	Acute myeloid leukemia		[111–112]
Oligomannose coated liposomes (OML)	Dipalmitoylphosphatidylethanolcholine, cholesterol and mannosetriose-dipalmitoylphosphatidylethathanolamine at 10:10:1	(1) Ovalbumin A (2) Soluble leishmanial antigen, soluble lysates of *Toxoplasmsa gondii*, *Trypanosoma brucei gambiense*, *Babesia rodhaini* or *Neospora caninum* (2) Soluble leishmanial antigen, soluble lysates of *Toxoplasmsa gondii*, *Trypanosoma brucei gambiense*, *Babesia rodhaini* or *Neospora caninum* (4) Cryj1 allergen	OMLs are recognized by Mannose binding lectin receptors (MBLRs) on Antigen presenting Cell (APC) and enter into cell. After antigen processing and presentation, they are flagged by Class I and Class II MHC molecules, OMLs, activate, mature APC with enhanced expression of costimulatory like CD40, CD80, CD86 and MHC II and proinflammatory cytokines, IL1, IL6. OMLs coated with allergen induce IFN gamma by Th1 cells inhibits Th2 response. Cry J entrapped OML prevents elevation of serum IgE level, control Type I allergy	Infection, cancer, allergy	Induce antigen-specific CD8+ Cytotoxic T Lymphocytes (CTLs), stimulation of invariant natural killer T cells leading to strong cellular immunity	[65, 113–119]

(continued)

Table 9.1 (continued)

Liposome type	Formulation	Agent transported	Mechanism	Disease/cell targeted/agent	Immunological response	References
		(5) OML coated with Human Papilloma virus type 16 E6 and E7 with OML coated papilloma virus (6) HLA-A80201 restricted HTLV-1 (7) Tax peptide				
Polysaccharide anchored liposomes	Cholesterol succinyl chitosan anchored liposomes	Epirubicin	Amphiphilic cholesterol succinyl chitosan (CHCS) conjugates were used as the anchoring materials to coating on the liposome surface by the incubation method. Compared with plain liposomes and chitosan-coated liposomes (CCLs), Epirubicin as a model drug, was effectively loaded into CALs. Drug release sustained both in phosphate buffer solution (PBS, pH 7.4) and 1 % (V/V) aqueous fetal bovine serum (FBS)	Drug release tested on Phosphate buffer and FBS	ND	[120]
	SPC-Cholesterol-R6H4-C18-HA	Paclitaxel (PTX)	SPC and Chol (20:1, w:w) were dissolved in chloroform. PTX with 3.3 % (w:w), C6 with 0.033 % (w:w) and NIRD15 with 3.3 % (w:w) of the total lipids weight were added to prepare drug-loaded nanoliposomes, Cell-penetrating peptides(CPP)-modified liposomes (R6H4-L, R8-L) were obtained by using stearyl CPP (R6H4-C18, R8-C18) as the hydration medium with 2.5 mol% of total lipids weight. HA-coated CPP-modified liposomes (HA-R6H4-L, HA-R8-L) were obtained by adding R6H4-L and R8-L into 1 mg/mL HA solution (1:2, v:v)	Tumor-targeted drug delivery. HA allowed protection and accumulates at the tumor site. pH sensitivity enhances the cellular uptake of the liposomes, cell penetrating effect of R6H4 enables endosomal/lysosomal escape	HepG2 and A549 cells, Heps tumor xenograft models	[121]

Liposome type	Formulation	Agent transported	Mechanism	Disease/cell targeted/agent	Immunological response	References
	O-palmitoyl amylopectin (OPA), a modified polysaccharide, anchored to drug, dipyridamole (DIP) liposomes	Hydrophobic model drug, dipyridamole (DIP), used as a coronary vasodilator and antiplatelet agent in medicine	EPC, cholesterol, and DIP (mass ratio = 100:12:2.25) were dissolved in chloroform. A thin film of dry lipid was hydrated at 40 °C with pH 7.4 PBS. Then, the OPA (formed by esterification of amylopectin in DMSO) anchored DIP liposomes (ODL) were prepared by anchoring O-palmitoyl amylopectin on the surface of PDL	OPA anchored liposomes increases DIP stabilities against irradiation, disperse medium, biofluid, and long-term storage as compared with only liposomes	Male S–D rats	[122]
Phospholipid -based nanoparticle formulation	DOX-Loaded GAGs. Lipids	Glycosaminoglycan hyaluronan, the major ligand of CD44, which is upregulated and undergoes different splice variations in many types of cancer cells	Lipids 1,2-dilauroyl-sn-glycero-3-phosphoethanolamine (DLPE) and 1,2-dilauroyl-sn-glycero-3-phosphoglycerol (DLPG) at a mole ratio of 9:1, respectively, were dissolved in ethanol. After drying and hydrating by MES buffer (0.1 M, pH 5.5) with or without DOX (drug loading was 1:12 drug/lipid (mol/mol). Activated HA (lipid/HA 10:1 (w/w) was linked DSPE-PEG 2000 added to phosphate buffer (pH 6.5, 7.4 or 8.0)/normal saline (pH 4.5–7.0), to form micelles. PP in phosphate buffers or normal saline was added to the SSM dispersion and allowed to equilibrate for 2 h in the dark at 25 °C	Monolayers of NCI/ ADR-RES (NAR) (P-gp-overexpressing human ovarian adenocarcinomaDOX-resistant), OVCAR-8 - (DOX-sensitive cell line), RAW264.7 (mouse macrophages) cells.	DOX-GAG formulation could bypass the P-gp-mediated drug resistance pumps.	[50, 123]
	Pancreatic polypeptide (PP) in sterically stabilized micelles (SSM)	Pancreatic polypeptide (PP)		SK-N-MC (human neuroblastoma) cells which endogenously express PP specific Y4 receptors, treatment of diabetes, obesity	PP-SSM confers stability to the PP against proteolytic digestion	[124]
	Liposomes (100 or 400 nm) were PEGylated and subsets of liposomes were either coated with the fluorescent dye	Dexamethasone	Prepared by the film method. A mixture of chloroform/methanol (10:1 volumetric ratio) containing Dipalmitoyl phosphatidylcholine (DPPC), PEG-(2000)-distearoyl phosphatidylethanolamine	Human primary blood leukocytes, Rheumatoid arthritis,	Reduced cytotoxicity in human fibroblasts, macrophages compared	[125]

(continued)

Table 9.1 (continued)

Liposome type	Formulation	Agent transported	Mechanism	Disease/cell targeted/agent	Immunological response	References
	NBD-PE or additionally loaded with dexamethasone		PEG-(2000)-DSPE, NBD-PE and cholesterol was prepared at a molar ratio of 1.85:0.15:0:1. A fluorescent marker 1 mol % of (*N*-(7-Nitrobenz-2-oxa-1,3-diazol-4-yl)-1,2-dihexadecanoyl-snglycero-3-phosphoethanolamine, triethyl-ammonium salt) NBD-PE was added to the organic phase in relation to the total amount of lipid, including cholesterol. The lipid film was hydrated at 50 °C in an aqueous solution of dexamethasone phosphate in a concentration of 100 mg/mL at a phospholipid concentration of 100 mM	atherosclerosis, colitis, and cancer,	to drug alone. Altered Macrophage maturation and polarization markers. Induced proinflammatory cytokine secretion (TNF, IL1β, IL6) in unstimulated cells, but reduced in inflammatory conditions. Inhibited monocyte and macrophage migration	
	SSM-Indisulam were prepared by coprecipitation method	Indisulam solubilized in sterically stabilized micelles (SSMs)	Mixed micelles containing indisulam were prepared by mixing poly-(ethylene glycol-2000)-grafted distearoyl phosphatidylethanolamine (DSPE-PEG2000) and egg-phosphatidylcholine at optimal molar ratio (90:10) and coprecipitation along with indisulam	MCF-7 breast cancer cell line	SSM is a promising nanocarrier for indisulam	[125]

Liposome type	Formulation	Agent transported	Mechanism	Disease/cell targeted/agent	Immunological response	References
	Carboplatin-phospholipid formulation	Carboplatin, a cisplatin-derived anticancer drug	Carboplatin (5 mmol/L) in Milli-Q water solution was added to a dry film of phospholipids	Carboplatin nanocapsules formulation reduced toxicity of carboplastin alone by 1,000-fold on carcinoma cell lines, improved cytotoxicity in IGROV-1 cells, increased platinum accumulation in the cells, uptake of the formulation by endocytosis fasting diabetic rats	Ovarian Carcinoma cell lines; IGROV-1, OVCAR-3, A2780, NSC lung carcinoma:NCI-H522 and Renal cell carcinoma: A498 cell lines	[126]
	Freeze-drying of emulsions for efficient entrapment of drugs and controlled, effective drug release phospholipid	Aqueous phase containing insulin and oil phase containing phosphatidylcholine	The water in oil type emulsions was prepared by emulsification A mixture of cyclohexane/dichloromethane/diethyl ether (4:1:1, v/v/v) was used as the organic solvent for the emulsion outer oil phase (O). SPC Soybean phosphatidylcholine dissolved in O to give lipid solutions. Insulin was dissolved in water (0.5 mg/mL) forming inner aqueous phase (W)		A significant reduction in the plasma glucose level of fasting diabetic rats after oral administration	[135]

**EE*, entrapment efficiency; *DL*, drug loading; *NR*, not reported; *HPESO*, hydrolyzed polymer of epoxidized soybean oil; *MDR*, multi-drug resistant; *PLGA*, poly(lactic-co-glycolic acid); *DLPC*, dilinoleoylphosphatidylcholine; *DMPE-DTPA*, 1,2-ditetradecanoyl-sn-glycero-3-phosphoethanolamine-N-diethylenetriaminepentaacetic acid; *DSPE-PEG*, 1,2-distearoyl-sn-glycero-3-phosphoethanolamine-N-[amino(polyethylene glycol)]; *PMOXA-PDMSPMOXA*, poly(2-methyloxazoline)-block-poly(dimethylsiloxan)-block-poly(2-methyloxazoline); *DPPC*, dipalmitoylphosphatidylcholine; *PEI*, polyethyleneimine; *EPC*, 1,2-dimyristoleoyl-sn-glycero-3-ethylphosphocholine; PGA, poly(glutamic acid); *DPTAP*, 1,2-dipalmitoyl-3-trimethylammonium-propane; PLA,poly(lactic acid); *OQLCS*, octadecyl-quaternized lysine-modified chitosan; *DHA*, cis-4,7,10,13,16,19-docosahexanoic acid; *PBAE*, poly-(β-amino ester); *MTT*, 3-(4,5-dimethylthiazol-2-yl)-2,5-diphenyltetrazolium bromide; *MTS*, 3-(4,5-dimethylthiazol-2-yl)-5-(3-carboxymethoxyphenyl)-2-(4-sulfophenyl)-2H-tetrazolium; *ATP*, adenosine triphosphate; *EPA*, Eicosapentaenoic acid; PUFAs, Polyunsaturated fatty acids; *PE*, phosphatidyl ethanolamine; *PEG*, poly(ethylene glycol); *AmpB*, amphotericin B; dauno, daunorubicin; dox, doxorubicin HCl; *cisPT*, cisplatin; PGE1, prostaglandin E1; CL, cardiolipin; *Ch*, cholesterol; *PC*, phosphatidyl choline; *PG*, phosphatidyl glycerol [1–7, 103–135]

Table 9.2 Liposomal formulations approved for commercial applications and in clinical trials

Trade name with major usage	Chemical name	Formulations	Route of entry	Company	Status/clinical trial phase
Doxil/Caelyx (Breast and ovarian neoplasm, multiple myeloma, hepatocellular carcinoma)	Doxorubicin	Stealth liposomes (Block copolymeric micelles); MPEG, HSPC, Chol, DPPC, MSPC, and DSPE-PEG$_{2000}$	Suspension (Intravenous)	Ortho Biotech; Schring-Plough; Sequus, Menlo Park, CA, USA	On the market since 1995 (USA) and 1996 (Europe); Phase III
MiKasome (Complicated Urinary Tract Infections, acute infections in cystic fibrosis, nosocomial pneumonia)	Amikacin	Liposome injection	Intravenous	Liposome Co., Princeton, NJ, USA	Phase I
Arikace (Lung infection)	Amikacin	Liposome; DPPC, and chol	Inhaled as aerosol		Phase III
SP1-077, Lipoplatin (Anticancer, mesothelioma)	Cisplatin (cisPT) and its analog	Stealth liposomes; HSPC, cholesterol and DSPE-mPEG, SPC, DPPG, chol, and DSPE-mPEG	Intravenous	NeXstar, Boulder, CO, USA	Phase I/II and III
Stimuvax (Non-small-cell lung carcinoma)	BLP25 lipopeptide	Monophosphoryl lipid A, chol, DMPG, and DPPC	Subcutaneous		Phase III
Ambisome (Fungal infections)	Amphotericin B (AmpB)	Liposomes (PC/Chol/DSPG/HSPC)	Suspension (i.v.)	Gilead Sciences; NeXstar, Boulder, CO, USA	In the market since 1990 (Europe) and 1997 (USA)
DC99 (Breast and ovarian neoplasm, multiple myeloma)	Doxorubicin	Liposomes (Block copolymeric micelles); DPPC, MSPC and DSPE-PEG$_{2000}$	Suspension (i.v.)	Liposome Co., Princeton, NJ, USA	Phase III
DaunoXome (Kaposi's sarcoma)	Dunorubicin	Liposomes (emulsion) DSPC and Chol	Suspension (i.v.)	Gilead Sciences; NeXstar, Boulder, CO, USA	On the market since 1996 (USA and Europe)
CPX-351 (Acute myeloid leukemia II)	Cytarabine and daunorubicin	Liposomes; DSPC, DSPG and chol	Intravenous		Phase II

Trade name with major usage	Chemical name	Formulations	Route of entry	Company	Status/clinical trial phase
CPX-1 (Colorectal cancer)	Irinotecan HCL and floxuridine	Liposomes; DSPC, DSPG and cholesterol	Intravenous		Phase II
Ventus (Acute respiratory distress syndrome)	Prostaglandin E1 (PGE1)	Liposomal PGE1	Mechanical ventilation	Liposome Co., Princeton, NJ, USA	Phase III not successful
Liposomal Nyotran (Fungal infection)	Nystatin	Liposomes	Intravenous	Aronex, The Woodlands, TX, USA	Phase III
Marqibo (Adult patients with acute lymphoblastic leukemia)	Vincristine sulfate	Liposomes; Egg sphingomyelin and chol	Liposomal injections (i.v.)	Spectrum Pharmaceuticals; Inex, Vancouver, BC, Canada	Phase I
Liposomal Annamycin (Antibreast cancer activity)	Annamycin	Liposomes; DSPC, DSPG and tween	Intravenous	Aronex, The Woodlands, TX, USA	Phase I/II
Epaxal (Hepatitis A)	Hepatitis-A vaccine	Liposomes; LECIVA-S70	Intravenous	Cruceal; Swiss Serum Institute, Bern, Switzerland	On Swiss market since 1994
Liposomal Trivalent influenza vaccine/Inflexal Berna V (Influenza)	Trivalent influenza vaccine	Liposomes	Intravenous	Swiss Serum Institute, Bern, Switzerland	Phase III
Atragen (Antirenal cell carcinoma)	Retinoic acid	Liposomes, DMPC, and soybean oil	Intravenous	Aronex, The Woodlands, TX, USA	Phase I/II
Hepuman-Berna (Hepatitis B vaccine)	Hepatitis B vaccine	Liposomes	Intravenous	Swiss Serum Institute, Bern, Switzerland	Phase I
Diphtheria, tetanus, influenza, and hepatitis-A liposomes	Diphtheria, tetanus, influenza, and hepatitis-A	Liposomes	Intravenous	Swiss Serum Institute, Bern, Switzerland	Phase I
ELA-Max (Pain management in children)	Lidocaine	Liposomal lidocaine	Topical anesthetics for venipuncture	Biozone Labs, Pittsburgh, CA, USA	In the US market since 1998

(continued)

Table 9.2 (continued)

Trade name with major usage	Chemical name	Formulations	Route of entry	Company	Status/clinical trial phase
ECOVAX 0157 (*Escherichia coli* vaccine)	*Escherichia coli* vaccine	Synthetic liposomes	Oral	Novavax, Rockville, MD, USA	Phase I
Shigella flexneri 2a oral vaccine (Shigella flexneri vaccine)	*Shigella flexneri* vaccine	Synthetic liposomes	Oral	Novavax, Rockville, MD, USA	Phase I
Depocyt/Liposomal Ara-C (Neoplastic meningitis and lymphomatous meningitis)	Cytarabine	Liposomes (DOPC/DSPC/DSPG/Chol/and triolein)	Suspension (intrathecal)/Spinal	Pacira (formerly SkyePharma); DepoCyt® (Almac Pharma Services Ltd, Craigavon, UK)	FDA regulated usage from 2007
Abelcet (Severe fungal infections)	Amphotericin B (Amp B)	Liposomes (Lipid complex)	Suspension (i.v.)	Enzon; Sigma-Tau Pharmaceuticals, Inc., Gaithersburg, MD, USA	More than 10 years in market
Amphotec (Severe fungal infections)	Amphotericin B (AmpB)	Liposomes (Lipid complex)	Suspension (i.v.)	Ben Venue Laboratories, Inc., Bedford, OH 44146, USA	Effective from 1996
Myocet (Combination therapy in breast cancer)	Doxorubicin	Liposome (powder) (Block copolymeric micelles) EPC, DPPC, MSPC, and DSPE-PEG_{2000}	Suspension (i.v.)	Zeneus; Enzon Pharmaceuticals, Piscataway, NJ, USA)	Not yet approved by the FDA for use in the United States
Visudyne (Age related molecular degeneration/photodermal therapy)	Verteporfin	Liposome, EPG and DMPC	Suspension (i.v.)	Novartis	Rx for US only
DepoDur (Pain management, post surgical analgesia)	Morphine sulfate	Liposome; DOPC, DPPG, chol and triolein	Epidural	Skype Pharma; Pacira Pharmaceuticals, Inc	FDA approved in 2004
LEP-ETU/Taxol, ENDO-TAG1 (Ovarian, breast and lung cancers, Intravenous antiangiogenesis, breast, and pancreatic caners)	Paclitaxel	Synthetic liposomes; DOTAP and DOPC	Intravenous	NeoPharm, Inc. Paclitaxel	Phase I and II

Trade name with major usage	Chemical name	Formulations	Route of entry	Company	Status/clinical trial phase
LEM-ETU (Leukemia, breast, stomach, liver and ovarian cancers)	Mitoxantrone	Liposome; DOPC, cholesterol, and cardiolipin	Intravenous		Phase I
INX-0076 (Advanced solid tumors)	Topotecan	Liposome; Egg sphingomyelin, and chol	Intravenous		Phase I
INX-0125 (Breast, colon and lung cancers)	**Vinorelbine**	**Liposome; Egg sphingomyelin and chol**	**Intravenous**		**Phase I**
OSI-211 (Ovarian, head and neck cancers)	Lurtotecan	Liposome; HSPC and cholesterol	Intravenous		Phase II
Inflexal V (Infuenza)	Liposomal IRIV vaccine	Liposome, LECIVA-S90	Intravenous	BERNA Biotech	

There are several lipid-based commercially available drug formulations whose structure is micelar, mixed micellar, a liquid-crystalline suspension or a lipid-stabilized emulsion. Several liposome–DNA complexes are also in various phases of clinical trials for different diseases. *IV*, intravenous; *DSPC*, 1,2-distearoyl-sn-glycero-3-phosphocholine; *DSPE*, 1,2-distearoyl-sn-glycero-3-phosphoethanolamine; *DSPG*, 1,2-distearoyl-sn-glycero-3-phosphoglycerol; *IP*, intraperitoneal; *PC*, phosphatidylcholine; *PEG*, polyethylene glycol; *CHOL*, Cholesterol [1–9].

Anionic liposomes (AL) complexes with folate (F-AL-Ad5) significantly improve the gene transduction ability of adenovirus vector Ad5 in airway epithelial cells which finds importance in delivering therapeutic genes to the airway epithelia and is promising in clinical medicine. Folate complexation makes F-AL-Ad5 more potent in gene transduction efficiency than naked AL-Ad5 [8]. The spherically constructed AL into chitosan/DNA (CS/DNA) complexes, AL/CS/DNA nanoparticles were shown to deliver the anti-caries DNA vaccine pGJA-P/VAX into nasal mucosa. The pH-dependant surface charge of the AL/CS/DNA enabled the intracellular release of loaded DNA with higher transfection efficiency and longer residence time at nasal mucosal surface. An AL/CS/DNA unique structural feature helps to significantly induce higher level of secretory IgA imposing a long term mucosal immunity and minimal cytotoxicity thereby enabling efficient elicitation of mucosal immunity mediated by improved DNA vaccine packaging and delivery nanoliposomes [2, 9].

Liposomes can fuse with hydrophilic particles thereby generating simple vectors of nanomedicine. Anionic silica particle fused with a cationic lipid (DOTAP) on a loading anionic fluorescent dye (calcein) could be efficiently taken up and released by Chinese hamster ovary cells under reduced pH condition [10].

Removal of salt has been reported to increase the efficiency of liposomes. Salt-free cationic and anionic (catanionic) lamellar vesicles made of tetra-decyl-trimethyl ammonium hydroxide and lauric acid reveal a much higher saturation binding with DNA thus implying high transfection efficiency while retaining its native stretched state during the interaction process. These catanionic vesicles are of great relevance in gene delivery of undisturbed and extended DNA molecules into the target cells [11].

pH sensitization of liposomes generates efficient liposomes as compared to pH insensitive liposomes. pH-sensitive liposomes composed of cholesteryl hemisuccinate (CHEMS) and DOPE revealed greater transfection efficiency of the pDNA as compared to pH insensitive ones [12].

Cationic lipid-based liposomes (CLP) can easily complex with negatively charged nucleic acids by electrostatic interactions, resulting in nanocarriers with biocompatibility, low toxicity, and easy large-scale production required for in vivo clinical trials. CLP-DNA complexes (lipoplexes) used for the delivery of plasmid DNA was modulated with the use of targeting ligands like epidermal growth factor and transferrin. They can target cancer cells over-expressing the receptors for these ligands specially [13].

The characteristics of liposomes (i.e., encapsulation efficacy, zeta potential, size, membrane fluidity, aggregation in serum, and uptake) with different surface potentials vary based on the enhanced permeation and retention (EPR) effect [14].

Neutral liposomes (NLP) like DOX-loaded NLPs and polyethylene glycol (PEG)-modified NLPs (NLP-PEG) preferentially localized in solid tumors whereas CLP like PEGylated CLPs (CLP-PEG) has a propensity for localizing in newly formed tumor vessels. In the presence of serum, cationic lipids help in liposome aggregation with unaltered membrane fluidity. The CLP reveal rapid uptake by rat aortic endothelial cells with reduced circulation times and poor distribution in tumors whereas NLP-PEG exhibited the highest anticancer efficacy [14].

The surfactant charge (anionic, neutral, and cationic), surfactant carbon chain length (C4, C12, and C16), and surfactant content (10 %, 20 %, and 29 %) greatly influences the morphology, physicochemical properties, stability, and permeability of liposome [15].

Liposomes derivatized with lipid-terminated polymer (TAT peptide-liposomes) were capable of anchoring in the liposomal membrane and increases the uptake of DNA by liposomes and transfection efficiency in dendritic cell cultures [16].

Gold (Au) nanoshells were synthesized using nanoliposome templates and then grown on the outer surface of the silica shells. DOX drug incorporated liposome/silicon-di-oxide (SiO_2)/Au killed cancer cells has higher therapeutic efficacy when irradiated with near-infrared light. This type of liposome has synergistic effect in delivering both DOX chemotherapy and photothermal therapy as compared to other Au nanoshells that cannot carry enough drugs [17].

Medium-chain fatty acids (MCFAs) nanoliposomes prepared by high-pressure micro-fluidization had a strong easy energy supply property, small average diameter, narrow size distribution, great entrapment efficiency with drug loading, and good stability as compared with crude MCFA liposomes. MCFAs-nanoliposomes can thus be used as a potential drug candidate for an easy energy supply [18].

Soya, rapeseed, and salmon lecithin liposome were stable than other drugs or liposomes made of pure phospholipids, with respect to their effect on cell proliferation and metabolic activity on bone marrow stem cells. Lecithin liposome holds great promise in the future as drug delivery systems for tissue engineering [19].

Different techniques like microarray-based immunoassays, enzyme-linked immunosorbent assay (ELISA), and antibody array are evolving for optimization of nanoformulations in which the signal-enhancing efficacy is highly increased by horseradish peroxidase (HRP)-loaded liposomes [20].

The response surface methodology (design) was used as an optimal preparation condition for preparing polyunsaturated fatty acids (docosahexaenoic acid and eicosapentaenoic acid) nanoliposomes having efficient entrapment efficiency in pharmaceuticals [21].

Emulsion liposomes (eLiposomes) composed of a perfluoropentane nanodroplet and an aqueous DPPC (di-palmitoyl phosphatidyl choline) liposome interior was used for carrying anti-cancer drug DOX. The drug delivery release of Dox is ultrasound sensitive and provides more effective ultrasound therapy for effective carrier of a chemotherapeutic agent in tumor treatment as it has the potential to reduce the side effects of cardiotoxicity caused by DOX [22].

The lactoferrin nanoliposomes prepared by reverse-phase evaporation method has better encapsulation efficiency, size distribution, more cellular uptake, higher stability, and portrays promises for oral administration of lactoferrin, as lectoferrin is well known for its antioxidant and anticarcinogenic properties [23].

DepoCyt (Almac Pharma Services Ltd, Craigavon, UK), a commercially available liposomal formulation of cytarabine indicated for the treatment of meningeal neoplasms. The specific multivesicular liposome technology like DepoFoam (Pacira Pharmaceuticals, Inc.) consists of multiple nonconcentric aqueous chambers within a single bilayer lipid membrane and a foam-like appearance. This multivesicular liposome has greater drug entrapment efficiency, robust structural stability with sustained release of the drug for targeting the meninges with least cytotoxicity [5].

9.1.2 Nanoliposomes and Agents Delivered

Encapsulating both hydrophobic and hydrophilic compounds, liposomes serve as effective platforms for intracellular drug delivery. Liposomes can serve as potent delivery agent for cancer therapeutics, vaccine, analgesic drugs, and many others. The triggered release of liposomes can be designed using external stimuli such as pH, ultrasound, and temperature. Moreover, using different compounds and preparation parameters, the vesicle size, surface properties, and surface charge can be easily modified. Long-circulating liposomes were prepared by adding polymers like polyethylene glycol (PEG) to the liposomal surface (PEGylation) to evade the RE system, longer body stay, and extended release of the encapsulated drug over time. For diagnostic imaging and targeted therapy, attaching antibodies and other markers to liposome surfaces serves as better tool [5–7].

For the safe delivery and entrapment of 5-fluorouracil (5-FU), nanoliposome carriers composed of endogenously occurring di-palmitoyl phosphatidyl choline (DPPC), negatively charged dicetyl phosphate (DCP), cholesterol (CHOL), and glycerol (3 %, v/v) were found to be effective. The release rate of 5-FU was enhanced by adding high ratio of drug: lipid to the liposomes [24].

Imaging of neovasculature is a promising new field in tumor diagnosis. Paramagnetic nanoliposomes targeted with RGD10, F56, and K237 peptides can bind to Integrin αvβ3 and VEGFR (vascular endothelial growth factor receptor)-1, VEGFR-2, respectively, in tumor neovasculature. The imaging was developed as potential tumor-specific Magnetic Imaging Resonance (MRI) contrast agents for detecting small tumors in animal models. RGD10 and K237 targeted paramagnetic nanoliposomes are helpful for tumor detection [25].

Hyaluronic acid-modified docetaxel-loaded liposomes (LMWHA-LPs) were found to significantly increase the docetaxel recovery in lymph nodes, after subcutaneous injection site, and plasma, after subcutaneous administration, with longer retention time compared to unmodified liposomes in vivo. LMWHA-LPs have been observed to be a highly promising lymphatic targeting carrier for chemotherapeutic drugs and diagnostic fluorescence agents [26].

Tea polyphenols, with biological functions reveal instability in oxygen and alkaline environments. Dynamic high-pressure micro-fluidization technique is used to prepare tea polyphenol nanoliposome (TPN) for effective delivery of tea polyphenols. TPN with higher entrapment efficiency, antioxidant activities, good sustained release property, and improved stability in alkaline solution, proved a better option for delivery of tea polyphenols [27].

Targeting of Toll-like receptor (TLR), TLR7 ligand, imiquimod, in the interior, and the TLR4 ligand, GLA, into the lipid bilayer of an anionic liposomal formulation synergistically trigger Th1 biased immunological responses in human. This finds importance in treatment that needs Th1-based responses where both TLR4 and TLR7 ligands would be coadministered with vaccine antigens [28].

The better targeting of curcumin (diferuloylmethane) for cancer treatment is effective by encapsulating curcumin in a nanoliposome to increase its absorption, bioavailability, and in vitro cytotoxicity effects. Salmon's lecithin liposome as compared to rapeseed and soya lecithins liposomes has also improved curcumin bioavailability [29, 30].

The efficient delivery of cinnamic acid in salmon lecithin nanoliposomes increases its entrapment efficiency, membrane permeability, electrophoretic mobility, elasticity, and membrane fluidity and thus shows promises as a better active therapeutic molecule delivery system [31].

9.1.3 Nanoliposomes as Vaccines

The integration of viral membrane proteins or peptide antigens into liposomes has the potential to evoke both cell mediated and humoral immune responses. These liposomes generate solid and durable immunity against the pathogen. The reconstituted viral liposomes, virosomes, have no viral genetic information and thus incapable of replication or cause infection and thus offers promises in preparation of vaccines. The lipid layers of virosomes, composed of DOPE and dioleoyl phosphatidyl Choline (DOPC) are used to mimic viral membrane for effective vaccine delivery. The virosome-based antigen delivery system for commercial applications uses both Epaxal and Inflexal V as vaccine products [5–7, 9].

Self-replicating RNA sequences have been utilized for correcting the aberrant function of the disease-causing gene and finds tremendous application in the vaccine technology. However, to function efficiently within the host body, delivery systems for such RNA replicons need to ensure their protection from nucleases, digestion by lysosomal enzymes on being internalized by receptor-mediated endocytosis, systemic circulation, therefore being safe and tolerated, even after multiple administrations. Lipid-based liposomal delivery devices are being designed for successful and effective delivery of such RNA molecules [32].

AL complexes with CS/DNA can efficiently deliver the anticaries DNA vaccine pGJA-P/VAX into nasal mucosa for efficient elicitation of mucosal immunity by inducing production of higher level of secretory IgA. This nanoliposome used for

DNA vaccine packaging offers a promising strategy to cross various physiological barriers for eliciting a definite protective mucosal immunity via intranasal administration [9].

Hydrophilic and hydrophobic fluorescent dyes as model drugs tagged with liposomes are found to be more effective in penetrating depth into the hair follicles, as compared to a standard, nonliposomal formulation [33].

The effective vaccination against *Leishmania major* infection was done by 1,2-dioleoyl-3-trimethylammonium-propane (DOTAP) nanoliposomes complexed with immune adjuvant for soluble Leishmania antigens (SLA) for higher encapsulation efficiency of SLA and low parasitic burden. The intrinsic adjuvanticity of the DOTAP-nanoliposomes as an antigen delivery system induces appropriate Th1 type of immune response and protection against *L. major* infection as compared to free SLA or free liposome [34].

Intravenous, intradermal, and eventually subcutaneous injections of liposomes provide an important agent in delivery of mRNA for vaccination. Liposomes containing mRNA of influenza virus nucleoprotein induce anti-influenza cytotoxic T lymphocytes, activating the T cell mediated immune responses in the host identical to those obtained in vivo with infectious virus. Liposomes combined with polymers to form lipopolyplexes were also useful for vaccination with nucleic acids [9, 32].

Many studies revealed the delivery of self-amplifying RNA via liposomes for gene therapy. CLP has shown efficient gene transfection and biocompatibility in in vitro studies, and has a few therapeutic applications in clinical trials, but critically its usage is limited for its in vivo instability. Hydrophobic CHOL, nonionic surfactants, or polyethylene glycol (PEG) has been widely used to increase the in vivo stability of CLP. However, Liposome/DNA complexes, injected intravenously sometimes form large aggregates with blood cells and these aggregates get entrapped in the lung capillary bed [9, 32].

Liposomes are promising tool for nonviral gene delivery. Several liposome-based vectors were used as vaccines and assayed in clinical trials. Allovectin-7® (Vical, San Diego, CA, USA), a pDNA carrying β2-microglobulin genes and major histocompatibility complex, class 1, B (HLA-B) genes were complexed with 1,2-dimyristyloxy-propyl-3-dimethyl-hydroxy ethyl ammoniumbromide/dioleoyl-phosphatidyl ethanolamine (DMRIE/DOPE) liposomes, for its assessment for safety and efficacy in Phase I and II clinical trials [5–7, 9, 32].

9.1.4 Liposomal Nanomedicine and Disease Targeting

Liposomes, at present, are widely applied as drug carriers in clinic, since the first liposomal pharmaceutical product, Doxil (formulation of DOX), received FDA approval in 1995. Till now, varied important types of liposomes, like CLP

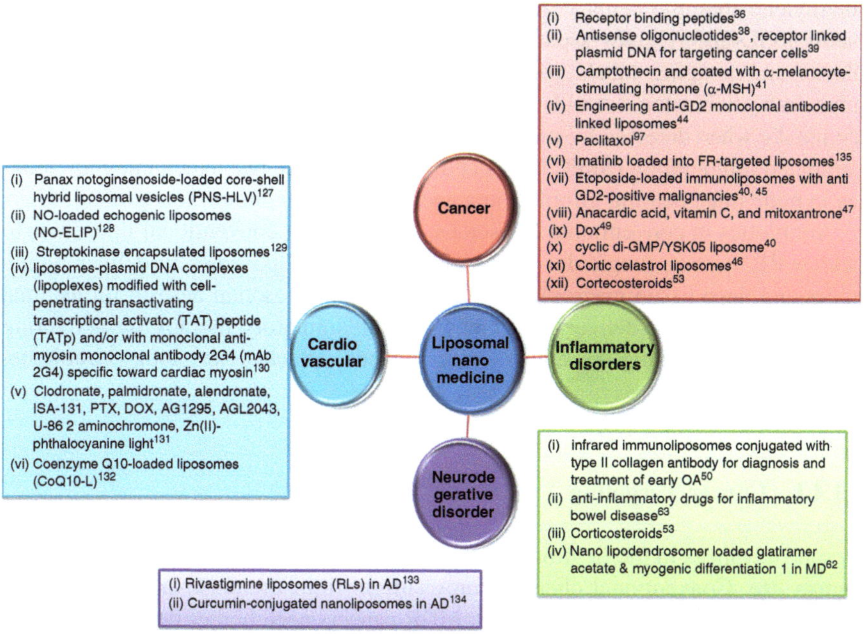

Fig. 9.3 Disease targeting by liposomal nanomedicine and its usage

(EndoTAG1-1) and temperature-sensitive liposomes (ThermoDox), have been extensively applied for clinical use. The medical applications of liposomes in various diseases in their preclinical and clinical stages were numerous (Table 9.2, Fig. 9.3). However, the application of liposomes in different human diseases remains the chief area of concern. In many cases, effective chemotherapy is not applied and it is severely restricted by the toxic side effects of the drugs. The unwanted toxic side effects of a drug are reduced by its encapsulation in a liposome and also its efficacy is also increased. The possible side effects related to a drug may be reduced by liposomal formulations by its extended plasma circulation time and varied drug distribution in body. Liposome encapsulation can alter the temporal and spatial distribution of the encapsulated drug molecules inside the body. Although many effective drugs have been studied in preclinical conditions in numerous veterinary disease models, but for human therapy, liposomal therapeutics, at present are widely used mostly in cancer therapy and systemic fungal infections. We need to exploit, in preventative conventional medicine or in nanomedicine, the liposome-based vaccines as future prospective in many diseases [1–9].

Biocompatible nanoliposomes carrying therapeutic agents find wide-scale application in targeting of a wide array of diseases (Table 9.2, Fig. 9.3). They also find application in tumor imaging. Formulation of lipids, PEG building blocks, DTPA (diethylene triamine pentaacetic acid)-derivatized lipids for radio-labeling, lipid-based RGD(arginylglycylaspartic acid, a tripeptide), and substance P building blocks and imaging labels combined in defined molar ratios targeted in nude mice with glioblastoma could successfully reveal their biodistribution and moderate uptake when detected by micro-single photon emission computed tomography/computed tomography (SPECT/CT) and MRI (magnetic resonance imaging) studies in nude mice bearing glioblastoma and/or melanoma tumor xenografts [35].

Antibiotics in conventional liposomes, fungicides in conventional liposomes to deliver RE system, fungicides in sterically stabilized liposomes to deliver lymph nodes and deep tissue macrophages, glucocorticoid liposomal aerosol for asthma, other drugs for lung delivery, vaccines, anti-inflammatory therapy by delivering drug or by blocking receptors with ligands attached to long circulating liposomes [5].

9.1.4.1 Cancer Targeting

Despite advances in the field of cancer research, it still remains the major threat to human life. Conventional chemotherapy suffers from the disadvantage of recurrence and multidrug resistant (MDR). MDR cancers need higher dose of apoptotic drugs, aerobic glycolysis, hypoxia, and increased number and elevated activity of MDR efflux transporter protein. MDR transporters enable cancer stem cells (CSCs) to evade the host immune surveillance and chemotherapy, thus leading to relapse of cancer. To circumvent this problem in MDR cancers, multifunctional engineered stimuli responsive nanoparticles carrying diverse therapeutic agents like drugs, genes with targeted ligands are being researched as promising delivery devices in cancer targeting. Quite a number of molecules including integrins, receptors like growth factor receptors (GFRs) and G-protein coupled receptors (GPCRs) have been researched as potential agents for coating nanoliposomes targeting peptide biomarkers either specifically expressed or over-expressed on cancer cell and physiology. Targeted receptors most studied in cancer include somatostatin receptors, cholecystokinin receptors, receptors associated with the Bombesin like peptides family, luteinizing hormone-releasing hormone receptors, and neurotensin receptors [36]. At present, most of the medical applications of liposome therapeutics in cancer therapy are in the level of preclinical trial studies. Liposome mediated targeting of cancer cell is gaining importance [37].

Targeting ASOs against drug efflux proteins including MDR1, MDR-associated protein (MRP)1, MRP2, and/or BCL-2/BCL-xL to suppress and reverse MDR transporters and induce apoptosis is a strategy employed to target MDR cancer. CLP (100 nm) composed of N-[1-(2,3-dioleyloxy) propyl]-n,n,n-trimethylammonium chloride and dioleoyl phosphotidyl ethanolamine core surrounded by a PEG shell enabled successful delivery of ASOs and/or anticancer drug epirubicin, on one hand down-regulated the MDR1 promoter activity and mRNA expression of human MDR1, MRP1, MRP2, and BCL-2, and on the other hand up-regulated mRNA expressions of p53 and proapoptotic proteins including BAX, caspase-3, -8, and -9, enhanced apoptosis, reduced viability in human colon adenocarcinoma Caco-2 cells [38].

Cationic lipids, employed in the delivery of genes suffer from the disadvantage of nonspecific gene expression in lung and liver tissue. To circumvent the problem of nonspecificity, a dual delivery system using NLP composed of plasmid DNA (pDNA) condensed with cationic PEG modified poly-lysine (PL/DNA) and neutral Her-2 targeting liposome conjugated Listeriolysin O (LLO) protein was used to target BCs expressing human epidermal growth factor receptor 2 (Her-2) cancer marker [39].

Cyclic dinucleotides di-guanosine monophosphate (c-di-GMP) are known to act as second messenger playing role in the signal transduction of cytosolic DNA immune responses and stimulates the immune system. Synthetic, pH sensitive lipid YSK05 with high fusogenicity property has been used as a carrier to transport c-di-GMP into the cytosol which could induce innate immunity response mediated by Interferon (IFN)-β production in Raw264.7 cells by the activation of STING-TBK1 pathway, up-regulated expression of CD80, CD86, and MHC class I, facilitated antigen-specific cytotoxic T cell (CTL) activity, and the inhibited tumor growth in a mouse model [40].

Functional liposomes composed of phosphatidyl choline (PC), CHOL, and stearyl amine encapsulating camptothecin and coated with α-melanocyte-stimulating hormone (α-MSH) with their unique property of antigen specific targeting and controlled drug delivery were reported to overshoot the antimelanoma activity of camptothecin when administered alone [41].

1,2-dipalmitoyl-*sn*-glycero-3-phosphocholine (DPPC) liposomes with enhanced drug retention and release property at 41 °C offers a novel approach of temperature controlled drug release overcoming the problems of drug leakage associated with liposomes [42].

Fusogenic liposome for delivery of chemotherapeutic drug DOX designed with protein–DNA complex core containing an ATP-responsive DNA scaffold has been reported to release drug through an ATP-mediated process triggered by liposomal fusion suggesting their potential application as an efficient anti-cancer agent [43]. Disialo-ganglioside GD-2 is a tumor marker expressed in many cancers and targeting cancer cells by antibodies against GD2 is a major strategy employed in cancer therapy [44]. Immuno-liposomes coated with anti-GD2 antibody, encapsulating anticancer drug etoposide targeting GD2 expressing cancer cells, has been reported to enter targeted cells via clathrin-coated pits, thereby causing reduction in cell proliferation in in-vitro studies and offers advantage over conventional chemotherapeutics in specific cancer cell targeting [45].

Anticancer activity of hydrophobic natural product Celastrol, derived from *Tripterygium wilfordii*, herb prepared as PEGylated celastrol liposomes has been shown to reduce prostate cancer cell viability [46].

Synergistic effects of anacardic acid, vitamin C, and mitoxantrone loaded to liposomes has been reported to have enhanced anticancer effects [47].

Curcumin, loaded on liposomes, and micelles, lipid nanoparticles, and emulsions have shown promising anticancer effects [30, 48].

Preformed vesicle (PFV), a type of liposome, PFVYLI with a hydrophobic penetration peptide (HPP), loaded with anticancer drug DOX (PFV-SSLs-DOX) targeted against different BCs cell revealed better accumulation in the tumor, growth inhibition, and lower systemic and cardiac toxicity thus proving a promising delivery device for therapeutic or imaging agents to tumors [49].

9.1.4.2 Inflammatory Diseases

Infections and inflammations lead to leaky vascular system just like tumor infiltration. Sterically stabilized liposomal formulations may therefore be an effective delivery tool for the transport of appropriate drugs in such circumstances. Many deep tissue residing infectious microbes are practically inaccessible, or escape the chemotherapy by conventional drugs as they reside in deep-tissue macrophages. Some liposomal encapsulations were effectively phagocytosed by the macrophages whereas some liposomes end up in the skin. The potential of long-circulating liposomes in inflammatory diseases should be the new field of research [5–7].

Corticosteroid, dexamethasone-loaded liposomes have been reported to specifically reduce cytotoxicity in human fibroblasts and macrophages, with downregulation of proinflammatory cytokines like tumor necrosis factor (TNF), interleukin (IL)1β, reduced migration of monocytes and macrophages in inflamed conditions thereby suggesting their efficient applicability and nontoxic role in patients suffering from rheumatoid arthritis, atherosclerosis, colitis, and cancer [50]. Nanomedicine-based strategies are being employed in targeting of atherosclerosis [51].

For the effective low-dose treatment of chronic inflammatory human skin disease, acne, vesicular (liposomes and niosomes) were used as a nanotechnological approaches along with colloidal drug delivery systems (micro-emulsion and nanoemulsion), particulate (solid lipid nanoparticles and microspheres), and miscellaneous systems (aerosol foams and micro-sponges) for acne therapy [52].

Glucocorticoids (GC), being potent immunosuppressive and anti-inflammatory agent is used extensively for the treatment of many different inflammatory diseases. Liposomal GC formulations for the treatment of rheumatoid arthritis, asthma, multiple sclerosis, and cancer, through the targeted delivery of GC to pathological sites enhance the biodistribution and the target site accumulation of GC to balance between its efficacy and their toxicity. Liposomal GC were better off as clinical therapeutics than prolonged and/or high-dose GC therapy with respect to lesser side effects, enhanced biodistribution, and lesser target site accumulation of GC to reduce toxicity [53].

The role of C6-ceramide as an antiproliferative and proapoptotic sphingolipid metabolite in regulating inflammatory responses was determined by nanoliposome formulation (Lip-C6) in corneal inflammation. Lip-C6 as compared to control liposomes significantly inhibited proinflammatory reactions in corneal cells after stimulation with UV-killed *Staphylococcus aureus* or lipopolysaccharides. Lip-C6 may act as a potential anti-inflammatory, nontoxic, therapeutic agent for delivering short-chain ceramide [54].

The potential first-line therapy for severe and life-threatening complication of childhood systemic inflammatory disorders like macrophage activation syndrome (MAS) includes liposome-incorporated dexamethasone (dexamethasone palmitate, DexP) than conventional free corticosteroids (methyl prednisolone) in effectively inhibiting inflammation in MAS patients for the management of the disease [55].

9.1.4.3 AIDS

One of the biggest challenges of twenty-first century remains the successful treatment and control of HIV/AIDS as more than 33 million individuals worldwide are infected with HIV and more than two million new cases of HIV infection has been reported. To control the situation, development of effective prevention strategies needs to be addressed. Anti-retroviral drugs and siRNA have been explored for HIV prophylaxis as assayed by clinical trials in absence of effective HIV vaccine. Liposomes may improve the efficacy of various modalities available for HIV prophylaxis in diagnosis, treatment, and prevention of this disease [56].

The viral protein αEnv from HIV virus coated on to liposomes linked to phosphatidyl serine triggers HIV-virus-like particles phagocytosis in macrophages offering promise to AIDS therapy [57]. Nanoliposomes with the property of crossing blood–brain barrier enable targeting of central nervous system diseases [58]. Liposomes, dendrimers, polymeric nanoparticles, micelles, and solid lipid nanoparticles are being tested for treatment of neuro AIDS [59].

Broadly neutralizing monoclonal antibodies (bNAbs), 2F5 and 4E10, bind to the membrane proximal external region (MPER) of gp41 of HIV virus to cross react with phospholipids. Chemical modifications on MPER adjacent to 2F5 and 4E10 epitopes increase immune responses to poorly immunogenic antigens, by breaking tolerance of inflammation-associated post-translational modifications to induce 2F5- and 4E10-like bNAbs. Tyrosine, serine, and threonine residues were modified with sulfate, phosphate, or nitrate moieties and presented in liposomes for rabbit immunizations protocol to explore it as a HIV-1 vaccine strategy by developing sera with strong anti-gp140 titers [60].

9.1.4.4 Other Diseases

As a result of their special properties, liposomes are widely studied for the treatment of various diseases and for numerous vaccination applications. The application includes steroid and nonsteroid anti-inflammatory drugs, insulin for diabetes, antiviral therapy, prostaglandins, antibiotics, and examples from many other diseases. The liposomal administration routes being also varied ranges from pulmonary, topical, and other parenteral routes. For the meaningful clinical applications of liposomes in different curable and noncurable diseases, a lot more basic research will have to be performed to end up in a promising approach [53, 61–66].

An aqueous formulation of liposome prepared by encapsulating lipophilic and photosensitive Vitamin K1 (Vit K1) has been reported to be administered successfully on the skin by nebulization wherein Vit K1 accumulates in the dermis and epidermis region of skin, enabling easy, efficient, and multiple administrations as compared to conventional semisolid preparations [67]. Octyl methoxycinnamate (OMC) liposomal nanosystem (liposome/OMC) was used successfully as a safe and effective sunscreen formulation [68].

The chronic inflammatory human skin disease, acne is a must encountered problems of the adolescent populations. Vesicular (liposomes and niosomes) has an enormous opportunity as a novel, low dose, and effective treatment systems in acne therapy [52]. Carrier-based delivery systems like liposomes, niosomes, solid lipid nanoparticles have the potential in treatment of acne vulgaris with simplified dosing regimen to improve treatment outcomes as large populations of adolescents suffer at some point in their lifetime from this multifactorial disease [69].

The P85-coated pluronic liposomes for the oral delivery of breviscapine, a hydrophobic drug used for treating cardiovascular disease, prepared by the thin film hydration technique effectively enhance the bioavailability of breviscapine, increases encapsulation efficiency to more than 90 %, increased absorption and in vitro drug release of breviscapine in Caco-2 cells. Thus, P85-modified liposomes can be therapeutically used in the treatment of cardiovascular diseases [70].

DexP used for first-line therapy for severe and potential life-threatening complication of childhood systemic inflammatory disorders like macrophage activation syndrome (MAS) is more potent than conventional free corticosteroids (methylprednisolone) in effectively inhibiting inflammation in MAS patients to manage their diseases [55].

The neurotrophic and neuroprotective effect of intranasal glial cell line-derived neurotrophic factor (GDNF) in the intact substantia nigra (SN) dopamine neurons were demonstrated using CLP as compared to Phosphate buffered saline (PBS) liposomes in the unilateral 6-hydroxydopamine (6-OHDA) model of Parkinson's disease (PD). GDNF loaded CLP has the potential efficacy in the treatment for PD [71].

The multifunctional envelope-type nanodevice for pancreatic β cells (β-MEND) liposomes encapsulating a 2′-O-Methyl RNA complementary to a micro RNA to suppress insulin secretion and the knockdown of the targeted micro RNA with an up-regulation of insulin secretion were observed in pancreatic β cells line, MIN6. Thus, the β-MEND has the potential as an efficient nucleic acids delivering system targeted for efficient internalization in MIN6 for application in research and therapy based treatments of diabetes [72].

Multifunctional nanosized liposomes incorporating the lipid-PEG-curcumin derivative with a Blood–Brain Barrier (BBB) transport mediator (anti-Transferin antibody) showed a high affinity for the amyloid deposits in brains and thus were targeted for treatment and diagnosis of Alzheimer Disease affected individuals [73].

The applicability of the nanocarrier systems *viz.* liposomes, nanoparticles, and nanoemulsion acting as nanotherapy for posterior eye diseases to defy ocular barriers, target retina, controlled release, and act as permeation enhancers is highly appreciable for treatment of blindness, the second most dreaded disease across the world [74].

The AeroEclipse II nebulizer for the delivery of pDNA complexed with the CLP like GL67A (pDNA/GL67A) formulations has efficacy to deliver aerosol slowly, generate aerosol only during the inspiratory phase, and biological function of pDNA/GL67A formulations was retained after nebulization. Thus this device was more efficient than other control devices in lung gene therapy for generating aerosol available for cystic fibrosis and for a range of acute and chronic diseased patients' inhalation [75].

Nanoliposomes complexes reduce the risk of systemic toxicity and increase the patient compliance in fatal respiratory infectious diseases like tuberculosis as compared to oral conventional anti-tuberculosis therapy [76].

For intrinsic adjuvanticity of the immunoadjuvant for SLA, 1,2-dioleoyl-3-trimethylammonium-propane (DOTAP) nanoliposomes containing SLA are appropriate delivery systems with high encapsulation efficiency and low parasite burden against *Leishmania major* infection. To induce a Th1 type of immune response and protection against *Leishmania major* infection, DOTAP liposomes act as a suitable adjuvant and delivery system as compared to free liposome or free SLA [34].

9.1.5 Liposomal Nanomedicine and Stem Cell

Liposomes composed of soya, rapeseed, and salmon lecithin have been reported to improve cell proliferation on rat bone marrow stem cells (rBMSC) [76]. Genetically modified stem cell therapy with reduced tumorigenic risks and safer applicability finds wide-scale application in regenerative medicine in the correction of several diseases. Nanoliposomal delivery of genes for therapeutic and corrective purposes to stem cells with minimum toxicity offers advantage over viral and nonviral delivery devices, providing a new arena in the field of cell-based therapeutics and serves as a promising tool for stem cell therapy. Noncovalently modified polyplexes with lysinylated and histidylated CHOL lipids have been reported for their improved efficiency and reduced toxicity for gene delivery in bone marrow stem cells [19]. Nanoformulation with either PEG-lipid membrane or PEG-Polylactic acid, PEG-PLA polymer membrane was being employed in designing of artificial red blood cells (RBC) [77]. PEG-grafted polyethylenimine (PEG-PEI) with easy construction, better transfection efficiency, and capability of potentially targeting modification have been reported to be promising tools in gene delivery to mesenchymal stem cells (MSCs) [78].

The cell proliferation and metabolic activity of rBMSCs were highly improved by the use of three stable lecithins (soya, rapeseed, salmon) liposomes than other phospholipids liposomes. These nanoliposomes used in pharmaceutics have better bioefficiency as drug delivery systems for tissue engineering [19].

The characteristics of veno-occlusive disease in pediatric hematopoietic stem cell (HSC) transplant recipients, their prophylactic regimen, risk factors, treatment modalities on severity and outcome were evaluated using liposomal amphotericin B (LAmB), vancomycin treatment as compared to oral busulphan use. LAmB vancomycin treatment and total parenteral nutrition played a fair role in low incidence of pediatric veno-occlusive disease in individuals with high-risk features [79].

Liposome protamine/DNA lipoplex (LPD) assembled from CLP and an anionic complex of protamine, DNA to target various peptides (VTAMEPGQ) that can home to rat mesenchymal stem cells (rMSCs). The transfection action of LPD was executed by the synergetic effect of rMSC-targeting peptide and nuclear localization signal (NLS) peptide for LPD for internalization inside the cells, leading to increased gene expression, guide sleeping Beauty (SB) transposon into nuclei with no obvious cell toxicity and influences the differentiation potential of rMSCs. Therefore, for promising nonviral gene delivery vector in stem cell therapy, the integration of SB transposon through LPD system is quite effective [80].

Mouse induced pluripotent stem (iPS) cell-derived Flk-1(+) cells were incubated with magnetic nanoparticle-containing liposomes (MCLs) and these cells were mixed with diluted extracellular matrix (ECM) to accelerate revascularization of ischemic hind limbs. This novel magnetite tissue engineering technology using MCLs represents a promising new modality for therapeutic angiogenesis [81].

The differentiation of fetal liver stem/progenitor cells (FLSPCs) into hepatocyte-like cells for in vivo therapeutic investigation, treatment of FLSPCs with Hepatocyte growth factors (HGF) with di-methyl sulfoxide (DMSO) for liposome-mediated transfection, GFP gene was introduced into the system. The self-renewal property of FLSPCs in soft agar culture and effectively differentiation into hepatocyte-like cells was mediated by GFP traced liposome-mediated transfection continued over successive generations [82].

Conventional chemotherapy may efficiently kill the differentiated drug sensitive breast cancer stem cells (BCSCs), but not the multidrug resistance (MDR) self-renewable BCSCs, leading to enrichment of the MDR-BCSCs. BCSCs play critical roles in self-renewal, differentiation, and generation of secondary tumors. BCSCs have high rate of uninhibited clathrin-independent and caveolin-independent endocytosis than the differentiated BC cells and thus open the possibilities for delivering therapeutic agents directly into the MDR-BCSCs through aptamer-coated liposomes [83].

Intravenous LAmB, the antifungal prophylaxis in pediatric patients undergoing hematopoietic stem cell transplantation (HSCT) was less efficient than Caspofungin (CAS), a broad-spectrum echinocandin, with lower nephrotoxicity than LAmB. LAmB treatment lowers potassium levels below normal values in patients and had more drug-related side effects [84].

The efficacy, homing/engraftment, and retention of bone marrow hematopoietic stem/progenitor cells (HSPCs) transplantation and gene therapy protocols were increased by safer and nonviral transfection of CXC chemokine receptor 4 (CXCR4) on hematopoietic stem/progenitor cells (HSPC) by using the CLP agent IBAfect. IBAfect is a better therapeutic agent than commercially available Lipofectamine 2000 and 1,2-dioleoyl-3-trimethylammonium-propane (DOTAP) liposomes [85].

To realize the safer quality control of regenerative medicine, hybrid liposomes (HL), HL23 (DMPC/10 mol%C(12)(EO)(23)) was used to selectively eliminate transformed hepatic stem cells, the forerunner of neoplastic cell transformation for future reduction of the risk of tumorigenesis after the cell transplantation [86].

The administration of clodronate-loaded liposomes recapitulated the loss of endosteal osteoblasts due to granulocyte colony-stimulating factor (G-CSF) mobilization. G-CSF mobilization suppresses endosteal bone formation and decreases expression of factors required for HSCs retention and self-renewal [87].

The safety and pharmacokinetics of the combination therapy of LAmB and CAS were same as compared to individual monotherapy when investigated in adult allogeneic hematopoietic stem cell recipients (aHSCRs) with granulocytopenia and refractory fever [84].

9.1.6 Liposomal Nanomedicine: Preclinical and Clinical Studies

In clinical and preclinical studies, liposomes improved pharmacokinetics and biodistribution of therapeutic drugs as compared to conventional drug formulations and thus minimize cytotoxicity by their accumulation at the target tissue [5–7].

At present, there are about 12 liposome-based drugs approved for clinical use and more drugs were undergoing various stages of clinical trials. For intravenous application, Doxil and Myocet™ (GPPharm, Barcelona, Spain) were approved;

Epaxal® (Berna Biotech Ltd, Berne Switzerland) and Inflexal® V (Berna Biotech España SA, Madrid, Spain) has been approved for intramuscular delivery [1–9]. The potential for liposome breakdown following exposure to bile salts makes it troublesome for oral delivery.

Globally research has led to the development of nanoliposomes that are currently applied in clinical practice. Several micelle designed by encapsulating anti-cancer drugs like paclitaxel (Micelle NK105), cisplatin (NK6004), and epirubicin (NK-6300) carrying paclitaxel are currently under Phase 3 trial [88].

DepoDur (Almac Pharma Services Ltd), an extended-release injectable liposomal formulation of morphine, exploiting DepoFoam drug formulation technology (Pacira Pharmaceuticals, Inc.) was highly indicated for epidural administration for the treatment of superior pain control following major surgery. The usage of DepoDur in clinical trials decreases the dosage of systemically administered analgesics like un-encapsulated morphine or intravenous administration of an opioid, needed for the first 1–2 postoperative pain management [5].

The synergistic delivery of TLR4/TLR7 adjuvant into an anionic liposome to trigger Th1 biased responses was pharmaceutically acceptable and will have a straightforward path into human clinical trials [28].

The role of curcumin (diferuloyl methane) for cancer treatment has been limited in its clinical use due to its low absorption and poor bioavailability. Encapsulation of curcumin by nanoliposome or in a salmon's lecithin liposome improved its bioavailability as a therapeutic agent [29].

AeroEclipse II nebulizer formulated with pDNA complexed CLP like GL67A (pDNA/GL67A) for slower aerosol delivery only during the inspiratory phase was used as efficient lung gene therapy to many acute and chronic diseases, including cystic fibrosis as part of phase IIa/b clinical studies [75].

The liposomal formulation of mifamurtide (immunomodulator with antitumor effects mediated by activation of monocytes and macrophages) was Mepact (Takeda Ireland Ltd, Wicklow, Ireland/Takeda Italia Farmaceutici S.p.A, Cerano, Italy) used in children, adolescents, and young adults for the treatment of high-grade, non-metastatic osteosarcoma after macroscopically complete surgical resection. Mepact was intravenously administered in conjunction with postoperative multiagent chemotherapy to significantly increase overall survival of patients in clinical trial [5, 89, 90].

The DexP, in comparison to commonly used corticosteroids, effectively inhibited inflammation in childhood systemic inflammatory disorders like MAS and was used as first-line therapy for MAS [55].

During clinical trials with gene therapy, safety is mainly concerned with the immunogenicity of the viral carriers and their random integration into the host genome. To improve clinical transplantation of hematopoietic stem/progenitor cells (HSPCs) and CXC chemokine receptor 4 (CXCR4) gene delivery therapies in leukemia, nonviral, and safer CLP agent IBAfect was used [85].

With safety as a primary concern for use of liposomes, the safety and pharmacokinetics of the combination therapy of LAmB and CAS in comparison to conventional mono-therapy were investigated in a randomized, risk-stratified, multicenter phase II clinical trial in immune-compromised adult aHSCRs with granulocytopenia and refractory fever [84, 91]. Liposome encapsulated therapeutic AmBisome (Gilead Sciences Ltd) has been clinically approved over AmB to overcome the associated side effects like infusion-related reactions and nephrotoxicity on chronic usage of AmB. The increased uptake of AmB-loaded liposomes by the RE system is noted for the small size and negative charge of the liposomes [91].

The clinical trial of the usage of DepoFoam bupivacaine compared to conventional bupivacaine in randomized, controlled, double-blind trials in patients used in therapeutic doses was well tolerated, had a higher safety margin, and showed a favorable safety profile in postoperative pain control [92].

The acute toxicity and chronic toxicity trials of DOX-Hydrochloride Nanoliposome (DHNP) prepared by ethanol injection-pH gradient method possesses the advantage of uniform distribution size, high encapsulation efficiency, big drug loading rate, and its toxicity is lower than free DOX [93].

9.1.7 Risks and Safety Issues

Liposomes formulated in aqueous solution generally face physical and chemical instabilities after long-term storage. Other common causes of liposome instabilities are hydrolysis and oxidation of phospholipids and liposome aggregation. Lyophilization, freezing, and spraying drying are the many methods developed for the stabilization of liposomes.

AL into CS/DNA complexes, AL/CS/DNA can efficiently deliver the anticaries DNA vaccine pGJA-P/VAX into nasal mucosa with longer term mucosal immunity and with minimal cytotoxicity [9].

The antifungal prophylaxis by intravenous LAmB in pediatric patients undergoing aHSCT has much lower potassium than normal values and higher nephrotoxicity. The drug-related side effects of LAmB in patients treated were aggravated by increased need for oral supplementation with potassium, sodium bicarbonate, and calcium upon discharge in immunocompromised pediatric patients, who underwent high-dose chemotherapy and HSCT [91].

The CLP agent IBAfect-mediated in vitro gene delivery to over-express chemokine stromal cell-derived factor (SDF)-1α/CXCL12 and its receptor CXC chemokine receptor 4 (CXCR4) on hematopoietic stem/progenitor cells (HSPCs) is a safe, nonviral, and efficient technique. It has great potential for improving the transfection efficiency of homing/engraftment and retention of HSPCs in the bone marrow for HSPC transplantation and gene therapy protocols for leukemia. In parallel, commercially available CLP (Lipofectamine 2000 and DOTAP), failed to deliver the CXCR4 gene into cells under the same conditions and has lesser efficiency [85].

The treatment of hybrid liposomes (HL), i.e., HL23 (DMPC/10 mol%C (12) (EO)(23)), could selectively eliminate the transformed hepatic stem cells using hepatoblast. The transformed cells in turn would have produced abnormal prothrombin, the clinical marker for hepatoma, and also formed colonies during neoplastic cell transformation. Thus, HL23 therapeutics would be a useful quality control culture method for the stem cells to reduce a risk of tumorigenesis after the cell transplantation [86].

The combination therapy of LAmB and CAS in immunocompromised adult aHSCRs patients' with granulocytopenia and refractory fever was as safe as monotherapy with CAS or LAMB and had similar plasma pharmacokinetics. Thus, this supports further investigations of the combination therapy in the management of patients with invasive opportunistic mycoses with poor prognosis [91].

The effects of extended drug delivery system using DepoFoam bupivacaine in place of bupivacaine single injection or continuous local infusion techniques for postoperative pain management is a promising and safer drug formulation that may significantly improve postoperative care and pain control in surgical patients [92].

The novel gene delivery system was developed using MLV liposomes composed of DOTAP and oligoamines-modified CHOL. The best transfection efficiency and the noncytotoxicity of the prepared lipoplexes in Neuro2A cells was tested using pDNA encoding Renilla luciferase. These nanocarriers are able to transfer the DNA into Neuro2A cells with low or no toxicity [93].

The cationic modification agent, polyethylenimine (PEI) of the chemotherapeutic drug, cisplatin or CDDP (*cis*-diamminedichloroplatinum(II))-encapsulated liposome reduced the nephrotoxicity associated with CDDP, had no system toxicity, efficient and safe for anti-tumor drug delivery [94].

The DOX-Hydrochloride Nanoliposome (DHNP) prepared by ethanol injection-pH gradient method possesses the advantage of uniform distribution size, high encapsulation efficiency, big drug loading rate, and its toxicity is lower than free DOX [95].

The amount of cytotoxicity of the anti-cancer drug, artemisinin compared to standard drug is increased by pegylated nanoliposomal formulation. Similarly, the cytotoxicity effect of nanoliposomal paclitaxel, an anti-breast cancer drug, is more than that of the standard form [96, 97].

9.2 Discussions

Currently, the chief priorities of the field of liposomal nanotherapeutics are to direct further research works toward development of competitive and translatable product to ensure a well-equipped regulatory environment. Thus, the key to success is an interdisciplinary and open-minded approach blended with the knowledge exchange between academia, industry, and drug regulatory agencies. So, the rapid and successful translation of emerging and under trial liposomal nanotherapeutics in the future can be derived from the well-defined quality-by-design approach. The aspiring motto of nanomedicine should revolve around scientific research and to translate this strength into commercially viable products. However, numerous recent initiatives in the field of nanomedicine are aimed at pushing forward with the best translatable concepts using funding and clinical proof of concept. Such an approach is expected to generate new interests in nanomedicinal technology with macro or micro impacts on patient benefits and therapeutic treatment options.

Liposomes have greater applicability in molecular and optical imaging, vascular targeting, and tumor cell membrane targeting, triggered drug release. As drug delivery device, it has the advantages of reduced volume of distribution of drug when packed within nanoliposome, advantages of reduced or nil side effects, nil toxicity, better pharmacokinetics, and tumor extravasations, improved bioavailability, targeting and accumulation.

Liposomal formulations have been developed to prolong drug circulating lifetime, enhance drug efficacy by increasing drug deposition, and reduce drug toxicity by avoiding critical normal tissues. Despite the clinical or preclinical approval of numerous liposome-based chemotherapeutics, challenges remain in the formulation and clinical deployment of micro- and nanoparticulate liposomes, as well as combining these novel liposomes with conventional drugs and standard-of-care therapies. Factors requiring optimization during drug delivery system include control of drug biodistribution, release rates of the encapsulated drug, and uptake by target cells. Quantitative mathematical calculations of the performance of the formulation can provide an important approach for understanding drug route, uptake, and disposition processes, as well as their therapeutic outcomes [98].

Doxil, the first liposomal pharmaceutical product, received Food and Drug Administration (FDA) approval in 1995, and now it has been widely applied as drug carriers in clinic. For clinical use, until now, varied important types of liposomes, like PEGylated liposomes (Doxil and Lipo-Dox), temperature-sensitive liposomes (Thermo-Dox), CLP (EndoTAG1-1), and virosomes (Expal and Inflexal V), have been investigated. But for clinical trials, more focus has been given on the types of delivered drugs (BLP25lipopeptide, Cisplatin, Grb2 antisense oligodeoxynucleotide, Bacteriophage T4 endonuclease 5), therapeutic applications (from topical delivery systems to portable aerosol delivery systems), and biological role (extended blood circulation time, various drug distribution in the body, reduced side effects). Though in clinical trials of PEGylated liposomes (Doxil and Lipo-Dox), significant incidence of stomatitis was observed due to PEGylation. In clinical trials, some of the new generation liposomes showed poor therapeutic efficiency, weaker accumulation in tumors, reduced anti-tumor activity, increased toxicity in tissues as compared to conventional free drug. Thus for clinical trials, formulation of liposomes must be regulated based on its efficacy, clinical therapeutic effects, preclinical trials in animal models, and toxic side effects of liposomal lipid composition [6].

The major challenges in the field of research in liposomal nanomedicine revolve around strategies of increased extravasation by manipulating architecture of tumor vasculature, improved bioavailability by increasing cell-specific internalization of liposomal nanoparticles (LN), modulating tumor cell membrane permeability and triggering drug release, targeted drug delivery by targeting intracellular receptors thereby promoting drug delivery and drug delivery by vasculature specific liposomes.

Research to generate better and effective LN is in process across the globe. Recombinant anti-human antibodies are being generated to tag with the LN for targeting specifically tumor cells and tumor vasculature, strategies for generating RGD (a tri-peptide) and NGR (Asparaginen-Glycine-Argine peptide motif) based delivery systems are being exploited. Short chain sphingolipid (SSC) analogs have been reported to improve delivery of amphiphilic drugs and increased cellular drug uptake by modulation of tumor membrane lipids without increasing toxicity. SSC could enhance DOX uptake [1–9]. Codelivery of C8-Glucosyl Ceramide (GC), DOX and PEGylated liposomal doxorubicin, PLD (GC-DOX-PLD) enabled enhanced uptake of DOX by the B16 melanoma cell with improved antitumor activity [53]. 8-GluCer-Dox-liposomes developed by Ceronco Biosciences (www.ceronco.nl) is currently undergoing Phase 1 clinical trial in 2014. Hyperthermia is used to improve liposomal chemotherapy and has paved its path in the generation of lipo-sensitive liposomal formulations like Lyso-Thermosensitive liposomes (TSL)/Thermodox®, OG-TSL [99–101]. Moreover, newly developed different liposomal combinations/formulations and liposomal vaccines have shown promising results in clinical studies and in commercial applications. In all, liposomal formulations are still important and potential tool for its clinical application of nanomedicines [45, 50, 61, 63, 65, 102–135].

This book chapter identifies several relevant pharmacokinetic/pharmacodynamic liposomal models that incorporate key physical, biochemical, and physiological processes involved in delivery of varied drugs by liposomal formulations. The observed data lend insight into factors determining overall therapeutic response, and in some cases, predict risk and safety conditions for optimizing drug delivery that include nanoparticulate drug carriers like liposomes as compared to conventional drug delivery system.

References

1. GaoW, Hu CM, Fang RH, Zhang L (2013) Liposome-like nanostructures for drug delivery. J Mater Chem B Mater Biol Med. doi:10.1039/C3TB21238F
2. Sadozai H, Saeidi D (2013) Recent developments in liposome-based veterinary therapeutics. ISRN Vet Sci 2013:167521
3. Markman JL, Rekechenetskiy A, Holler E, Ljubimova JY (2013) Nanomedicine therapeutic approaches to overcome cancer drug resistance. Adv Drug Deliv Rev 65(13–14):1866–1879
4. Deshpande PP, Biswas S, Torchilin VP (2013) Current trends in the use of liposomes for tumor targeting. Nanomedicine (Lond) 8(9):1509–1528. doi:10.2217/nnm.13.118
5. Hafner A, Lovrić J, Lakoš GP, Pepić I (2014) Nanotherapeutics in the EU: an overview on current state and future directions. Int J Nanomedicine 9:1005–1023
6. Chang H-I, Yeh M-K (2012) Clinical development of liposome-based drugs: formulation, characterization, and therapeutic efficacy. Int J Nanomedicine 7:49–60. doi:10.2147/IJN.S26766

7. Dua JS, Rana AC, Bhandari AK (2012) Liposome: methods of preparation and applications. Int J Pharma Stud Res 3(2):14–20
8. Zhong Z, Wan Y, Han J, Shi S, Zhang Z, Sun X (2011) Improvement of adenoviral vector-mediated gene transfer to airway epithelia by folate-modified anionic liposomes. Int J Nanomedicine 6:1083–1093. doi:10.2147/IJN.S19745
9. Chen L, Zhu J, Li Y, Lu J, Gao L, Xu H, Fan M, Yang X (2013) Enhanced nasal mucosal delivery and immunogenicity of anti-caries DNA vaccine through incorporation of anionic liposomes in chitosan/DNA complexes. PLoS One 8(8):e71953. doi:10.1371/journal.pone.0071953
10. Liu J, Stace-Naughton A, Jiang X, Brinker CJ (2009) Porous nanoparticle supported lipid bilayers (protocells) as delivery vehicles. J Am Chem Soc 131(4):1354–1355. doi:10.1021/ja808018y
11. Xu L, Feng L, Dong R, Hao J, Dong S (2013) Transfection efficiency of DNA enhanced by association with salt-free catanionic vesicles. Biomacromolecules 14(8):2781–2789. doi:10.1021/bm400616y
12. Chen Y, Sun J, Lu Y, Tao C, Huang J, Zhang H, Yu Y, Zou H, Gao J, Zhong Y (2013) Complexes containing cationic and anionic pH-sensitive liposomes: comparative study of factors influencing plasmid DNA gene delivery to tumors. Int J Nanomedicine 8:1573–1593. doi:10.2147/IJN.S42800
13. Düzgüneş N, de Ilarduya CT (2012) Genetic nanomedicine: gene delivery by targeted lipoplexes. Methods Enzymol 509:355–367. doi:10.1016/B978-0-12-391858-1.00018-6
14. Zhao W, Zhuang S, Qi XR (2011) Comparative study of the in vitro and in vivo characteristics of cationic and neutral liposomes. Int J Nanomedicine 6:3087–3098. doi:10.2147/IJN.S25399
15. Duangjit S, Pamornpathomkul B, Opanasopit P, Rojanarata T, Obata Y, Takayama K, Ngawhirunpat T (2014) Role of the charge, carbon chain length, and content of surfactant on the skin penetration of meloxicam-loaded liposomes. Int J Nanomedicine 9:2005–2017. doi:10.2147/IJN.S60674
16. Pappalardo JS, Langellotti CA, Di Giacomo S, Olivera V, Quattrocchi V, Zamorano PI, Hartner WC, Levchenko TS, Torchilin VP (2014) In vitro transfection of bone marrow-derived dendritic cells with TATp-liposomes. Int J Nanomedicine 9:963–973. doi:10.2147/IJN.S53432
17. Wu C, Yu C, Chu M (2011) A gold nanoshell with a silica inner shell synthesized using liposome templates for doxorubicin loading and near-infrared photothermal therapy. Int J Nanomedicine 6:807–813. doi:10.2147/IJN.S16701
18. Liu W, Liu WL, Liu CM, Liu JH, Yang SB, Zheng HJ, Lei HW, Ruan R, Li T, Tu ZC, Song XY (2011) Medium-chain fatty acid nanoliposomes for easy energy supply. Nutrition 27(6):700–706. doi:10.1016/j.nut.2010.06.010
19. Arab Tehrany E, Kahn CJ, Baravian C, Maherani B, Belhaj N, Wang X, Linder M (2012) Elaboration and characterization of nanoliposome made of soya; rapeseed and salmon lecithins: application to cell culture. Colloids Surf B Biointerfaces 95:75–81. doi:10.1016/j.colsurfb.2012.02.024
20. Ruktanonchai U, Nuchuchua O, Charlermroj R, Pattarakankul T, Karoonuthaisiri N (2012) Signal amplification of microarray-based immunoassay by optimization of nanoliposome formulations. Anal Biochem 429(2):142–147. doi:10.1016/j.ab.2012.07.012
21. Rasti B, Jinap S, Mozafari MR, Abd-Manap MY (2013) Optimization on preparation condition of polyunsaturated fatty acids nanoliposome prepared by Mozafari method. J Liposome Res Jun;24(2):99–105. doi:10.3109/08982104.2013.839702. Epub 2013 Oct 7
22. Lin CY, Javadi M, Belnap DM, Barrow JR, Pitt WG (2014) Ultrasound sensitive eLiposomes containing doxorubicin for drug targeting therapy. Nanomedicine 10(1):67–76. doi:10.1016/j.nano.2013.06.011

23. Guan R, Ma J, Wu Y, Lu F, Xiao C, Jiang H, Kang T (2012) Development and characterization of lactoferrin nanoliposome: cellular uptake and stability. Nanoscale Res Lett 7 (1):679. doi:10.1186/1556-276X-7-679
24. Elmeshad AN, Mortazavi SM, Mozafari MR (2014) Formulation and characterization of nanoliposomal 5-fluorouracil for cancer nanotherapy. J Liposome Res 24(1):1–9. doi:10.3109/08982104.2013.810644
25. Xu QH, Shi JY, Zhang J, Sun YF, Chang AH, Zhao YM, Cai WJ, Liu D, Zhou CC, Fan LH, Su B (2014) Comparison of tumor neovasculature-targeted paramagnetic nanoliposomes for MRI in mice xenograft models. Clin Transl Oncol. doi:10.1007/s12094-013-1091-4
26. Tiantian Y, Wenji Z, Mingshuang S, Rui Y, Shuangshuang S, Yuling M, Jianhua Y, Xinggang Y, Shujun W, Weisan P (2014) Study on intralymphatic-targeted hyaluronic acid-modified nanoliposome: influence of formulation factors on the lymphatic targeting. Int J Pharm 471(1–2):245–257. doi:10.1016/j.ijpharm.2014.05.027
27. Zou LQ, Liu W, Liu WL, Liang RH, Li T, Liu CM, Cao YL, Niu J, Liu Z (2014) Characterization and bioavailability of tea polyphenol nanoliposome prepared by combining an ethanol injection method with dynamic high-pressure microfluidization. J Agric Food Chem 62(4):934–941. doi:10.1021/jf402886s
28. Fox CB, Sivananthan SJ, Duthie MS, Vergara J, Guderian JA, Moon E, Coblentz D, Reed SG, Carter D (2014) A nanoliposome delivery system to synergistically trigger TLR4 AND TLR7. J Nanobiotechnol 12(1):17. doi:10.1186/1477-3155-12-17
29. Hasan M, Belhaj N, Benachour H, Barberi-Heyob M, Kahn CJ, Jabbari E, Linder M, Arab-Tehrany E (2014) Liposome encapsulation of curcumin: physico-chemical characterizations and effects on MCF7 cancer cell proliferation. Int J Pharm 461(1–2):519–528. doi:10.1016/j.ijpharm.2013.12.007
30. Ghosh S, Pal S, Prusty S, Girish KVS (2012) Curcumin and cancer: recent developments. J Res Biol 3:251–272
31. Bouarab L, Maherani B, Kheirolomoom A, Hasan M, Aliakbarian B, Linder M, Arab-Tehrany E (2014) Influence of lecithin-lipid composition on physico-chemical properties of nanoliposomes loaded with a hydrophobic molecule. Colloids Surf B Biointerfaces 115:197–204. doi:10.1016/j.colsurfb.2013.11.034
32. Rodríguez-Gascón A, del Pozo-Rodríguez A, Solinís MÁ (2014) Development of nucleic acid vaccines: use of self-amplifying RNA in lipid nanoparticles. Int J Nanomedicine 9:1833–1843. doi:10.2147/IJN.S39810
33. Jung S, Patzelt A, Otberg N, Thiede G, Sterry W, Lademann J (2009) Strategy of topical vaccination with nanoparticles. J Biomed Opt 14(2):021001. doi:10.1117/1.3080714
34. Firouzmand H, Badiee A, Khamesipour A, Heravi Shargh V, Alavizadeh SH, Abbasi A, Jaafari MR (2013) Induction of protection against leishmaniasis in susceptible BALB/c mice using simple DOTAP cationic nanoliposomes containing soluble Leishmania antigen (SLA). Acta Trop 128(3):528–535. doi:10.1016/j.actatropica.2013.07.021
35. Rangger C, Helbok A, Sosabowski J, Kremser C, Koehler G, Prassl R, Andreae F, Virgolini IJ, von Guggenberg E, Decristoforo C (2013) Tumor targeting and imaging with dual-peptide conjugated multifunctional liposomal nanoparticles. Int J Nanomedicine 8:4659–4671. doi:10.2147/IJN.S51927
36. Accardo A, Aloj L, Aurilio M, Morelli G, Tesauro D (2014) Receptor binding peptides for target-selective delivery of nanoparticles encapsulated drugs. Int J Nanomedicine 9:1537–1557. doi:10.2147/IJN.S53593
37. Ghosh S, Saha S, Sur A (in press) Polymeric nanoparticles: application in cancer management. Encyclopedia of biomedical polymers and polymeric biomaterials. Taylor & Francis
38. Lo YL, Liu Y (2014) Reversing multidrug resistance in Caco-2 by silencing MDR1, MRP1, MRP2, and BCL-2/BCL-xL using liposomal antisense oligonucleotides. PLoS One Mar 17;9(3):e90180.doi: doi:10.1371/journal.pone.0090180

39. Kullberg M, McCarthy R, Anchordoquy TJ (2014) Gene delivery to Her-2+ breast cancer cells using a two-component delivery system to achieve specificity. Nanomedicine pii: S1549-9634(14)00116-6. doi: 10.1016/j.nano.2014.02.013
40. Miyabe H, Hyodo M, Nakamura T, Sato Y, Hayakawa Y, Harashima H (2014) A new adjuvant delivery system 'cyclic di-GMP/YSK05 liposome' for cancer immunotherapy. J Control Release 184:20–27. doi:10.1016/j.jconrel.2014.04.004
41. Lin CH, Al-Suwayeh SA, Hung CF, Chen CC, Fang JY (2013) Camptothecin-loaded liposomes with α-melanocyte-stimulating hormone enhance cytotoxicity toward and cellular uptake by melanomas: an application of nanomedicine on natural product. J Tradit Complement Med 3(2):102–109. doi:10.4103/2225-4110.110423
42. Tejera-Garcia R, Parkkila P, Zamotin V, Kinnunen PK (2014) Principles of rational design of thermally targeted liposomes for local drug delivery. Nanomedicine pii: S1549-9634(14) 00131-2. doi: 10.1016/j.nano.2014.03.013
43. Mo R, Jiang T, Gu Z (2014) Enhanced anticancer efficacy by ATP-mediated liposomal drug delivery. Angew Chem Int Ed Engl. doi:10.1002/anie.201400268
44. Ahmed M, Cheung NK (2014) Engineering anti-GD2 monoclonal antibodies for cancer immunotherapy. FEBS Lett 588(2):288–297. doi:10.1016/j.febslet.2013.11.030
45. Brown BS, Patanam T, Mobli K, Celia C, Zage PE, Bean AJ, Tasciotti E (2014) Etoposide-loaded immunoliposomes as active targeting agents for GD2-positive malignancies. Cancer Biol Ther 15(7):851–861
46. Wolfram J, Suri K, Huang Y, Molinaro R, Borsoi C, Scott B, Boom K, Paolino D, Fresta M, Wang J, Ferrari M, Celia C, Shen HJ (2014) Evaluation of anticancer activity of celastrol liposomes in prostate cancer cells. J Microencapsul 31(5):501–507
47. Legut M, Lipka D, Filipczak N, Piwoni A, Kozubek A, Gubernator J (2014) Anacardic acid enhances the anticancer activity of liposomal mitoxantrone towards melanoma cell lines—in vitro studies. Int J Nanomedicine 9:653–668. doi:10.2147/IJN.S54911
48. Naksuriya O, Okonogi S, Schiffelers RM, Hennink WE (2014) Curcumin nanoformulations: a review of pharmaceutical properties and preclinical studies and clinical data related to cancer treatment. Biomaterials 35(10):3365–3383. doi:10.1016/j.biomaterials.2013.12.090
49. Cai D, Gao W, He B, Dai W, Zhang H, Wang X, Wang J, Zhang X, Zhang Q (2014) Hydrophobic penetrating peptide PFVYLI-modified stealth liposomes for doxorubicin delivery in breast cancer therapy. Biomaterials 35(7):2283–2294. doi:10.1016/j.biomaterials.2013.11.088
50. Bartneck M, Peters FM, Warzecha KT, Bienert M, van Bloois L, Trautwein C, Lammers T, Tacke F (2014) Liposomal encapsulation of dexamethasone modulates cytotoxicity, inflammatory cytokine response, and migratory properties of primary human macrophages. Nanomedicine 10(6):1209–1220. doi:10.1016/j.nano.2014.02.011
51. Schiener M, Hossann M, Viola JR, Ortega-Gomez A, Weber C, Lauber K, Lindner LH, Soehnlein O (2014) Nanomedicine-based strategies for treatment of atherosclerosis. Trends Mol Med 20(5):271–281
52. Garg T (2014) Current nanotechnological approaches for an effective delivery of bio-active drug molecules in the treatment of acne. Artif Cells Nanomed Biotechnol 20:1–8
53. Ozbakir B, Crielaard BJ, Metselaar JM, Storm G, Lammers T (2014) Liposomal corticosteroids for the treatment of inflammatory disorders and cancer. J Control Release 190:624–36. doi:10.1016/j.jconrel.2014.05.039
54. Sun Y, Fox T, Adhikary G, Kester M, Pearlman E (2008) Inhibition of corneal inflammation by liposomal delivery of short-chain, C-6 ceramide. J Leukoc Biol 83(6):1512–1521. doi:10.1189/jlb.0108076
55. Nakagishi Y, Shimizu M, Kasai K, Miyoshi M, Yachie A (2014) Successful therapy of macrophage activation syndrome with dexamethasone palmitate. Mod Rheumatol Apr 23. [Epub ahead of print]
56. Date AA, Destache CJ (2013) A review of nanotechnological approaches for the prophylaxis of HIV/AIDS. Biomaterials 34:6202–6228. doi:10.1016/j.biomaterials.2013.05.012

57. Gramatica A, Petazzi RA, Lehmann MJ, Ziomkowska J, Herrmann A, Chiantia S (2014) αEnv-decorated phosphatidylserine liposomes trigger phagocytosis of HIV-virus-like particles in macrophages. Nanomedicine pii: S1549-9634(14)00043-4
58. Alyautdin R, Khalin I, Nafeeza MI, Haron MH, Kuznetsov D (2014) Nanoscale drug delivery systems and the blood–brain barrier. Int J Nanomedicine 9:795–811. doi:10.2147/IJN.S52236
59. Sagar V, Pilakka-Kanthikeel S, Pottathil R, Saxena SK, Nair M (2014) Towards nanomedicines for neuroAIDS. Rev Med Virol 24(2):103–124
60. Venditto VJ, Wieczorek L, Molnar S, Teque F, Landucci G, Watson DS, Forthal D, Polonis VR, Levy JA, Szoka FC Jr (2014) Chemically modified peptides based on the membrane proximal external region of HIV-1 envelope induce high titer, epitope specific non-neutralizing antibodies in rabbits. Clin Vaccine Immunol pii: CVI.00320-14
61. Cho H, Stuart JM, Magid R, Danila DC, Hunsaker T, Pinkhassik E, Hasty KA (2014) Theranostic immunoliposomes for osteoarthritis. Nanomedicine 10(3):619–627. doi:10.1016/j.nano.2013.09.004
62. Afzal E, Zakeri S, Keyhanvar P, Bagheri M, Mahjoubi P, Asadian M, Omoomi N, Dehqanian M, Ghalandarlaki N, Darvishmohammadi T, Farjadian F, Golvajoee MS, Afzal S, Ghaffari M, Cohan RA, Gravand A, Ardestani MS (2013) Nanolipodendrosome-loaded glatiramer acetate and myogenic differentiation 1 as augmentation therapeutic strategy approaches in muscular dystrophy. Int J Nanomedicine 8:2943–2960. doi:10.2147/IJN.S43219
63. Zhang JX, Wang K, Mao ZF, Fan X, Jiang DL, Chen M, Cui L, Sun K, Dang SC (2013) Application of liposomes in drug development—focus on gastroenterological targets. Int J Nanomedicine 8:1325–1334. doi:10.2147/IJN.S42153
64. Sun M, Su X, Ding B, He X, Liu X, Yu A, Lou H, Zhai G (2012) Advances in nanotechnology-based delivery systems for curcumin. Nanomedicine (Lond) 7(7):1085–1100. doi:10.2217/nnm.12.80
65. Kozako T, Arima N, Yoshimitsu M, Honda SI, Soeda S (2012) Liposomes and nanotechnology in drug development: focus on oncotargets. Int J Nanomedicine 7:4943–4951. doi:10.2147/IJN.S30726
66. Agrati C, Marianecci C, Sennato S, Carafa M, Bordoni V, Cimini E, Tempestilli M, Pucillo LP, Turchi F, Martini F, Borioni G, Bordi F (2011) Multicompartment vectors as novel drug delivery systems: selective activation of Tγδ lymphocytes after zoledronic acid delivery. Nanomedicine 7(2):153–161. doi:10.1016/j.nano.2010.10.003
67. Campani V, Marchese D, Pitaro MT, Pitaro M, Grieco P, De Rosa G (2014) Development of a liposome-based formulation for vitamin K1 nebulization on the skin. Int J Nanomedicine. doi:10.2147/IJN.S58365
68. Mota Ade C, de Freitas ZM, Ricci Júnior E, Dellamora-Ortiz GM, Santos-Oliveira R, Ozzetti RA, Vergnanini AL, Ribeiro VL, Silva RS, dos Santos EP (2013) In vivo and in vitro evaluation of octyl methoxycinnamate liposomes. Int J Nanomedicine 8:4689–4701. doi:10.2147/IJN.S51383
69. Vyas A, Kumar Sonker A, Gidwani B (2014) Carrier-based drug delivery system for treatment of acne. Scientific World Journal 2014:276260. doi:10.1155/2014/276260
70. Zhou Y, Ning Q, Yu DN, Li WG, Deng J (2014) Improved oral bioavailability of breviscapine via a Pluronic P85-modified liposomal delivery system. J Pharm Pharmacol 66:903–911. doi:10.1111/jphp.12215
71. Migliore MM, Ortiz R, Dye S, Campbell RB, Amiji MM, Waszczak BL (2014) Neurotrophic and neuroprotective efficacy of intranasal GDNF in a rat model of Parkinson's disease. Neuroscience pii: S0306-4522(14)00411-4. doi: 10.1016/j.neuroscience.2014.05.019
72. Yamada Y, Tabata M, Yasuzaki Y, Nomura M, Shibata A, Ibayashi Y, Taniguchi Y, Sasaki S, Harashima H (2014) A nanocarrier system for the delivery of nucleic acids targeted to a pancreatic beta cell line. Biomaterials 35(24):6430–6438. doi:10.1016/j.biomaterials.2014.04.017

73. Mourtas S, Lazar AN, Markoutsa E, Duyckaerts C, Antimisiaris SG (2014) Multifunctional nanoliposomes with curcumin-lipid derivative and brain targeting functionality with potential applications for Alzheimer disease. Eur J Med Chem 80:175–183. doi:10.1016/j.ejmech.2014.04.050
74. Kaur IP, Kakkar S (2014) Nanotherapy for posterior eye diseases. J Control Release. May 24. pii: S0168-3659(14)00334-4. doi:10.1016/j.jconrel.2014.05.031. [Epub ahead of print]
75. Davies LA, Nunez-Alonso GA, McLachlan G, Hyde SC, Gill DR (2014) Aerosol delivery of DNA/liposomes to the lung for cystic fibrosis gene therapy. Hum Gene Ther Clin Dev. doi:10.1089/humc.2014.019
76. Song H, Wang G, He B, Li L, Li C, Lai Y, Xu X, Gu Z (2012) Cationic lipid-coated PEI/DNA polyplexes with improved efficiency and reduced cytotoxicity for gene delivery into mesenchymal stem cells. Int J Nanomedicine 7:4637–4648
77. Chang TM (2012) From artificial red blood cells, oxygen carriers, and oxygen therapeutics to artificial cells, nanomedicine, and beyond. Artif Cells Blood Substit Immobil Biotechnol 40 (3):197–199. doi:10.3109/10731199.2012.662408
78. Chen XA, Zhang LJ, He ZJ, Wang WW, Xu B, Zhong Q, Shuai XT, Yang LQ, Deng YB (2011) Plasmid-encapsulated polyethylene glycol-grafted polyethylenimine nanoparticles for gene delivery into rat mesenchymal stem cells. Int J Nanomedicine 6:843–853
79. Gökce M, Kuskonmaz B, Cetin M, Uckan Cetinkaya D, Tuncer M (2013) Coexisting or underlying risk factors of hepatic veno-occlusive disease in pediatric hematopoietic stem cell transplant recipients receiving prophylaxis. Exp Clin Transplant 11(5):440–446
80. Ma K, Wang DD, Lin Y, Wang J, Petrenko V, Mao C (2013) Synergetic targeted delivery of sleeping-beauty transposon system to mesenchymal stem cells using LPD nanoparticles modified with a phage-displayed targeting peptide. Adv Funct Mater 23(9):1172–1181
81. Kito T, Shibata R, Ishii M, Suzuki H, Himeno T, Kataoka Y, Yamamura Y, Yamamoto T, Nishio N, Ito S, Numaguchi Y, Tanigawa T, Yamashita JK, Ouchi N, Honda H, Isobe K, Murohara T (2013) iPS cell sheets created by a novel magnetite tissue engineering method for reparative angiogenesis. Sci Rep 3:1418. doi:10.1038/srep01418
82. Liu WH, Liu ZC, You N, Zhang N, Wang T, Gong ZB, Liu HB, Dou KF (2012) Several important in vitro improvements in the amplification, differentiation and tracing of fetal liver stem/progenitor cells. PLoS One 7(10):e47346. doi:10.1371/journal.pone.0047346
83. Palaniyandi K, Pockaj BA, Gendler SJ, Chang XB (2012) Human breast cancer stem cells have significantly higher rate of clathrin-independent and caveolin-independent endocytosis than the differentiated breast cancer cells. J Cancer Sci Ther 4(7):214–222
84. Döring M, Hartmann U, Erbacher A, Lang P, Handgretinger R, Müller I (2012) Caspofungin as antifungal prophylaxis in pediatric patients undergoing allogeneic hematopoietic stem cell transplantation: a retrospective analysis. BMC Infect Dis 12:151
85. Gul-Uludag H, Xu P, Marquez-Curtis LA, Xing J, Janowska-Wieczorek A, Chen J (2012) Cationic liposome-mediated CXCR4 gene delivery into hematopoietic stem/progenitor cells: implications for clinical transplantation and gene therapy. Stem Cells Dev 21 (10):1587–1596. doi:10.1089/scd.2011.0297
86. Oshikata A, Miyazaki R, Matsushita T, Ueoka R (2011) Selective elimination of the transformed hepatic stem cells using hybrid liposomes. Yakugaku Zasshi 131(5):757–763
87. Winkler IG, Sims NA, Pettit AR, Barbier V, Nowlan B, Helwani F, Poulton IJ, van Rooijen N, Alexander KA, Raggatt LJ, Lévesque JP (2010) Bone marrow macrophages maintain hematopoietic stem cell (HSC) niches and their depletion mobilizes HSCs. Blood 116(23):4815–4828. doi:10.1182/blood-2009-11-253534
88. Matsumura Y (2014) The drug discovery by nanomedicine and its clinical experience. Jpn J Clin Oncol Jun;44(6):515–25. doi:10.1093/jjco/hyu046. Epub 2014 Apr 21
89. Nardin A, Lefebvre ML, Labroquère K, Faure O, Abastado JP (2006) Liposomal muramyl tripeptide phosphatidylethanolamine: targeting and activating macrophages for adjuvant treatment of osteosarcoma. Curr Cancer Drug Targets 6(2):123–133

90. Anderson PM, Tomaras M, McConnell K (2010) Mifamurtide in osteosarcoma—a practical review. Drugs Today (Barc) 46(5):327–337
91. Groll AH, Silling G, Young C, Schwerdtfeger R, Ostermann H, Heinz WJ, Gerss J, Kolve H, Lanvers-Kaminsky C, Vieira Pinheiro JP, Gammelin S, Cornely OA, Wuerthwein G (2010) Randomized comparison of safety and pharmacokinetics of caspofungin, liposomal amphotericin B, and the combination of both in allogeneic hematopoietic stem cell recipients. Antimicrob Agents Chemother 54(10):4143–4149. doi:10.1128/AAC.00425-10
92. Portillo J, Kamar N, Melibary S, Quevedo E, Bergese S (2014) Safety of liposome extended-release bupivacaine for postoperative pain control. Front Pharmacol 5:90
93. Malaekeh-Nikouei B, Malaekeh-Nikouei M, Oskuee RK, Ramezani M (2009) Preparation, characterization, transfection efficiency, and cytotoxicity of liposomes containing oligoamine-modified cholesterols as nanocarriers to Neuro2A cells. Nanomedicine 5 (4):457–462. doi:10.1016/j.nano.2009.02.001
94. Sun X, Chen J, Gu X, Liang W, Wang J (2014) Efficacy and toxicity of cisplatin liposomes modified with polyethylenimine. Pharmazie 69(4):281–286
95. Xie M, Chen Y, Wu L (2013) Preparation of doxorubicin-hydrochloride nanoliposomes by ethanol injection-pH gradient method and their safety evaluation. J Nanosci Nanotechnol 13 (1):216–221
96. Dadgar N, Alavi SE, Esfahani MK, Akbarzadeh A (2013) Study of toxicity effect of pegylated nanoliposomal artemisinin on breast cancer cell line. Indian J Clin Biochem 28 (4):410–412. doi:10.1007/s12291-013-0306-3
97. Esfahani MK, Alavi SE, Movahedi F, Alavi F, Akbarzadeh A (2013) Cytotoxicity of liposomal paclitaxel in breast cancer cell line mcf-7. Indian J Clin Biochem 28 (4):358–360. doi:10.1007/s12291-013-0296-1
98. Ait-Oudhia S, Mager DE, Straubinger RM (2014) Application of pharmacokinetic and pharmacodynamic analysis to the development of liposomal formulations for oncology. Pharmaceutics 6(1):137–174. doi:10.3390/pharmaceutics6010137
99. Needham D, Anyarambhatla G, Kong G, Dewhirst MW (2000) A new temperature-sensitive liposome for use with mild hyperthermia: characterization and testing in a human tumor xenograft model. Cancer Res 60(5):1197–1201
100. Lindner LH, Eichhorn ME, Eibl H, Teichert N, Schmitt-Sody M, Issels RD, Dellian M (2004) Novel temperature-sensitive liposomes with prolonged circulation time. Clin Cancer Res 10 (6):2168–2178
101. Langer R (1998) Drug delivery and targeting. Nature 392(6679 Suppl):5–10
102. Deepa K, Singha S, Panda T (2014) Doxorubicin nanoconjugates. J Nanosci Nanotechnol 14(1):892–904
103. Li L, ten Hagen TL, Schipper D, Wijnberg TM, van Rhoon GC, Eggermont AM, Lindner LH, Koning GA (2010) Triggered content release from optimized stealth thermosensitive liposomes using mild hyperthermia. J Control Release 143(2):274–279. doi:10.1016/j.jconrel.2010.01.006
104. Chan JM, Zhang L, Yuet KP, Liao G, Rhee JW, Langer R, Farokhzad OC (2009) PLGA-lecithin-PEG core-shell nanoparticles for controlled drug delivery. Biomaterials 30(8):1627–1634. doi:10.1016/j.biomaterials.2008.12.013
105. Ning M, Guo Y, Pan H, Chen X, Gu Z (2005) Preparation, in vitro and in vivo evaluation of liposomal/niosomal gel delivery systems for clotrimazole. Drug Dev Ind Pharm 31(4–5):375–383
106. Haley B, Frenkel E (2008) Nanoparticles for drug delivery in cancer treatment. Urol Oncol 26(1):57–64. doi:10.1016/j.urolonc.2007.03.015
107. Choi SJ, Oh JM, Choy JH (2010) Biocompatible nanoparticles intercalated with anticancer drug for target delivery: pharmacokinetic and biodistribution study. J Nanosci Nanotechnol 10(4):2913–2916
108. Nobs L, Buchegger F, Gurny R, Allémann E (2004) Current methods for attaching targeting ligands to liposomes and nanoparticles. J Pharm Sci 93(8):1980–1992

109. Stathopoulos GP, Boulikas T, Vougiouka M, Rigatos SK, Stathopoulos JG (2006) Liposomal cisplatin combined with gemcitabine in pretreated advanced pancreatic cancer patients: a phase I-II study. Oncol Rep 15(5):1201–1204
110. Mandal B, Bhattacharjee H, Mittal N, Sah H, Balabathula P, Thoma LA, Wood GC (2013) Core-shell-type lipid-polymer hybrid nanoparticles as a drug delivery platform. Nanomedicine 9(4):474–491. doi:10.1016/j.nano.2012.11.010
111. Hu CM, Zhang L, Aryal S, Cheung C, Fang RH, Zhang L (2011) Erythrocyte membrane-camouflaged polymeric nanoparticles as a biomimetic delivery platform. Proc Natl Acad Sci U S A 108(27). 10980-5. doi:10.1073/pnas.1106634108. Epub 2011 Jun 20
112. Aryal S, Hu CM, Fang RH, Dehaini D, Carpenter C, Zhang DE, Zhang L (2013) Erythrocyte membrane-cloaked polymeric nanoparticles for controlled drug loading and release. Nanomedicine (Lond) 8(8):1271–1280. doi:10.2217/nnm.12.153
113. Tanaka S, Kuroda Y, Ihara F, Nishimura M, Hiasa J, Kojima N, Nishikawa Y (2014) Vaccination with profilin encapsulated in oligomannose-coated liposomes induces significant protective immunity against Toxoplasma gondii. Vaccine 32(16):1781–1785. doi:10.1016/j.vaccine.2014.01.095
114. Senchi K, Matsunaga S, Hasegawa H, Kimura H, Ryo A (2013) Development of oligomannose-coated liposome-based nasal vaccine against human parainfluenza virus type 3. Front Microbiol 4:346. doi:10.3389/fmicb.2013.00346
115. Kojima N, Ishii M, Kawauchi Y, Takagi H (2013) Oligomannose-coated liposome as a novel adjuvant for the induction of cellular immune responses to control disease status. Biomed Res Int 2013:56292. doi:10.1155/2013/562924
116. Ishii M, Kojima N (2013) Effective stimulation of invariant natural killer T cells by oligomannose-coated liposomes. Int Immunopharmacol 15(4):685–692. doi:10.1016/j.intimp.2013.03.009
117. Kawakita A, Shirasaki H, Yasutomi M, Tokuriki S, Mayumi M, Naiki H, Ohshima Y (2012) Immunotherapy with oligomannose-coated liposomes ameliorates allergic symptoms in a murine food allergy model. Allergy 67(3):371–379
118. Ishii M, Koyama A, Iseki H, Narumi H, Yokoyama N, Kojima N (2010) Anti-allergic potential of oligomannose-coated liposome-entrapped Cry j 1 as immunotherapy for Japanese cedar pollinosis in mice. Int Immunopharmacol 10(9):1041–1046. doi:10.1016/j.intimp.2010.06.003
119. Kozako T, Hirata S, Shimizu Y, Satoh Y, Yoshimitsu M, White Y, Lemonnier F, Shimeno H, Soeda S, Arima N (2011) Oligomannose-coated liposomes efficiently induce human T-cell leukemia virus-1-specific cytotoxic T lymphocytes without adjuvant. FEBS J 278 (8):1358–1366. doi:10.1111/j.1742-4658.2011.08055.x
120. Wang Y, Tu S, Li R, Yang X, Liu L, Zhang Q (2010) Cholesterol succinyl chitosan anchored liposomes: preparation, characterization, physical stability, and drug release behavior. Nanomedicine 6(3):471–477. doi:10.1016/j.nano.2009.09.005
121. Jiang T, Zhang Z, Zhang Y, Lv H, Zhou J, Li C, Hou L, Zhang Q (2012) Dual-functional liposomes based on pH-responsive cell-penetrating peptide and hyaluronic acid for tumor-targeted anticancer drug delivery. Biomaterials 33(36):9246–9258. doi:10.1016/j.biomaterials.2012.09.027
122. Cheng J, Zhu JB, Wen N, Xiong F (2006) Stability and pharmacokinetic studies of O-palmitoyl amylopectin anchored dipyridamole liposomes. Int J Pharm 313(1–2):136–143
123. Cohen K, Emmanuel R, Kisin-Finfer E, Shabat D, Peer D (2014) Modulation of drug resistance in ovarian adenocarcinoma using chemotherapy entrapped in hyaluronan-grafted nanoparticle clusters. ACS Nano 8(3):2183–2195. doi:10.1021/nn500205b
124. Banerjee A, Onyuksel H (2012) Human pancreatic polypeptide in a phospholipid-based micellar formulation. Pharm Res 29(6):1698–1711. doi:10.1007/s11095-012-0718-4
125. Cesur H, Rubinstein I, Pai A, Onyüksel H (2009) Self-associated indisulam in phospholipid-based nanomicelles: a potential nanomedicine for cancer. Nanomedicine 5(2):178–183. doi:10.1016/j.nano.2008.09.001

126. Hamelers IH, van Loenen E, Staffhorst RW, de Kruijff B, de Kroon AI (2006) Carboplatin nanocapsules: a highly cytotoxic, phospholipid-based formulation of carboplatin. Mol Cancer Ther 5(8):2007–2012
127. Zhang J, Han X, Li X, Luo Y, Zhao H, Yang M, Ni B, Liao Z (2012) Core-shell hybrid liposomal vesicles loaded with panax notoginsenoside: preparation, characterization and protective effects on global cerebral ischemia/reperfusion injury and acute myocardial ischemia in rats. Int J Nanomedicine 7:4299–4310. doi:10.2147/IJN.S32385
128. Kim H, Britton GL, Peng T, Holland CK, McPherson DD, Huang SL (2014) Nitric oxide-loaded echogenic liposomes for treatment of vasospasm following subarachnoid hemorrhage. Int J Nanomedicine 9:155–165. doi:10.2147/IJN.S48856
129. Holt B, Gupta AS (2012) Streptokinase loading in liposomes for vascular targeted nanomedicine applications: encapsulation efficiency and effects of processing. J Biomater Appl 26(5):509–527. doi:10.1177/0885328210374778
130. Ko YT, Hartner WC, Kale A, Torchilin VP (2009) Gene delivery into ischemic myocardium by double-targeted lipoplexes with anti-myosin antibody and TAT peptide. Gene Ther 16 (1):52–59. doi:10.1038/gt.2008.135
131. Brito L, Amiji M (2007) Nanoparticulate carriers for the treatment of coronary restenosis. Int J Nanomedicine 2(2):143–161
132. Verma DD, Hartner WC, Thakkar V, Levchenko TS, Torchilin VP (2007) Protective effect of coenzyme Q10-loaded liposomes on the myocardium in rabbits with an acute experimental myocardial infarction. Pharm Res 24(11):2131–2137
133. Ismail MF, Elmeshad AN, Salem NA (2013) Potential therapeutic effect of nanobased formulation of rivastigmine on rat model of Alzheimer's disease. Int J Nanomedicine 8:393–406. doi:10.2147/IJN.S39232
134. Lazar AN, Mourtas S, Youssef I, Parizot C, Dauphin A, Delatour B, Antimisiaris SG, Duyckaerts C (2013) Curcumin-conjugated nanoliposomes with high affinity for Aβ deposits: possible applications to Alzheimer disease. Nanomedicine 9(5):712–721. doi:10.1016/j.nano.2012.11.004
135. Ye P, Zhang W, Yang T, Lu Y, Lu M, Gai Y, Ma X, Xiang G (2014) Folate receptor-targeted liposomes enhanced the antitumor potency of imatinib through the combination of active targeting and molecular targeting. Int J Nanomedicine 9:2167–2178
136. Wang T, Wang N, Hao A, He X, Li T, Deng Y (2010) Lyophilization of water-in-oil emulsions to prepare phospholipid-based anhydrous reverse micelles for oral peptide delivery. Eur J Pharm Sci. 2010 Mar 18;39(5):373–9. doi: 10.1016/j.ejps.2010.01.006. Epub 2010 Jan 20

Part III
Characterization and Detection Methods for Nanoparticles

Chapter 10
Identification of Nanoparticle in Organic Matrices

Antonietta M. Gatti and Stefano Montanari

Abstract The chapter discusses the ways to identify micron- and sub-micron-sized particles in a variety of matrices and verify, during manufacturing processes and in the final products, the presence and distribution of the engineered nanoparticles deliberately used or those in any case present as unwelcome side-products of human activities, particularly high-temperature ones.

A brief introduction is presented on the use of an Environmental Scanning Electron Microscope and of the X-ray microprobe of an Energy Dispersive Spectroscopy that can give information on the elemental composition of the nanoparticles trapped in the products.

10.1 Introduction

Micro- and, in particular, nanosized particles are a largely new type of environmental pollution whose importance has been underestimated at least until very recently. In most cases they are generated by human activities, but their deliberate addition as engineered nanoparticles in many different products is growing more and more widespread. The novel diagnostic approach offered by nanopathology was able to demonstrate their invasiveness and pathogenicity. It is therefore necessary to evaluate their presence in the environment (air, earth, water) and in all the matrices in which they can penetrate (e.g., the organism) or where they can be laid or drop. It is also important to notice that not infrequently raw materials are already polluted by that particularly fine dust and the sole analysis of the final products where a number of ingredients, and, obviously, of different raw materials, are mixed does not supply all the information needed to track the origin of pollution and, eventually, to solve the problem. But in more than one instance analyzing the

A.M. Gatti (✉)
ISTEC, National Council of Research of Italy, via Granarolo, 64, 48018 Faenza, RA, Italy
e-mail: gatti@nanodiagnostics.it

S. Montanari
NANODIAGNOSTICS srl, via E.Fermi 1/L, 41057 San Vito, MO, Italy

M. Lungu et al. (eds.), *Nanoparticles' Promises and Risks*,
DOI 10.1007/978-3-319-11728-7_10

proper samples can directly address the problem of the origin of a pathogenic pollution.

10.2 The Technique

It is particularly useful to have a technique that preserves the sample and allows the investigation to be repeatable, but it is mandatory that the technique or the preservation process do not add laboratory pollution.

The pieces of equipment we use to observe our specimens are environmental scanning electron microscopes (ESEM Quanta 200 or FEG ESEM Quanta 250, both by FEI Company, The Netherlands). The choice is due to their possibility to work not in low vacuum the way SEMs do, but, thanks to a particular air trap placed in the column, at environmental conditions, so that no preparation of the sample (dehydration, Carbon or Gold/Palladium coating to make the sample electroconductive) is necessary. This is a measure to prevent the formation of artifacts.

An EDS microprobe (Energy Dispersion System by EDAX, USA) equipping the ESEM collects the X-rays generated by the sample's surface when hit by the electron beam, sorts and plots them by energy, and allows to identify the elements, each of which produces peaks characteristic to that particular element in this energy distribution. Suitable software identifies the elemental composition of the materials imaged of all elements with an atomic number greater than Boron's. Most elements can be detected at concentrations of the order of 0.1 %.

For the morphological observations and for the chemical analyses the sample is placed on an Aluminum stub which supports an adhesive Carbon substrate whose contribution in terms of Carbon and Oxygen does not affect the analytical result, since the particles to be studied are inorganic.

The samples can be observed in different modalities: in low vacuum, from 20 to 30 kV, in secondary and in backscattered electron mode, with spots from 3 to 5. The most common mode of detection is by secondary electrons emitted by the atoms of the sample excited by the electron beam. That way the surface of the specimen is shown. Backscattered electrons are electrons emitted by the beam that are reflected from the sample. The intensity of that signal is strongly related to the atomic number of the element. The backscattered mode allows to see materials with a higher atomic density than the biological matrix that contains them.

The analyses on biological specimens we usually carry out are either on fresh or on preserved materials. Common fresh samples are fluids like blood, sperm, and sweat, while preserved ones, generally in paraffin blocks, are tissues coming from postmortem or biopsic takings. When the sample relates to an infectious disease, a fixation in formalin is performed beforehand. Unlike fresh samples that need no special preparation, those stored in paraffin blocks must be cut into 5–10-micron sections with a microtome and freed from paraffin with xylol. For the various preparation steps, please see the book Nanopathology [1].

10.3 Dry Samples

Samples like dry or fresh leaves (among them also spices, tea, etc.), flour, lyophilized food (powder), potato chips, etc., are particularly easy to observe. They can just be put on the adhesive Carbon disc of the stub and studied as such or, when they are relatively large and their surface is irregular, they can be cut. In that case, both sides of the sections must be examined. If incidental or engineered nanosized particles are on the samples or embedded in it, they can be spotted by difference of atomic density with the surrounding matrix. The morphology of fresh samples (for instance, a leaf) must be observed at low energy, down to 5 keV (Fig. 10.1). On the y-axis of the EDS spectra there is the intensity of the signal in counts while on the x-axis is the energy in keV.

Some problems can be met with some industrial snack foods because of their very high content of Sodium chloride, a compound that dwarfs all the particles that may be present. So, much patience is required by the operator to tell kitchen salt from other particles. Some relevant results are presented in Figs. 10.2 and 10.3.

Another difficulty is the analysis of rubbery matrices like chewing gum. If the analysis is carried out using a high-energy beam (at 30 keV), the sample is spoilt and the results nullified. It is therefore necessary to work at lower voltage, for instance 10–15 kV. But such a low energy is not enough to get the elemental spectrum because of the underestimation of the elements with the peaks at high energy. With the combination of the two modes, information on both morphological and chemical nature of the particles can be obtained.

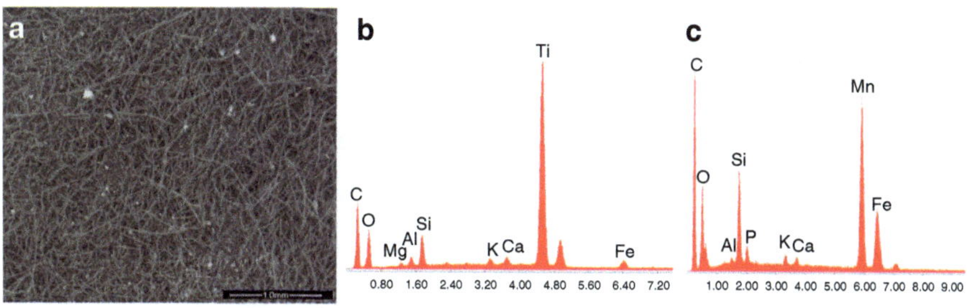

Fig. 10.1 The SEM image shows the skin of a peach with small debris coming from environmental pollution. The debris are metallic alloys composed of: (**b**) Titanium, Silicon, Aluminum, Calcium, Magnesium, Potassium, Iron; and (**c**) Manganese, Silicon, Iron, Phosphorus, Aluminum, Potassium, Calcium. Potassium and Calcium are elements belonging to the skin, while the other elements belong to industrial environmental pollution deposited on the peach's skin

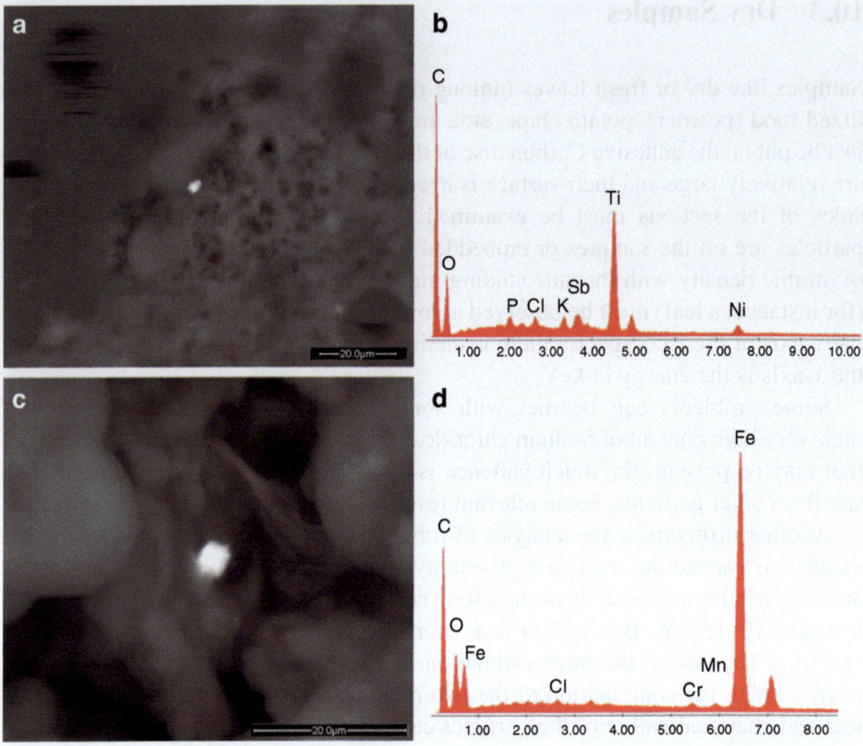

Fig. 10.2 SEM images of inorganic debris identified in an industrial cake for children. The metallic particulate matter it contains (Titanium, Nickel, and Antimony (**b**) and (**d**) stainless steel: Iron, Chromium and Nickel) is probably related to the industrial preparation process

10.4 Fatty and Oily Samples

In some cases, like yogurt or ketchup, the sample can be smeared in a thin layer directly on the Carbon disc with a plastic spatula not to run the risk of introducing metal pollutants coming from a metal tool.

To analyze pasty and semisolid fatty and oily specimens, the particles, dispersed inside the matrix as they are, must be put in a condition to be seen (Figs. 10.4, 10.5, 10.6, 10.7, and 10.8). In some circumstances we must filter the sample and analyze the filter after dehydration (Millipore, pore: 0.45 μm, White Gridded HAWG, diameter 25 mm (No. HAWG02500, Lot No. H6AN39726).

To do that, in a number of cases (e.g., chocolate (Fig. 10.4), butter, etc.) the specimens must be heated. It may happen that a thin, fatty layer clogs the filter and even trying to force somewhat the filtration with a vacuum pump may prove useless. This, for example, is the case with whole milk. So, it is the supernatant layer that, after having been left to desiccate at 40 °C for an hour, is set on the Carbon disc and observed. That is also the case of mortadella (a type of Italian

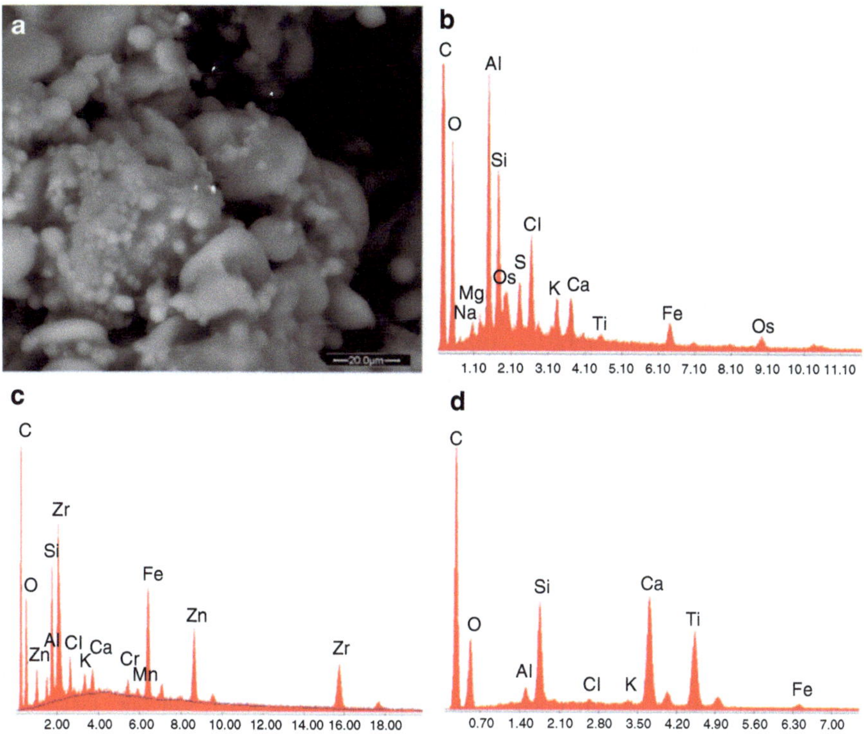

Fig. 10.3 SEM image of unintentional contamination of a biscuit by nanoparticles of (**b**) Aluminum, Silicon, Chlorine, Potassium, Calcium, Iron, Osmium, Sodium, Magnesium, Titanium; (**c**) Zirconium, Silicon, Iron, Zinc, Aluminum, Chlorine, Potassium, Calcium, Chromium, Manganese; (**d**) Calcium, Silicon, Titanium, Aluminum, Iron, Chlorine, Potassium

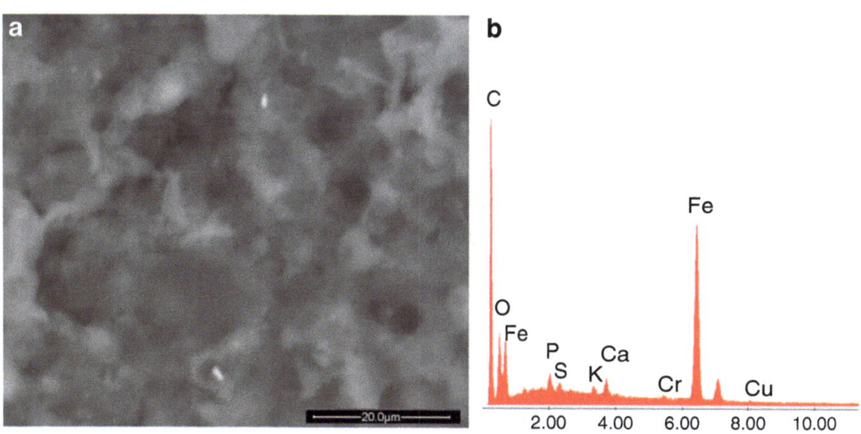

Fig. 10.4 SEM image of a chocolate samples with submicronic debris of Iron, Chromium, Phosphorus, Sulfur, Potassium, Calcium, Copper

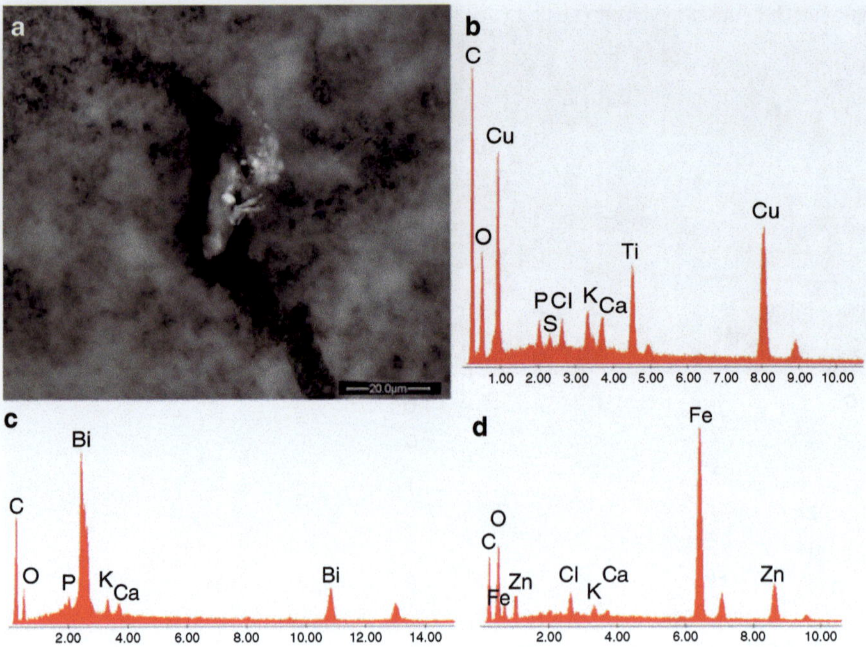

Fig. 10.5 SEM image of an Italian yogurt where submicronic and nanosized inorganic particles were detected. Particles of (**b**) Copper, Titanium, Phosphorus, Sulfur, Chlorine, Potassium, Calcium; (**c**) Bismuth, Phosphorus, Potassium, Calcium; (**d**) Iron, Zinc, Chlorine, Calcium

sausage also known in English as Bologna) that is a mix of fatty cubes embedded in homogenized pork meat.

There are a few differences in the preparation of fatty and oily samples according to their being solid or fluid. Usually, the fluid specimen is spread on a cellulose filter, dehydrated in an oven for 1 h and observed under the microscope. In the case of chocolate and yogurt (Fig. 10.5), they are heated, in some cases diluted, and finally deposited on a filter. In some cases, for instance milk or fruit juice, it is necessary to make a preventive centrifugation before the deposition on the support for the ESEM investigation.

In the case we mentioned above of mortadella, the presence of particulate matter was immediately suspected at the preliminary inspection of the specimens because of the presence of grayish spots barely visible to the naked eye in the pink areas of the meat and better visible through an optical microscope. In that case, under a stereomicroscope, we cut only those blots and smeared the matter on the Carbon disc of the stub in order to get a thin film. Embedded in that matrix we spotted irregularly shaped stainless steel debris whose corrosion had caused the discoloration of the meat. The origin of those debris was the wear of the blades used to mince the frozen pork meat (Fig. 10.8).

The identification of inorganic particulate matter, not biodegradable, so biopersistent and not biocompatible, is quite a novelty in the field of food [2–8]

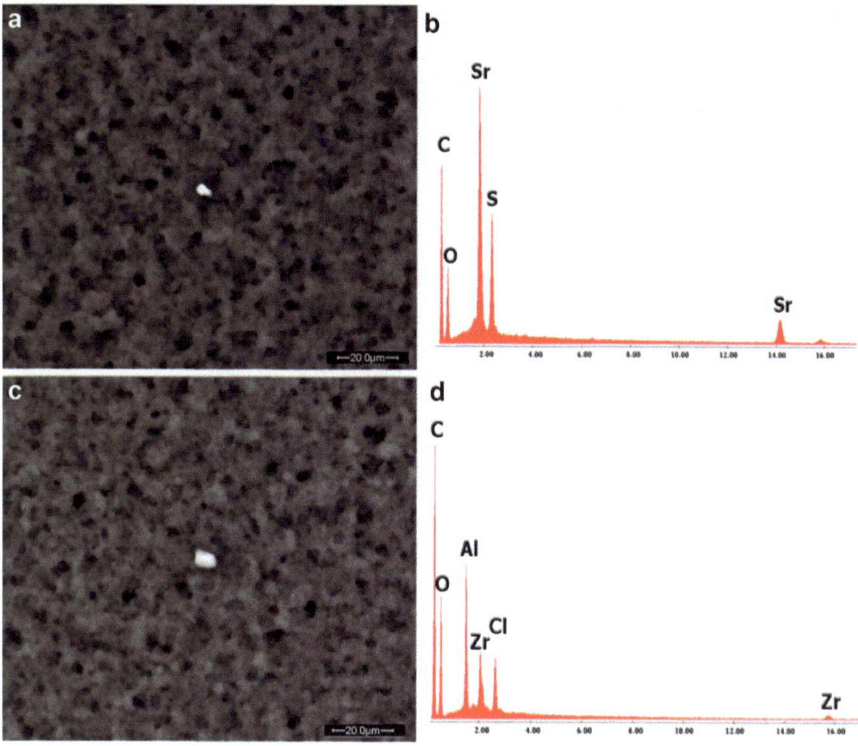

Fig. 10.6 SEM images of an artificial frozen cream with inorganic contamination. The debris are composed of (**b**) Strontium, Sulfur; (**d**) Aluminum, Zirconium, Chlorine. Such chemical compositions have never been reported in any material handbook

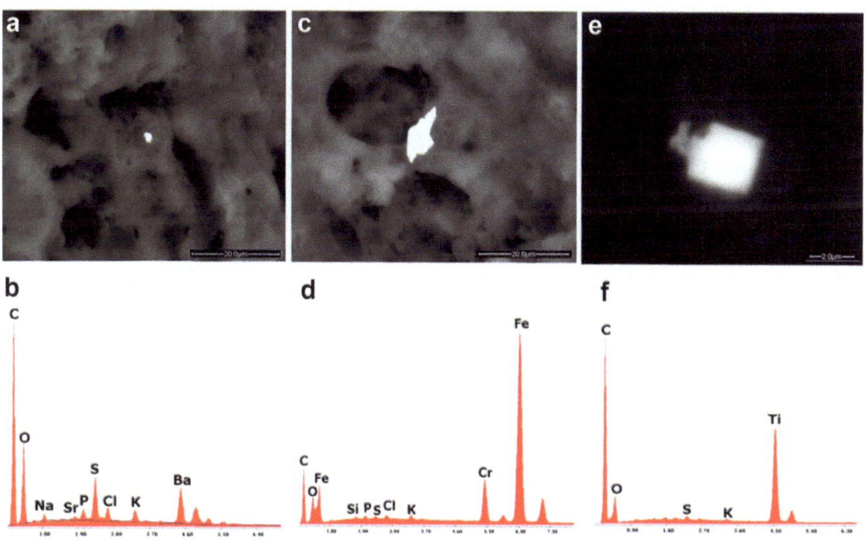

Fig. 10.7 SEM image of inorganic contamination of a homogenized baby food. The particles found are of (**b**) Barium, Sulfur; (**d**) stainless steel; (**f**) and Titanium

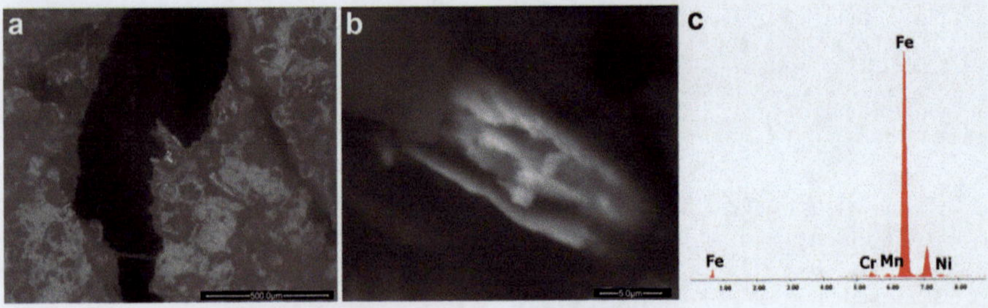

Fig. 10.8 The SEM images show inorganic debris embedded in the meat

and there is no legislation that imposes a control for such a contamination. Their presence could be related to allergic symptoms but also to other pathologies (see Chap. 5).

Acknowledgements The authors are indebted to Ms. Lavinia Nitu and Dr. Federico Capitani for the invaluable technical work and human support in writing this chapter.

References

1. Gatti AM, Montanari S (2008) Nanopathologies: the health impact of nanoparticles. In: PanStanford (ed) Singapore, Appendix pp 287–290
2. Charbonneau JE (2001) Investigation of foreign substances in food. Scanning 23:51–57
3. Charbonneau JE (2001) Investigation of corrosion and container integrity in metal food containers using scanning electron microscopy-x-ray microanalysis. Scanning 23:198–203
4. Chin HB, Kimball JR, Allen B, Charbonneau JE (1984) Corrosion in canned light-colored fruits. ITRI Publication 660:465–470
5. Tiede K, Boxall AB, Tear SP, Lewis J, David H, Hassellov M (2008) Detection and characterization of engineered nanoparticles in food and the environment. Food Addit Contam Part A Chem Anal Control Expo Risk Assess 25(7):795–821
6. Das M, Ansari KM, Tripathi A, Dwivedi PDJ (2011) Need for safety of nanoparticles used in food industry. J Biomed Nanotechnol 7(1):13–16
7. Luo P, Morrison I, Dudkiewicz A, Tiede K, Boyes E, O'Toole P, Park S, Boxall AB (2013) Visualization and characterization of engineered nanoparticles in complex environmental and food matrices using atmospheric scanning electron microscopy. J Microsc 250(1):32–41
8. Gatti AM, Tossini D, Gambarelli A, Montanari S, Capitani F (2008) Inorganic micro- and nanosized contaminants in bread and biscuits. Crit Rev Food Sci Nutr 49(3):36–43

Chapter 11
Gold Nanoparticles in Biosensing Analyses

Subash C.B. Gopinath, Marimuthu Citartan, Thangavel Lakshmipriya, Thean-Hock Tang, and Yeng Chen

Abstract Monitoring biomolecular interactions has become a constituent element of various medical and clinical applications. The importance of measuring the biomolecular interaction is evident as a potpourri of sensing strategies is currently and continuously devised. One of the strategies that can immensely enhance the sensitivity of the interaction is the one mediated by nanoparticles. Among the various nanoparticles, gold nanoparticle (AuNP) is opted for due to its capacity to improve the sensitivity and selectivity of biomolecular interactions. In addition to this, AuNP is easy to be manufactured, chemically inert, have uniform dispersibility and are able to react with chemically modified biomolecules. In this overview, the roles adopted by AuNP in various biosensing applications are discussed.

11.1 Introduction

Gold nanoparticles (AuNPs) possess excellent features that make them a magnificent material for chemical and biomolecular sensing applications [1–9]. AuNPs have features such as inert, have compatibility with surface functionalization and can make in different sizes. Another interesting characteristic of AuNPs is the absorption of the AuNPs which is at 520 nm as a result of surface plasmon excitation [10–12]. Due to this feature, AuNPs have become a platform for the

S.C.B. Gopinath (✉)
Department of Oral Biology & Biomedical Sciences and OCRCC, Faculty of Dentistry, University of Malaya, 50603 Kuala Lumpur, Malaysia

Advanced Medical & Dental Institute (AMDI), Universiti Sains Malaysia, 13200 Kepala Batas, Penang, Malaysia
e-mail: gopis11@gmail.com

M. Citartan • T. Lakshmipriya • T.-H. Tang
Advanced Medical & Dental Institute (AMDI), Universiti Sains Malaysia, 13200 Kepala Batas, Penang, Malaysia

Y. Chen
Department of Oral Biology & Biomedical Sciences and OCRCC, Faculty of Dentistry, University of Malaya, 50603 Kuala Lumpur, Malaysia

M. Lungu et al. (eds.), *Nanoparticles' Promises and Risks*,
DOI 10.1007/978-3-319-11728-7_11

development of sensors in a potpourri of applications [13–26]. The ease of preparing AuNPs also attributes to the vast application of AuNPs for sensing purpose [27].

11.2 Preparation of AuNP

The most common method for the preparation of AuNP is by sodium citrate reduction of hydrogen tetrachloroaurate ($HAuCl_4$), whereby the colour changes from pale yellow to deep red when synthesis is completed [28, 29]. Sodium citrate acting as the reducing agent that can provide AuNPs with desired sizes. Studies have shown that different ratios of gold (Au) salt to sodium citrate give rise to different sizes of AuNPs [29]. Other synthetic strategy include Brust-Schiffrin method for the preparation of thiol-protected AuNPs, which involves the two-phase synthetic strategy that depends on strong thiol-Au interactions to protect AuNPs with thiol ligands [30]. In this strategy, the transfer of $AuCl^{4-}$ from aqueous phase to toluene using the surfactant tetraoctylammonium bromide (TOAB) is then reduced by sodium borohydride ($NaBH_4$) in presence of dodecanethiol (colour change from orange to deep brown). Prior to using AuNPs in sensing application, types of biomolecules and the method adopted to immobilize the biomolecules on the surface of AuNPs must be taken into account, which engenders a potpourri of AuNP-based sensing assays.

11.3 Biosensing Configurations of AuNP

To create an ideal biosensing configurations on AuNP, two potential biomolecules, namely aptamer and antibody as the probes are very commonly in use [10–12, 26]. These molecules can be conjugated with AuNP through thiol-linker (Fig. 11.1).

11.3.1 Aptamer-Based Sensing

Isolated by a process known as Systematic Evolutions of Ligands by Exponential Enrichment (SELEX), aptamers are single stranded DNA or RNA that have high binding affinity and specificity against the target (Fig. 11.2). Due to its specificity against the target, aptamers are utilized as the detection probe in AuNP-based assay. The most common form of the assay is the differential aggregation of AuNPs in the presence or absence of the target. This AuNP-based biosensing system that can result in the visual transduction of the molecular event without the requirement of special equipments [31, 32]. In this platform, the controlled assembly and disassembly of aptamer on unmodified AuNP is the key strategy and the binding of target can be evaluated by colorimetric measurements as described below.

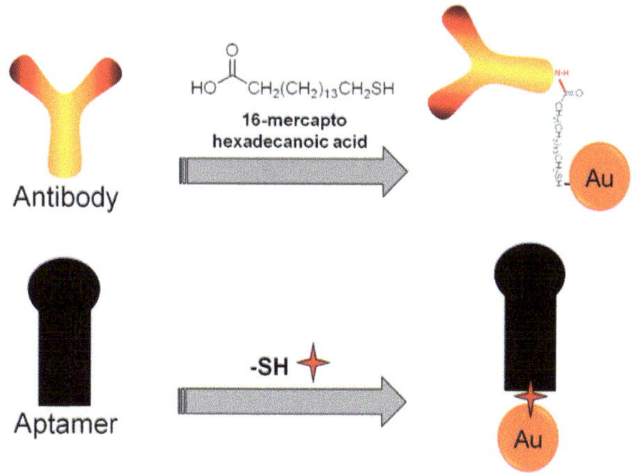

Fig. 11.1 Thiolation on probes. Both antibody and aptamer are shown as the probe. Chemical structure of 16-mercapto hexadecanoic acid is shown

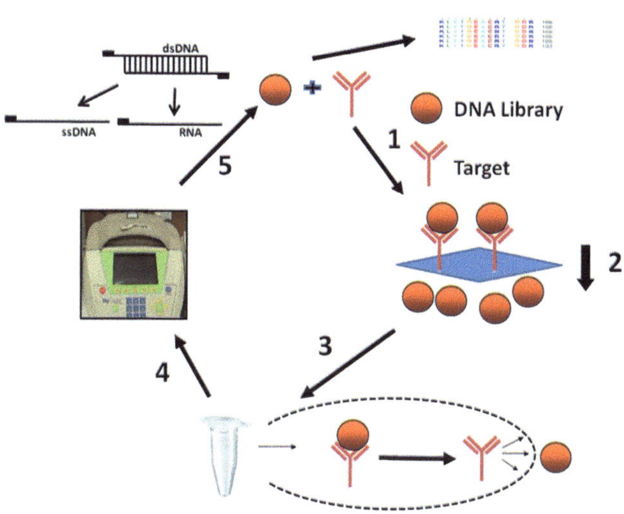

Fig. 11.2 Representation of SELEX cycle. SELEX process involves the following steps. (1) Incubation of random oligonucleotide pool with the target molecule; (2) Separation of bound from unbound nucleic acids; (3) Elution of the bound nucleic acids; (4) Amplification of the bound nucleic acids; (5) ssDNA or RNA generation for next cycle

11.3.1.1 Colorimetric Assay

The main feature of this AuNP-based colorimetric biosensing is the differential colour production in the presence or absence of the target. In the presence of the target that affects the interparticle distance, well-dispersed AuNP particles harness red colour while aggregated AuNP particles produce blue or purple colour (Fig. 11.3). Another advantage of AuNP is the ability of the Au atoms on the AuNPs' surface to interact with the nitrogenous bases of ssDNA/RNA, enabling

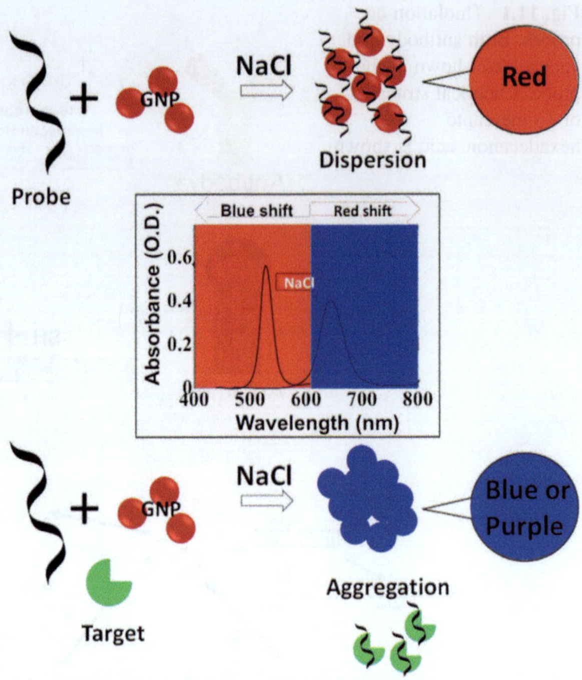

Fig. 11.3 Aptamer-based colorimetric assay. Observation of the dispersion (*red colour* indication) and aggregation (*blue* or *purple* colour indication) of AuNPs under salt induction are shown. Probe indicates the aptamer. Colour-based spectral changes are shown with *blue* and *red* shifts

immobilization of these nucleic acids on the AuNP surface [33, 34]. This spontaneous adsorption of the ssDNA/RNA onto the surface of AuNPs affects the interparticle distance of the AuNPs. Based on this feature, many aptamer-based assays were devised for the sensitive detection of the target molecule.

DNA aptamer against target molecule Hg^{2+} was used in the AuNP-based assay. In this assay, the presence of the target Hg^{2+}, T-rich single stranded oligonucleotide DNA aptamer forms duplex with Hg^{2+} and is not able to bind to the AuNP. Under this condition, NaCl will lessen the electrostatic repulsion between the AuNPs which results in aggregation and blue or purple colour formation.

In the absence of the Hg^{2+}, the ssDNA aptamer is spontaneously absorbed onto the surface of AuNP, which prevents the action of Na^+, stabilizing the nanoparticles against the salt-induced aggregation and gives red colour. In a different strategy, colorimetric aptasensor for the detection of *Staphylococcus aureus* using tyramine signal amplification (TSA) technology was developed. In this assay, aptamer specific against *S. aureus* was functionalized with biotin and immobilized on the surface of the microtiter plate. Subsequent to the addition of the target, biotinylated-aptamer-streptavidin-HRP conjugates, biotinylated tyramine, hydrogen peroxide and avidin-catalase were added consecutively. The addition of the Au (III) chloride trihydrate follows which changes the colour of the reaction and also the absorbance at 550 nm. The limit of detection is 9 CFU mL^{-1} [35]. In another format, an adenosine sensor system was designed using aptamer against adenosine. This construct that consists of two types of ssDNA-modified AuNPs and a linker DNA molecule that is anchored to anti-adenosine DNA aptamer. In the absence of the target, the aptamer, which is complementary to a part of the ssDNA results in the DNA hybridization, which aggregates the AuNPs (blue or purple). In the presence of the target adenosine, aptamer changes its structure and results in the disassembly of the AuNP aggregates (red) [36]. Compared to aptamer-based assays that is dependent on the aggregation and disaggregation of the AuNPs based on the spontaneous absorption of the aptamer, antibody-based assay involves the immobilization of the antibody on the surface of the AuNPs.

11.3.2 Antibody-Based Sensing

Antibodies are protein-based probes that recognize their target antigen with high affinity and specificity. They are produced using an immunized animal or via cloning/recombinant DNA technology. As glycoproteins belonging to the immunoglobulin group, antibodies have different isotypes with an amine (NH_2) terminus at antigen binding regions and the carboxyl (COOH) terminus at the stem. One strategy of immobilizing antibody on the surface of AuNP is by using amino-thiol coupling with the aid of 16-mercaptohexadecanoic acid (MHA). Conjugation of the antibody's amino group to the carboxyl group of the MHA allows immobilization of the antibody onto the Au surface via the thiol group at the other end (Fig. 11.1). Regardless of the type of immobilization, antibodies or target immobilized on the surface of the AuNPs (Fig. 11.4a) is able to form complex with their partner molecule that affects the degree of AuNP aggregation. Antibody complexed AuNP will cause drastic changes on its absorption (Fig. 11.4b). In addition, with increase in the sizes of AuNPs, there will be a concomitant increment in the sensitivity [26]. In a study, monoclonal anti-O157:H7 antibody immobilized on the surface of AuNPs was used for detection of *Escherichia coli*. The immobilization of the antibody on the surface of the AuNPs was carried out by the addition of the antibody into the oxidized MP suspension at a ratio of 0.3:1, mixed and incubated at 4 °C in the dark [37]. In another colorimetric assay, antibody functionalized AuNPs was used for the β-casein

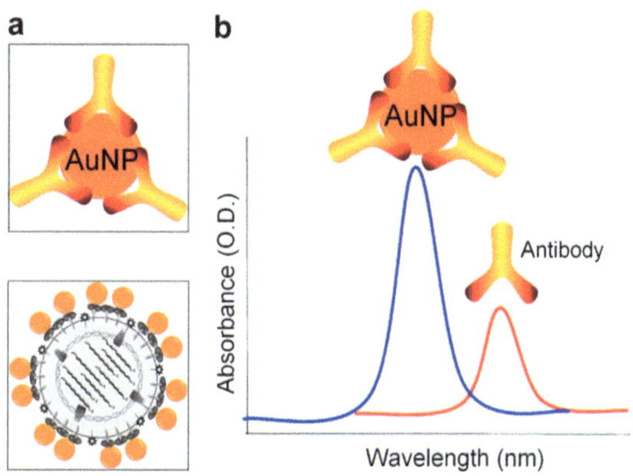

Fig. 11.4 AuNP and bioconjugates. (**a**) AuNP and antibody or virus conjugates (**b**) spectral changes for AuNP-antibody conjugates or antibody alone. There is a signal enhancement with AuNP conjugation compared to antibody or other biomolecules such as virus alone. Absorbance-based these changes are shown

detection in bovine milk samples. The linear dynamic range of 0.08–250 μg mL^{-1} and limit of detection of 0.03 μg mL^{-1} were achieved, respectively [38].

In a sandwich-based assay, bio-barcode assay employs two different antibodies being polyclonal capturing antibodies immobilized on the surface of magnetic bead while monoclonal antibodies are conjugated on the surface of AuNPs. In the presence of the target protein, immunosandwich is formed between the particles and as AuNPs carry hundreds of thiolated ssDNA that can hybridize with complementary DNA, these ssDNA will be released via melting and analysed by PCR [39]. Other than acting as the platform for antibody immobilization, AuNPs are also conjugated to antibody for signal production, as in the case of lateral flow test.

11.3.2.1 Lateral Flow Test

Lateral flow test or immunochromatography test (ICT) is designed to detect a particular target in a complex mixture [40–42]. ICT consists of a consecutive series of pads (sample pad, conjugate pad, capillary pad and absorption pad) that are able to transport the fluid from one pad to another. Sample pad that consists of a sponge can absorb the sample for flow initiation. The fluid is transported from one pad to then next until it reaches the capillary pad that contains the test line. For the detection of the specific target, the corresponding antibody against the target will be immobilized on the surface of the test lines. In the presence of the target, complex will form between the target, primary antibody and the secondary antibody conjugated to Au nanoparticles (e.g. anti-rabbit or anti-mouse). AuNPs adopts a role as the visible signal producer, as the aggregation of the AuNPs will result in the appearance of the coloured lines (Fig. 11.5).

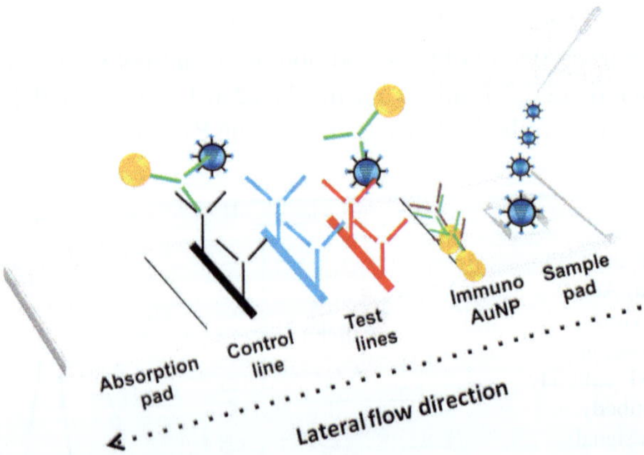

Fig. 11.5 Schematics for an immunochromatography test (ICT). AuNP-antibody conjugates play an important role. ICT consists of a consecutive sample, conjugate, capillary and absorption pads that are able to transport the fluid from one pad to another. For the detection of the specific target, the corresponding antibody against the target will be immobilized on the surface of the test lines. In the presence of the target, complex will form between the target and antibody conjugated to Au nanoparticles. The aggregation of the AuNPs will result in the appearance of the coloured lines

For the formation of sandwich complex, the antibodies must have different binding sites for the target to avert competition but in certain cases, sandwich can still form owing to the formation of trimer by HA molecules. AuNPs also function to enhance the signal amplification in ICT. Mori et al. [41] used 50 nm AuNPs and have amplified the signal using silver particles >10 μM, which improved the sensitivity up to 1,000-fold.

11.4 AuNP-Facilitated Label-Free Sensors

To enhance the sensitivity, AuNPs have been used in conjunction with many label-free sensors such as surface plasmon resonance, surface-enhanced Raman scattering (SERS) and waveguide mode. SERS is a phenomenon in which frequency of the incoming light is similar to the frequency of the scattered light (Fig. 11.6). The SERS effect can be enhanced in the presence of AuNP by up to 10^{14} orders of magnitude [43–47]. SERS was applied for the multiplex detection of *S. typhimurium* and *S. aureus* in complex food matrices. Using silica-coated magnetic probes as the platforms which are immobilized with antibodies, detection of bacteria was achieved via AuNPs functionalized with a Raman reporter with a detection limit of 10^3 CFU mL^{-1} [48].

Surface Plasmon is the excitation of the surface Plasmon by incident light which happens under the condition of total internal reflection [49]. The conjugation or interaction of the biomolecules with the corresponding target will change the frequencies of the free electrons on the surface, which can be monitored in real time. Due to the high dielectric constant of AuNPs and electromagnetic coupling between AuNPs and the metal film on the surface, the sensitivity of the surface

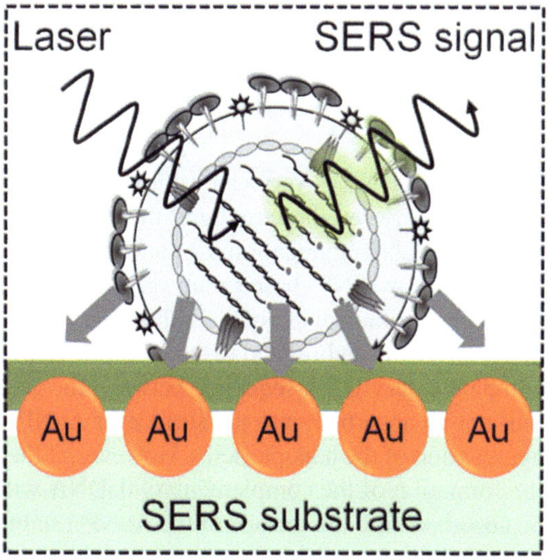

Fig. 11.6 AuNP in Surface-enhanced Raman scattering (SERS)-based detection. Specific probes immobilized on the sensor surface interact with the virus or other molecule, thereby shifting the SERS spectra

plasmon resonance (SPR)-based sensor can be increased. Several different materials are also used to support AuNPs such as optical fibres, ITO glass, sol-gel matrix for the detection of human IgG [50], streptavidin [51], interleukin-1β [52] and propanethiol [53]. Similarly, waveguide-mode sensor, which mimic the principle of SPR, has been generated with higher sensitive sensing using AuNP conjugation for the detection of influenza viruses [10–12].

11.5 Sensing Applications of AuNPs

11.5.1 Detection of Proteins

The state of a disease can be diagnosed by detecting the presence of certain biomarkers related to the disease. In this case, AuNP-based detection can play an important role for the detection of the target protein. In one example, AuNP was utilized for the detection of *Ricinis communis* agglutinin (RCA120), with a detection sensitivity of 1 ppm [54]. In a fluorescent-based competition analyses, the binding affinity of the protein (Lectin Con A) against lyconanoparticles (AuNPs conjugated with underivatized mono-, oligo- and polysaccharides) was determined [50]. A wide range of additional protein biomarkers have been detected besides cancer biomarkers, including cholera toxin [55], inter-leukin [56] vascular endothelial growth factors [57], Annexin II and MUC5AC antigens [58], antigens of *Schistosoma japonicum* [59], *Salmonella typhi* [60, 61], *Escherichia coli* O157:H7 [62] and osteoprotegerin [63]. AuNP-based electrochemical sandwich immunosensor was designed for the detection of fenolase (ENO1) antigen, a potential diagnostic marker for lung cancer, with a detection limit of 11.9 fg [64].

11.5.2 Detection of DNA

Detection of DNA is very amenable with the AuNP-based sensor, as the formation of duplex between the target and the probe DNA can also impart changes on the degree of AuNP aggregation. Lee and colleagues have designed a single nucleotide polymorphism (SNP)-based analysis for the genotyping of rs2131877 that is located at human chromosome 3q29. In this study, oligonucleotide complementary to the target DNA was hybridized to the PCR product from human DNA. Formation of the complementary DNA sequence between the oligonucleotide and the PCR product will not be absorbed onto the surface of AuNP. This causes the Na^+ to induce the aggregation of the nanoparticles. However, if there is a single-mismatch nucleotide, the formation of the complementary dsDNA will be prevented, causing the Na^+ to be absorbed onto the surface of the AuNPs, stabilizing the nanoparticles against the aggregation action Na^+. In another approach, S1 nuclease was adopted and this nuclease is only sensitive against the ssDNA. In the presence of the non-complementary/mismatched target, this nuclease will cleave the ssDNA, producing deoxynucleotide monophosphates (dNMPs) that stabilized the AuNPs against salt-induced aggregation of Na^+.

11.5.3 Detection of Whole Cell Bacteria or Virus

AuNPs are often used with antibodies and aptamers for the detection of bacteria. For example, the DNA probe complementary to the DNA or total RNA of the microbe is immobilized on the surface of AuNP. The aggregation or separation of the AuNPs which can result in different colours may indicate the presence or absence of the target bacteria. In one study of using paper-based assay to detect *Mycobacterium tuberculosis* complex, primer complementary to a specific region of the RNA polymerase beta subunit gene was conjugated to AuNP. The genomic DNA and the AuNP probe were mixed together and added into the wells that were printed on the surface of a paper which are filled with $MgCl_2$. The binding of the probe on the surface of the AuNPs with the genomic DNA will result in red colour while the absence of the target protein will result in blue or purple colour [65]. In another system, an antibody-based sensor using AuNPs was also developed for the detection of *Francisella tularensis* [66].

11.5.4 Detection of Cancers

The presence of tumour cells is often used as a method to diagnose the occurrence of cancer among patients. In one study, melanin, natural light absorber in melanoma cells, was used to induce photoacoustic waves for the detection of tumour cells. To induce optical contrast in the non-pigmented cancer cells, attachment of the AuNPs to the prostate cancer cell line was carried out. This enables the detection of the cells by photoacoustic flow meter [67]. In a different method of detection, after performing the selective labelling of cancer cells in suspension, interaction between the AuNP and the proteins expressed on the surface of the cell membrane was carried out [68]. This system enables the detection of up to 4×10^3 cancer cells. Imaging of cancer cells was also made possible by using anti-epidermal growth factor receptor (anti-EGFR) antibodies immobilized on the surface of AuNPs that form complex with the EGFR on the surface of the cancer cells [69].

11.5.5 Detection of Small Molecules, Toxic Chemicals and Drugs

The presence of large surface area to volume ratio, ability to reduce the overpotentials of many electroanalytical reactions and the ability to maintain reversibility of redox reaction have capacitated AuNPs for the application in detecting small molecules, toxic chemicals and drugs. These molecules are present in very minute amounts and the most common format of assay with AuNPs is electrochemical assay. Wang et al. [70] have designed electrochemical assay using self-assemble dithiothreitol(DTT)-dodecanethiol (DDT)-Au colloid modified Au-electrode. This assay immensely improved the detection of epinephrine at the Au electrode, with a detection limit of 60 nM. AuNPs are widely applied for the detection of glucose [71, 72], dopamine [73, 74], uric acid [75, 76], ascorbic acid [77–80], bisphenol A [81] and nitrite [82].

Conclusions

With the above applications in the sensing technology, it was proved in many instances that AuNP is the potential nanoparticle and play a pivotal role. Further ahead, the production of various shapes and sizes of the AuNPs and other Au-based nanostructures paved the way to generate a wide range of higher sensitive sensing platforms. In addition, the tailor-made commercial availability of AuNPs makes easier to end users. Apart from these, the visual nature of AuNPs under microscopic analyses creates the faithful way of direct visualization of the tiny bioanalytes conjugated AuNPs in the complex mixture. With these sensing progress with AuNPs, there will be several AuNP-based sensors created with interdisciplinary sciences with higher sensitive and specificity.

Acknowledgements T.H. Tang was supported by a Universiti Sains Malaysia (USM) Research Grant (Number: 1001/CIPPT/813043). Y. Chen was supported by UM.C/625/1/HIR/MOHE/MED/16/5. T. Lakshmipriya was supported by the Research fellowship from USM.

References

1. Daniel MC, Astruc D (2004) Gold nanoparticles: assembly, supramolecular chemistry, quantum-size-related properties, and applications toward biology, catalysis, and nanotechnology. Chem Rev 104:293–346
2. Zayats M, Baron R, Popov I, Willner I (2005) Biocatalytic growth of Au nanoparticles: from mechanistic aspects to biosensors design. Nano Lett 5:21–25
3. Haick HJ (2007) Chemical sensors based on molecularly modified metallic nanoparticles. J Phys D Appl Phys 40:7173–7186

4. Sperling RA, Rivera Gil P, Zhang F, Zanella M, Parak WJ (2008) Biological applications of gold nanoparticles. Chem Soc Rev 37:1896–1908
5. Zhao ML, Ni DD, Wang JW, Di JW, Tu YF (2008) Determination of nitrite at gold nanoparticles modified indium tin oxide electrode with direct electrodeposition. Chin J Anal Chem 36:1729–1731
6. Boisselier E, Astruc D (2009) Gold nanoparticles in nanomedicine: preparations, imaging, diagnostics, therapies and toxicity. Chem Soc Rev 38:1759–1782
7. Radwan SH, Azzazy HME (2009) Gold nanoparticles for molecular diagnostics. Expert Rev Mol Diagn 9:511–524
8. Bunz UHF, Rotello VM (2010) Gold nanoparticle-fluorophore complexes: sensitive and discerning "noses" for biosystems sensing. Angew Chem Int 49:3268–3279
9. Zeng SW, Yong KT, Roy I, Dinh XQ, Yu X, Luan F (2011) A review on functionalized gold nanoparticles for biosensing applications. Plasmonics 6:491–506
10. Gopinath SCB, Awazu K, Fujimaki M, Shimizu K (2013) Evaluation of Anti-A/Udorn/307/1972 antibody specificity to Influenza viruses using a waveguide mode sensor. PLoS One 8: e81396
11. Gopinath SCB, Awazu K, Fujimaki M, Shimizu K, Shima T (2013) Observations of immuno-gold conjugates on influenza viruses using waveguide-mode sensors. PLoS One 8:e69121
12. Gopinath SCB, Awazu K, Fujimaki M, Shimizu K (2013) Aptamers that bind to the hemagglutinin of recent pandemic H1N1 and inhibits efficiently the agglutination. Acta Biomater 9:5080–5087
13. Baldrich E, Restrepo A, O'Sullivan CK (2004) Aptasensor development: elucidation of critical parameters for optimal aptamer performance. Anal Chem 76:7053–7063
14. Varma MM, Nolte DD, Inerowicz HD, Regnier FE (2004) Spinning-disk self-referencing interferometry of antigen-antibody recognition. Opt Lett 29:950–952
15. Guo P (2005) RNA nanotechnology: engineering, assembly and applications in detection, gene delivery and therapy. J Nanosci Nanotechnol 5:1964–1982
16. Gronewold TMA, Glass S, Quandt E, Famulok M (2005) Monitoring complex formation in the blood coagulation cascade using aptamer-coated SAW sensors. Biosens Bioelectron 20:2044–2052
17. Odenthal KJ, Gooding JJ (2007) An introduction to electrochemical DNA biosensors. Analyst 132:603–610
18. Peng L, Varma MM, Cho W, Regnier FE, Nolte DD (2007) Adaptive interferometry of protein on a BioCD. Appl Opt 46:5384–5395
19. Wang X, Zhao M, Nolte DD (2008) Combined fluorescent and interferometric detection of protein on a BioCD. Appl Opt 47:2779–2789
20. Gopinath SCB, Awazu K, Kumar PKR, Tominaga J (2008) Monitoring biomolecular interactions on a digital versatile disc: a BioDVD platform technology. ACS Nano 2:1885–1895
21. Marquette CA, Blum LJ (2008) Electro-chemiluminescent biosensing. Anal Bioanal Chem 390:155–168
22. Song S, Wang L, Li J, Zhao J, Fan C (2008) Aptamer-based biosensors. Trends Anal Chem 27:108–117
23. Giljohann DA, Seferos DS, Daniel WL, Massich MD, Patel PC, Mirkin CA (2010) Gold nanoparticles for biology and medicine. Angew Chem Int Ed 49:3280–3294
24. Iliuk AB, Hu L, Tao WA (2011) Aptamer in bioanalytical applications. Anal Chem 83:4440–4452
25. Zanoli LM, D'Agata R, Spoto G (2012) Functionalized gold nanoparticles for ultrasensitive DNA detection. Anal Bioanal Chem 402:1759–1771
26. Gopinath SC, Lakshmipriya T, Awazu K (2014) Colorimetric detection of controlled assembly and disassembly of aptamers on unmodified gold nanoparticles. Biosens Bioelectron 51:115–123
27. Saha K, Agasti SS, Kim C, Li X, Rotello VM (2012) Gold nanoparticles in chemical and biological sensing. Chem Rev 112:2739–2779

28. Turkevich J, Stevenson PC, Hillier J (1951) A study of the nucleation and growth processes in the synthesis of colloidal gold. Discuss Faraday Soc 11:55
29. Frens G (1973) Controlled nucleation for the regulation of the particle size in monodisperse gold suspensions. Nat Phys Sci 241:20–22
30. Brust M, Walker M, Bethell D, Schiffrin DJ, Whyman RJ (1994) Synthesis of thiol-derivatised gold nanoparticles in a 2-phase liquid-liquid system. Chem Soc Chem Commun:801–802
31. Baptista P, Pereira E, Eaton P, Doria G, Miranda A, Gomes I, Quaresma P, Franco R (2008) Gold nanoparticles for the development of clinical diagnosis methods. Anal Bioanal Chem 391:943–950
32. Larguinho M, Baptista PV (2012) J. Gold and silver nanoparticles for clinical diagnostics—from genomics to proteomics. Proteomics 75:2811–2823
33. Li H, Rothberg L (2004) Colorimetric detection of DNA sequences based on electrostatic interactions with unmodified gold nanoparticles. Proc Natl Acad Sci U S A 101:14036–14039
34. Li H, Rothberg L (2004) Label-free colorimetric detection of specific sequences in genomic DNA amplified by the polymerase chain reaction. J Am Chem Soc 126:10958–10961
35. Yuan J, Wu S, Duan N, Ma X, Xia Y, Chen J, Ding Z, Wang Z (2014) A sensitive gold nanoparticle-based colorimetric aptasensor for Staphylococcus aureus. Talanta 127:163–168
36. Liu JW, Lu Y (2006) Fast colorimetric sensing of adenosine and cocaine based on a general sensor design involving aptamers and nanoparticles. Angew Chem Int Ed Engl 45:90–94
37. Shen Z, Hou N, Jin M, Qiu Z, Wang J, Zhang B, Wang X, Wang J, Zhou D, Li J (2014) A novel enzyme-linked immunosorbent assay for detection of Escherichia coli O157:H7 using immunomagnetic and beacon gold nanoparticles. Gut Pathog 6:14
38. Li YS, Zhou Y, Meng XY, Zhang YY, Song F, Lu SY, Ren HL, Hu P, Liu ZS, Zhang JH (2014) Gold nanoparticle aggregation-based colorimetric assay for β-casein detection in bovine milk samples. Food Chem 162:22–26
39. Nam JM, Thaxton CS, Mirkin CA (2003) Nanoparticle-based bio-bar codes for the ultrasensitive detection of proteins. Science 301:1884–1886
40. Hara M, Takao S, Fukuda S, Shimazu Y, Miyazaki K (2008) Evaluation of three immunochromatographic kits for rapid detection of influenza virus A and B. LabMedicine 39:603–610
41. Mori M, Katada J, Chiku H, Nakamura K, Oyamada T (2012) Development of highly sensitive immunochromatographic detection kit for seasonal influenza virus using silver amplification. Fujifilm Res Develop 57:5–11
42. Mitamura K, Kawakami C, Shimizu H, Abe T, Konomi Y, Yasumi Y, Yamazaki M, Ichikawa M, Sugaya N (2013) Evaluation of a new immunochromatographic assay for rapid identification of influenza A, B, and A(H1N1)2009 viruses. J Infect Chemother 19:633–638
43. Fleischm M, Hendra PJ, McQuilla AJ (1974) Raman-spectra of pyridine adsorbed at a silver electrode. Chem Phys Lett 26:163–166
44. Albrecht MG, Creighton JA (1977) Anomalously intense Raman spectra of pyridine at a silver electrode. J Am Chem Soc 99:5215–5217
45. Jeanmaire DL, Vanduyne RP (1977) Surface Raman spectroelectrochemistry: part I. Heterocyclic, aromatic, and aliphatic amines adsorbed on the anodized silver electrode. J Electroanal Chem 84:1–20
46. Kneipp K, Wang Y, Kneipp H, Perelman LT, Itzkan I, Dasari R, Feld MS (1997) Single molecule detection using surface-enhanced Raman scattering (SERS). Phys Rev Lett 78:1667–1670
47. Nie SM, Emory SR (1997) Probing single molecules and single nanoparticles by surface-enhanced Raman scattering. Science 275:1102–1106
48. Wang Y, Ravindranath S, Irudayaraj J (2011) Separation and detection of multiple pathogens in a food matrix by magnetic SERS nanoprobes. Anal Bioanal Chem 399:1271–1278
49. Kretschmann E, Raether H (1968) Radiative decay of nonradiative surface plasmons excited by light. Z Naturforsch 23:2135–2136
50. Wang X, Ramstrom O, Yan MD (2010) Quantitative analysis of multivalent ligand presentation on gold glyconanoparticles and the impact on lectin binding. Anal Chem 82:9082–9089

51. Lin YB, Zou Y, Mo YY, Guo JP, Lindquist RG (2010) E-beam patterned gold nanodot arrays on optical fiber tips for localized surface plasmon resonance biochemical sensing. Sensors 10:9397–9406
52. Chiang CY, Hsieh ML, Huang KW, Chau LK, Chang CM, Lyu SR (2010) Fiber-optic particle plasmon resonance sensor for detection of interleukin-1β in synovial fluids. Biosens Bioelectron 26:1036–1042
53. Briglin SM, Gao T, Lewis NS (2004) Detection of organic mercaptan vapors using thin films of alkylamine-passivated gold nanocrystals. Langmuir 20:299–305
54. Otsuka H, Akiyama Y, Nagasaki Y, Kataoka K (2001) Quantitative and reversible lectin-induced association of gold nanoparticles modified with alpha-lactosyl-omega-mercapto-poly (ethylene glycol). J Am Chem Soc 123:8226–8230
55. Loyprasert S, Hedstrom M, Thavarungkul P, Kanatharana P, Mattiasson B (2010) Sub-attomolar detection of cholera toxin using a label-free capacitive immunosensor. Biosens Bioelectron 25:1977–1983
56. Liang KZ, Mu WJ (2006) Flow-injection immuno-bioassay for interleukin-6 in humans based on gold nanoparticles modified screen-printed graphite electrodes. Anal Chim Acta 580:128–135
57. Kim GI, Kim KW, Oh MK, Sung YM (2010) Electrochemical detection of vascular endothelial growth factors (VEGFs) using VEGF antibody fragments modified Au NPs/ITO electrode. Biosens Bioelectron 25:1717–1722
58. Kim DM, Noh HB, Park DS, Ryu SH, Koo JS, Shim YB (2009) Immunosensors for detection of Annexin II and MUC5AC for early diagnosis of lung cancer. Biosens Bioelectron 25:456–462
59. Lei CX, Gong FC, Shen GL, Yu RQ (2003) Amperometric immunosensor for *Schistosoma japonicum* antigen using antibodies loaded on a nano-Au monolayer modified chitosan-antrapped carbon paste electrode. Sens Actuat B Chem 96:582–588
60. Dungchai W, Siangproh W, Chaicumpa W, Tongtawe P, Chailapakul O (2008) Salmonella typhi determination using voltammetric amplification of nanoparticles: a highly sensitive strategy for metalloimmunoassay based on a copper-enhanced gold label. Talanta 77:727–732
61. Yang GJ, Huang JL, Meng WJ, Shen M, Jiao XA (2009) A reusable capacitive immunosensor for detection of Salmonella spp. based on grafted ethylene diamine and self-assembled gold nanoparticle monolayers. Anal Chim Acta 647:159–166
62. Lin YH, Chen SH, Chuang YC, Lu YC, Shen TY, Chang CA, Lin CS (2008) Disposable amperometric immunosensing strips fabricated by Au nanoparticles-modified screen-printed carbon electrodes for the detection of foodborne pathogen Escherichia coli O157:H7. Biosens Bioelectron 23:1832–1837
63. Singh K, Rahman MA, Son JI, Kim KC, Shim YB (2008) An amperometric immunosensor for osteoproteogerin based on gold nanoparticles deposited conducting polymer. Biosens Bioelectron 23:1595–1601
64. Ho JAA, Chang HC, Shih NY, Wu LC, Chang YF, Chen CC, Chou C (2010) Diagnostic detection of human lung cancer-associated antigen using a gold nanoparticle-based electrochemical immunosensor. Anal Chem 82:5944–5950
65. Veigas B, Jacob JM, Costa MN, Santos DS, Viveiros M, Inácio J, Martins R, Barquinha P, Fortunato E, Baptista PV (2012) Gold on paper-paper platform for Au-nanoprobe TB detection. Lab Chip 12:4802–4808
66. Kleo K, Schäfer D, Klar S, Jacob D, Grunow R, Lis F (2012) Immunodetection of inactivated Francisella tularensis bacteria by using a quartz crystal microbalance with dissipation monitoring. Anal Bioanal Chem 404:843–851
67. Viator JA, Gupta S, Goldschmidt BS, Bhattacharyyal K, Kannan R, Shukla R, Dale PS, Boote E, Katti K (2010) Gold nanoparticle mediated detection of prostate cancer cells using photoacoustic flowmetry with optical reflectance. J Biomed Nanotechnol 2:187–191

68. Maltez-da Costa M, de la Escosura-Muñiz A, Nogués C, Barrios L, Ibáñez E, Merkoçi A (2012) Detection of circulating cancer cells using electrocatalytic gold nanoparticles. Small 8:3605–3612
69. Li N, Larson T, Nguyen HH, Sokolov KV, Ellington AD (2010) Directed evolution of gold nanoparticle delivery to cells. Chem Commun 46:392–394
70. Wang L, Bai JY, Huang PF, Wang HJ, Zhang LY, Zhao YQ (2006) Self-assembly of gold nanoparticles for the voltammetric sensing of epinephrine. Electrochem Commun 8:1035–1040
71. Kurniawan F, Tsakova V, Mirsky VM (2006) Gold nanoparticles in non-enzymatic electrochemical detection of sugars. Electroanalysis 18:1937–1942
72. Li Y, Song YY, Yang C, Xia XH (2007) Hydrogen bubble dynamic template synthesis of porous gold for nonenzymatic electrochemical detection of glucose. Electrochem Commun 9:981–988
73. Raj CR, Ohsaka TJ (2003) Voltammetric detection of uric acid in the presence of ascorbic acid at a gold electrode modified with a self-assembled monolayer of heteroaromatic thiol. J Electroanal Chem 540:69–77
74. Li MG, Gao F, Yang P, Wang L, Fang B (2008) Conveniently assembling dithiocarbamate and gold nanoparticles onto the gold electrode: a new type of electrochemical sensors for biomolecule detection. Surf Sci 602:151–155
75. Raj CR, Okajima T, Ohsaka TJ (2003) Gold nanoparticle arrays for the voltammetric sensing of dopamine. J Electroanal Chem 543:127–133
76. Zhang SJ, Xu ML, Zhang YZ (2009) Simultaneous voltammetric detection of salsolinol and uric acid in the presence of high concentration of ascorbic acid with gold nanoparticles/functionalized multiwalled carbon nanotubes composite film modified electrode. Electroanalysis 21:2607–2610
77. Hu GZ, Ma YG, Guo Y, Shao SJ (2008) Electrocatalytic oxidation and simultaneous determination of uric acid and ascorbic acid on the gold nanoparticles-modified glassy carbon electrode. Electrochim Acta 53:6610–6615
78. Kalimuthu P, John SAJ (2008) Size dependent electrocatalytic activity of gold nanoparticles immobilized onto three dimensional sol–gel network. J Electroanal Chem 617:164
79. Kannan P, John SA (2009) Determination of nanomolar uric and ascorbic acids using enlarged gold nanoparticles modified electrode. Anal Biochem 386:65–72
80. Lu LP, Lin XQ (2004) Glassy carbon electrode modified with gold nanoparticles and DNA for the simultaneous determination of uric acid and norepinephrine under coexistence of ascorbic acid. Anal Sci 20:527–530
81. Yin HS, Zhou YI, Ai SY, Han RX, Tang TT, Zhu LS (2010) Electrochemical behavior of bisphenol A at glassy carbon electrode modified with gold nanoparticles, silk fibroin, and PAMAM dendrimers. Microchim Acta 170:99–105
82. Zhao W, Brook MA, Li YF (2008) Design of gold nanoparticle-based colorimetric biosensing assays. Chembiochem 9:2363–2371

Chapter 12
Electron Magnetic Resonance (EMR) Technique and Nanoparticle Characterization

Ashutosh Kumar Shukla

Abstract This chapter summarizes the applications of Electron Magnetic Resonance (EMR) spectroscopy for characterization of nanoparticle systems. It covers review of reported recent research on EMR characterization of metallic nanoparticles and metal oxyhydroxide/metal oxide nanoparticles which are important for technological as well as biomedical applications. Photo generated charge carriers in nanocrystalline materials as studied by EMR are also included in this review. Efforts have been made to highlight open problems that need further investigations.

12.1 Basic Features of EMR

Spectroscopy may be subdivided in different branches depending on the energy involved [1]. Electron Magnetic Resonance (EMR) spectroscopy basically deals with microwave region of electromagnetic radiation and it is based on the interaction between electronic magnetic moments and magnetic field. Resonance absorption of microwave radiation by unpaired electron spins in a magnetic field is observed in EMR. One studies the direct transitions between electronic Zeeman levels in EMR spectroscopy. Zavoisky discovered EMR in 1945 for study of transition metal ions in salts [2]. EMR spectra are usually characterized by the spin Hamiltonian parameters and therefore rigorous EMR analysis requires quantum mechanical concepts. Theory of EMR and its applications in various disciplines have been discussed in many monographs and review articles. EMR can be used to study the samples with paramagnetic centers which are the sites with unpaired electrons. The various parameters of EMR spectrum, viz. intensity, line width, g value, and hyperfine splitting constant, provide variety of information about the paramagnetic specimen. EMR can provide spectroscopic data on unpaired

A.K. Shukla (✉)
Physics Department, Ewing Christian College, Allahabad, U.P. 211003, India
e-mail: drakshukla@gmail.com

M. Lungu et al. (eds.), *Nanoparticles' Promises and Risks*,
DOI 10.1007/978-3-319-11728-7_12

electronic spin states at both the atomic (conventional EMR) and the nanoscale (ferromagnetic resonance, i.e., FMR) [3]. Information obtained from EMR spectrum depends upon the nature and form of specimen.

12.2 EMR Characterization of Nanoparticles

From a fundamental point of view, the study of nanoparticles is very important since the physics of infinitely large systems is not applicable in the finite-size particles [4]. Nanomaterials exhibit different properties from their coarse grained materials. Specific properties lead to their technological importance. It is the reason that they have attracted researchers these days. Therefore, the preparation and characterization of nanoparticles are prime concerns of the scientific community. In this chapter, EMR spectroscopy of metallic nanoparticles and metal oxide nanoparticles along with some others have been discussed and we have tried to avoid repetition from earlier review [3] unless it is specifically within the general scope of the present article.

12.3 EMR of Metallic Nanoparticles

12.3.1 Silver Nanoparticles

EMR of silver nanoparticles embedded in SiO_2 matrix is reported by Mitrikas et al. [4]. The magnetic properties were investigated in the low temperature region between 4 and 300 K. CW EMR spectrum of the silver nanoparticles was found to exhibit a modest homogeneous broadening character [1]. Authors have reported the spin–lattice relaxation time was measured by pulsed EMR. However, good fit for their data could not be obtained using the approach of Khaliullin et al. [5] for spin–lattice relaxation of conduction electrons. This led to the need of detailed studies on silver nanoparticles to understand their spin–lattice relaxation mechanisms. However, they could conclude that factors such as the particle diameters and the matrix might play important roles in the spin–lattice relaxation.

12.3.2 Nickel Nanoparticles

Development of nickel nanoparticles has attracted attention of researchers because of their unique properties and potential applications in a variety of fields including electronics [6], magnetism [7, 8], energy technology [9], and biomedicine [10, 11]. Alonso et al. [12] have reported study on the characterization of the nickel nanoparticles, utilized in transfer hydrogenation reactions with isopropanol as the hydrogen donor, as well as a series of catalytic and kinetic experiments. Presence of nickel nanoparticles in the zero-valence state has been confirmed by EMR experiments.

12.3.3 Gold Nanoparticles

Ionita et al. [13] reported characterization of ligand dynamics attached to gold nanoparticles. EMR spectra of a series of spin labeled nanoparticles at variable temperature and at different frequencies were recorded. Series of Au nanoparticles protected by a monolayer of organic ligands and labeled them with disulfide functionalized nitroxides. The chain length of the spin labels and the surrounding ligands was systematically varied. Authors found that the rotational diffusion rate of the spin labels increases with the length of the linker connecting the nitroxide unit to the Au surface. They could further suggest that packing of the organic shell also influences the diffusion parameters.

12.4 EMR of Metal Oxyhydroxide and Metal Oxide Nanoparticles

Metal oxyhydroxide and metal oxide nanoparticles are the important class of materials because their optical, magnetic, and electrical properties find a wide range of technological and biomedical applications. The following section summarizes some of the EMR investigations of metal oxyhydroxide and metal oxide nanoparticles.

12.4.1 Nanoparticles of Oxyhydroxides of Iron

Akaganeite is one of the polymorphs of iron hydroxides. Fe^{3+} ion in Akaganeite nanoparticles is responsible for EMR signal. Parameshwari et al. [14] have performed systematic EMR study of peak-to-peak line width and spin–spin relaxation time for samples prepared with different dextrose concentration. The spin–spin relaxation process which accounts for the energy difference transferred to the neighboring electrons is important as for as the line widths are concerned. They have reported the spin–spin relaxation time for different concentration of dextrose used in coprecipitation and found that relaxation time decreases with increasing dextrose concentration. On the basis of their study of relaxation time they have inferred that akaganeite nanoparticles may be potential candidate for microwave applications. The superparamagnetic domains present in the akaganeite nanoparticles make them important for pharmaceutical and biomedical applications like targeted drug delivery.

12.4.2 Nanomaterials Based on ZnO

Nanomaterials based on ZnO have attracted significant attention due to their possible applications in electro-optical devices, transparent ultraviolet protection films, and spintronic devices [15, 16]. Zinc oxide exhibits ferromagnetism when doped with most of the transition metals [17]. Misra et al. [18] have investigated the mechanism of ferromagnetism in the $Zn_{1-x}Fe_xO$ system. As the nanosamples possess different surface structure/groups depending on the solvent used for synthesis, authors prepared two sets of iron-doped zinc oxide nanosamples using similar chemical hydrolysis methods, one in diethylene glycol, and another in denatured ethanol solutions and EMR spectra were recorded at X and Q-band. Authors have reported coexistence of ferromagnetic and paramagnetic states in diethylene glycol prepared samples. They could further discuss that ferromagnetism occurs in the regions, which are rich in oxygen vacancies whereas in some regions Fe ions are not bound together through exchange interactions due to lack of oxygen vacancies and such localized Fe ions are responsible for paramagnetic signal.

Sibera et al. [19] have performed EMR studies of the iron doped zinc oxide for different concentration of the dopant. Very intense and broad EMR line centered at effective g value near 2 is reported (Fig. 12.1). Resonance field and line width were not found to depend significantly on the concentration of Fe_2O_3, while the intensity strongly increases with increasing concentrations of Fe_2O_3. They could conclude that dipole–dipole interaction depends essentially on concentration of magnetic nanoparticles.

Antibacterial and anticancerous activities of ZnO nanoparticles have also been reported in literature. Lipovsky et al. [20] have performed EMR coupled with the spin-trapping technique to demonstrate the formation of hydroxyl radicals and singlet oxygen in a water suspension of ZnO nanoparticles. Authors reported that level of oxy radicals increased considerably when the suspension was irradiated with visible light (Fig. 12.2). They have suggested that the production of OH and singlet oxygen radicals by ZnO is responsible for the cellular damage leading to its bactericidal activity.

12.4.3 TiO_2 Nanoparticles

Titanium dioxide is a very important material, intensely investigated due to its various technological applications, especially in the nanoparticles form [21–

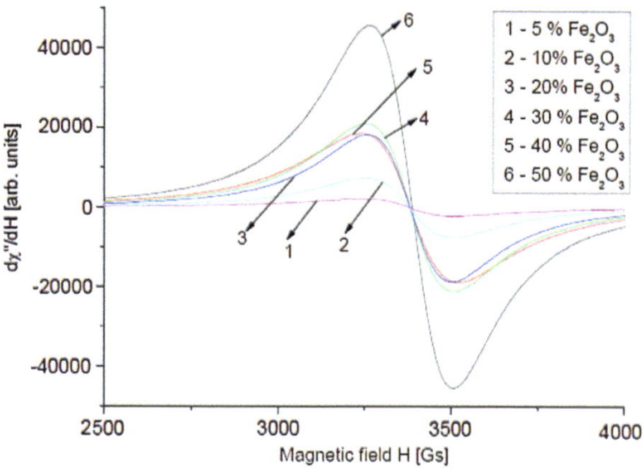

Fig. 12.1 EMR spectra of different concentrations of Fe_2O_3 in the samples. (Reproduced with permission)

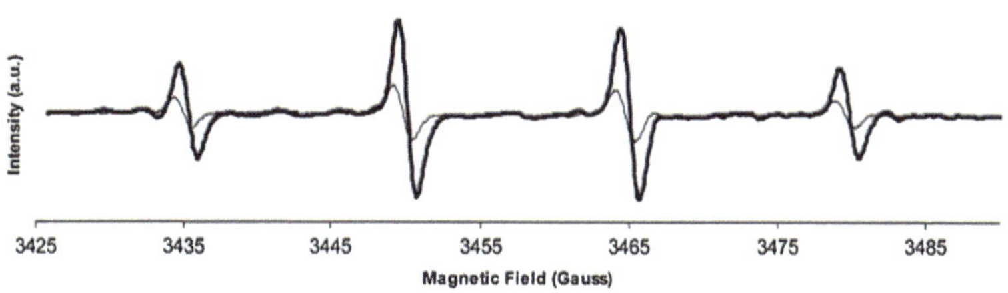

Fig. 12.2 OH production by ZnO nanoparticles before (*black*) and after (*bold black*) broadband visible light illumination using DMPO as a spin trap. (Reproduced with permission)

25]. Sroiraya et al. [26] have studied TiO_2 nanoparticles irradiated with ultraviolet light to detect hydroxyl radicals using EMR spectroscopy with spin trapping technique. They have analyzed the effect of particle size, concentration, and agglomeration effects. Authors have proposed a mechanism for the generation of hydroxyl radical that can be used to support the antibacterial effect of TiO_2 photocatalysis.

12.4.4 CuO Nanoparticles

Prasetyanto et al. [27] have reported the preparation of CuO nanoparticles onto SBA16 type mesoporous silica matrix using cyclam complex as precursor and also by impregnation method. EMR of samples prepared by both these methods were recorded. The spectrum of the CuO-SBA-16 (cyclam) could be described by an axial spin Hamiltonian. Spin Hamiltonian parameters have been reported and square planar symmetry for Cu(II) ions coordinated to four nitrogens has been proposed by the authors. From the intensity pattern of EMR, authors were able to compare the concentration of Cu(II) ions in the samples.

12.4.5 Fe_3O_4 Nanoparticles

Fe_3O_4 nanoparticles find wide applications in the biomedical field as anticancer agent [28, 29] and corrosion protective pigments in paints and coatings [30]. Fe_3O_4 nanoparticles are amongst the common ferrite materials which exhibit unique electrical and magnetic properties due to the transfer of electrons between Fe^{2+} and Fe^{3+} on octahedral sites [31]. Rahman et al. [32] have studied EMR spectra of nano ferrites. Authors have confirmed the superparamagnetic nature of nano ferrite from EMR studies. Their EMR study reveals a relatively weak transition other than the main strong feature of X-Band solid state EMR line. They have suggested that the weaker transition confirms the super paramagnetic nature of these nano ferrites.

Zhao et al. [33] designed and synthesized novel functional nanoparticles with nitroxide free radicals spin labeling in order to investigate their new applications EMR bioprobes. Their results of the EMR measurement of carboxylic acids modified Fe_3O_4 nanoparticles indicated that the spin-labeled nanoparticles with both the superparamagnetism of ferrimagnetic nanoparticles and the paramagnetism of nitroxide free radicals could act as a potential vehicle to deliver drugs.

Gamarra et al. [34] have studied the kinetics of the biodistribution and the elimination of the biocompatible Fe_3O_4 ferrofluid (Endorem) covered with dextran after intravenous administration to Winstar rats. They have used EMR to determine the iron content found in the liver and the blood samples of rats.

Rakhmatullin et al. [35] have recently reported EMR characterization of CeO_2: Er^{3+} nanoparticles. The obtained results are compared with appropriated characteristics of the Er-doped CeO_2 single crystal and the macroscopic powder of commercial pure cerium dioxide. They found that Er^{3+} ions in CeO_2 nanoparticles with average grain sizes of 22 and 300 nm occupy mainly cubic sites, similar to the single crystal with the same low concentration of these ions. However, the line widths observed in nanoparticles of CeO_2 reveal increased distortions with the decrease of the grain size. They have concluded that the vacancies are located in the next nearest neighbor positions. Authors have pointed out that these studies may lead to design and enhance catalytic properties of cerium dioxide.

12.5 EMR of Irradiated and Activated Nanoparticle Systems

12.5.1 Photo-Generated Charge Carriers in Nanocrystalline Powders

Paramagnetic intermediates produced in the irradiated titanium dioxide systems can be investigated using EMR spectroscopy [36]. Brezova et al. [37] have reported EMR study of species generated upon UV exposure of anatase and rutile nanocrystalline powders, along with their analogs containing up to 90 % of ^{17}O atoms. They prepared anatase and rutile nanoparticulate samples via $TiCl_4$ hydrolysis using either ordinary or ^{17}O-enriched water with subsequent heat treatment under vacuum. These samples were investigated in the temperature range 100–160 K. The presence of ^{17}O atoms in the anatase or rutile crystal lattice caused significant line broadening of the EMR signals monitored at 100 K. This was attributed to the electron and hole sites, due to the interaction of the unpaired electron with a ^{17}O nucleus. They have further reported that an increase in the temperature leads to a significant decline of the EMR signals and hence they could conclude that the stability of the photo-generated charge carriers is limited.

12.5.2 Fluorite Nanoparticles Activated by Er^{3+} Ions

Irisova et al. [38] have reported EMR of CaF_2 nanoparticles activated with Er^{3+} ions and attributed two intense lines to be originated from surface defects. The EMR line due to intrinsic CaF_2 surface defects can be used to estimate the grain size of the sample. Authors have reported most of the Er^{3+} centers to be nominally high symmetry cubic centers perturbed by the disorder near the surface.

Conclusions

This chapter is an attempt to review the applications of EMR spectroscopy in nanoparticle characterization. Mainly, EMR of metallic nanoparticles and oxide nanoparticles has been the subject matter of this chapter. As for as metallic nanoparticles are concerned, the EMR characterization is capable of providing very useful information on magnetic properties but the spin–lattice relaxation is not completely understood. In particular, the effect of particle diameters and the matrix on the spin–lattice relaxation in metallic nanoparticles requires series of investigations.

Further, many studies on magnetic nanoparticles have used classical ferromagnetic approach (FMR) to describe the dynamics of magnetization. However, this is still challenging to explain line shape and their temperature dependence. Noginova et al. [39] have proposed an alternative model where magnetic nanoparticles are considered as a giant spin and the EMR signal was assumed to be the sum of contributions of quantum transitions between energy levels associated with the projections of the giant spin onto the direction of the magnetic field. This model could explain some of the general tendencies observed in the intermediate size spin systems. Their work is concentrated about Fe_2O_3 system. More experimental and theoretical work is needed on different systems to understand the transition from quantum to classical case. Many of the metal oxides, in particular nano-TiO_2 and ZnO, have been studied by EMR spectroscopy due to their bactericidal and

(continued)

anticancerous properties. The biocompatible ferro-fluids based on oxide nanoparticles are of particular interest in clinical applications such as drug targeting [40]. At the same time, mechanism for their cytotoxic effect is still to be understood. Recently reported study of Er^{3+} doped CeO_2 nanoparticles [35] indicates the future scope for EMR of rare earth impurity doped oxide nanoparticles.

Acknowledgement Author thanks the referee for valuable encouraging remarks.

References

1. Poole CP (1983) Electron spin resonance: a comprehensive treatise on experimental techniques. Wiley, New York
2. Zavoisky E (1945) Relaxation of liquid solutions for perpendicular fields. J Phys USSR 9:211
3. Smirnov AI (2011) EPR studies of nanomaterials. In: Misra SK (ed) Multifrequency electron paramagnetic resonance: theory and applications. Wiley, New York, p 825
4. Mitrikas G, Deligiannakis Y, Trapalis CC, Boukos N, Kordas G (1998) CW and pulsed EPR study of silver nanoparticles in a SiO_2 matrix. J Sol–Gel Sci Technol 13:503
5. Khaliullin GG, Khusainov MG (1988) Theory of spin–lattice relaxation of conduction electrons in small metallic particles. Sov Phys JETP 67:524
6. Tseng W, Chen C (2006) Dispersion and rheology of nickel nanoparticle inks. J Mater Sci 41:1213
7. Hyeon T (2003) Chemical synthesis of magnetic nanoparticles. Chem Commun: 927
8. Gubin SP, Koksharov YA, Khomutov GB, Yurkov GY (2005) Magnetic nanoparticles: preparation, structure and properties. Russ Chem Rev 74:489
9. Karmhag R, Tesfamichael T, Wackelgard E, Nikalsson G, Nygren M (2000) Oxidation kinetics of nickel particles: comparison between free particles and particles in an oxide matrix. Sol Energy 68:329
10. Lee SI, Lee N, Park J, Kim BH, Yi Y, Kim T, Kim TK, Lee IH, Paik SR, Hyeon T (2006) Ni/NiO core/shell nanoparticles for selective binding and magnetic separation of histidine-tagged proteins. J Am Chem Soc 128:10658
11. Llamazares SR, Merchan J, Olmedo I, Marambio HP, Munoz JP, Jara P, Sturm JC, Chornik B, Pena O, Yutronic N, Kogan MJ (2008) Ni/Ni oxides nanoparticles with potential biomedical applications obtained by displacement of a nickel-organometallic complex. J Nanosci Nanotechnol 8:3820
12. Alonso F, Riente P, Sirvent JA, Yus M (2010) Nickel nanoparticles in hydrogen-transfer reductions: characterisation and nature of the catalyst. Appl Catal A: Gen 378:42
13. Ionita P, Wolowska J, Chechik V, Caragheorgheopol A (2007) Ligand dynamics in spin-labeled Au nanoparticles. J Phys Chem C 111:16717
14. Parameshwari R, Priyadarshini P, Chandrasekaran G (2011) Optimization, structural, spectroscopic and magnetic studies on stable akaganeite nanoparticles via co-precipitation method. Am J Mat Sci 1:18
15. Chen Y, Bagnall DM, Koh H, Park K, Higara K, Zhu Z, Yao T (1998) Plasma assisted molecular beam epitaxy of ZnO on c-plane sapphire: growth and characterization. J Appl Phys 84:3912
16. Nemeth J, Rodriguez-Gattorno G, Diaz A, Dekany I (2004) Synthesis of ZnO nanoparticles on a clay mineral surface in dimethyl sulfoxide medium. Langmuir 20:2855

17. Coey JMD, Venkatesan M, Fitzgerald CB (2005) Donor impurity band exchange in dilute ferromagnetic oxides. Nat Mater 4:173
18. Misra SK, Andronenko SI, Thurber A, Punnoose A, Nalepa A (2014) An X- and Q-band Fe^{3+} EPR study of nanoparticles of magnetic semiconductor $Zn_{1-x}Fe_xO$. J Magn Magn Mater 363:82
19. Sibera D, Narkiewicz U, Guskos N, Zolnierkiewicz G (2009) The preparation and EPR study of nanocrystalline $ZnFe_2O_4$. J Phys Conf 146:012014
20. Lipovsky A, Tzitrinovich Z, Friedmann H, Applerot G, Gedanken A, Lubart R (2009) EPR study of visible light-induced ROS generation by nanoparticles of ZnO. J Phys Chem C 113:15997
21. Fernandez A, Lassaletta G, Jimenez VM, Justo A, Gonzalezelipe AR, Herrmann JM, Tahiri H, Aitichou Y (1995) Preparation and characterization of TiO_2 photocatalysts supported on various rigid supports (glass, quartz and stainless steel). Comparative studies of photocatalytic activity in water purification. Appl Catal B Environ 7:49
22. Schierbaum KD, Kirner UK, Geiger JF, Gopel W (1991) Conductance, work function and catalytic activity of SnO_2-based gas sensors. Sens Actuators B 3:205
23. Linsebigler AL, Lu G, Yates JT Jr (1995) Photocatalysis on TiO2 surfaces: principles, mechanisms, and selected results. Chem Rev 95:735
24. Strukov DB, Sneider GS, Stewart DR, Williams RS (2008) The missing memristor found. Nature 453:80
25. Waser R, Dittmann R, Staikov G, Szot K (2009) Redox-based resistive switching memories—nanoionic mechanisms, prospects, and challenges. Adv Mater 21:2632
26. Sroiraya S, Triampo W, Morales NP, Triampo D (2008) Kinetics and mechanism of hydroxyl radical formation studied via electron spin resonance for photocatalytic nanocrystalline titania: effect of particle size distribution, concentration, and agglomeration. J Ceram Process Res 9:146
27. Prasetyanto EA, Sujandi, Lee S, Park S (2007) Highly dispersed Cuo nanoparticles on SBA-16 type mesoporous silica with cyclam SBA-16 as a precursor. Bull Korean Chem Soc 28:2359
28. Oh JK, Park JM (2011) Iron oxide-based superparamagnetic polymeric nanomaterials: design, preparation, and biomedical application. Prog Polym Sci 36:168
29. Pan D, Zhang H, Fan T, Chen J, Duan X (2011) Nearly monodispersed core–shell structural Fe_3O_4@DFUR–LDH submicro particles for magnetically controlled drug delivery and release. Chem Commun 47:908
30. Alam J, Riaz U, Ashraf SM, Ahmad S (2008) J Coat Technol Res 5:123
31. Chen R, Song G, Wei Y (2010) J Phys Chem C 114:13409
32. Rahman O, Mohapatra SC, Ahmad S (2012) Fe_3O_4 inverse spinal super paramagnetic nanoparticles. Mater Chem Phys 132:196
33. Zhao H, Zhang Z, Zhao Z, Yu R, Wan Y, Lan M (2011) Preparation and characterization of novel spin labeled magnetic nanoparticles. Adv Mat Lett 2:172
34. Gamarra LF, Pontuschka WM, Amaro E Jr, Costa-Filho AJ, Brito GES, Vieira ED, Carneiro SM, Escriba DM, Falleiros AMF, Salvador VL (2008) Kinetics of elimination and distribution in blood and liver of biocompatible ferrofluids based on Fe_3O_4 nanoparticles: an EPR and XRF study. Mater Sci Eng C 28:519
35. Rakhmatullin RM, Kurkin IN, Pavlov VV, Semashko VV (2014) EPR, optical, and dielectric spectroscopy of Er-doped cerium dioxide nanoparticles. Phys Status Solidi B 1–7. doi: 10.1002/pssb.201451116
36. Wang Z, Ma W, Chen C, Ji H, Zhao J (2011) Probing paramagnetic species in titania-based heterogeneous photocatalysis by electron spin resonance (ESR) spectroscopy—a mini review. Chem Eng J 170:353
37. Brezova V, Barbierikova Z, Zukalova M, Dvoranova D, Kavan L (2014) EPR study of ^{17}O-enriched titania nanopowders under UV irradiation. Catalysis Today 230:112
38. Irisova IA, Rodionov AA, Tayurskii DA, Yusupov RV (2011) EPR studies of the mechanochemically Er^{3+}-activated fluorite nanoparticles. J Phys Conf 324:012026

39. Noginova N, Chen F, Weaver T, Giannelis EP, Bourlinos AB, Atsarkin VA (2007) Magnetic resonance in nanoparticles: between ferro- and paramagnetism. J Phys Condens Matter 19:246208
40. Lubbe AS, Bergemann C, Huhnt W, Fricke T, Reiss H, Brock JW, Huhn D (1996) Preclinical experiences with magnetic drug targeting: tolerance and efficacy. Cancer Res 56:4694

Chapter 13
Nanoparticle Characterization Using Nanoparticle Tracking Analysis

Antoanetta Lungu, Mihai Lungu, Adrian Neculae, and Raluca Giugiulan

Abstract Characterization of nanoparticles in environmental samples involves determining their size, their chemical composition, and their concentrations in the bulk matrix. Environmental changes in their environment metallic nanoparticles as colloidal aggregates tend to be stable or dispersions. Filter size differential is most commonly used method to isolate nanoparticles in aqueous solutions. Microfiltration, nanofiltration, cross-flow filtration, and ultracentrifugation are commonly used to achieve the highest degree of separation. Chemical characterization of the nanoparticles has traditionally been performed using a transmission microscope/scanning electron (TEM/SEM), followed by spectroscopy, energy dispersive X-ray (EDS), and X-ray diffraction (XRD). Due to the inherent limitations these methods were combined. This review describes the current status and challenges of isolation, separation, and detection of nanoparticles in samples. Nanoparticle Tracking Analysis (NTA) is a common procedure for the analysis of nanoparticles in complex aqueous matrices.

13.1 Introduction

Nanoparticles include any type of particle with at least one dimension of less than 500 nm. Nanoparticles play an important role in a wide variety of fields including advanced materials, pharmaceuticals, and environmental detection and monitoring. They can generally be classified as one of two types: engineered or nonengineered. Engineered nanoparticles are intentionally designed and created with physical properties tailored to meet the needs of specific applications. They can be end products in and of themselves, as in the case of quantum dots or pharmaceutical drugs, or they can be components later incorporated into separate end products, as carbon black in rubber products. The particle's physical properties are important to their performance and the performance of any product into which they are

A. Lungu (✉) • M. Lungu • A. Neculae • R. Giugiulan
Department of Physics, West University of Timisoara, Blv. V. Parvan No. 4, 300223 Timisoara, Romania
e-mail: antoanettalungu@yahoo.com

M. Lungu et al. (eds.), *Nanoparticles' Promises and Risks*,
DOI 10.1007/978-3-319-11728-7_13

incorporated. Nonengineered nanoparticles are unintentionally generated or naturally produced, such as atmospheric nanoparticles created during combustion. For nonengineered nanoparticles, physical properties play an important role as they determine whether or not ill effects which will occur as a result of the presence of these particles [1].

Examples of industries with engineered and nonengineered nanoparticles:

- Some industries with engineered nanoparticles: chemical mechanical polishing, pharmaceuticals, performance chemicals, cosmetics, bio detection and labeling, food products, quantum dots, ceramics:
- Some industries with nonengineered particles: environmental monitoring controlled environments, environmental detection.

Depending on the application of interest, nanoparticles may be known by a number of alternative and trade-specific names, including particulate matter, aerosols, colloids, nanocomposites, nanopowders, and nanoceramics.

Nanoparticle characterization parameters include: hydrated surface analysis, surface area and porosity, particle size distribution, solubility, adsorption potential, zeta potential, aggregation, wettability, shape and size of interactive surface. Nanoparticles are classified based on their dimensionality, morphology, composition, uniformity, and agglomeration. It is necessary to make a clear distinction between nanostructured thin films or other fixed nanometer-scale objects (the circuits within computer microprocessors) and free nanoparticles [2].

The motion of free nanoparticles is not constrained, and they can easily be released into the environment leading to human exposure that may pose a serious health risk. In contrast are the many objects containing nanostructured elements that are firmly attached to a larger object, where the fixed nanoparticles should pose no health risk when properly handled. An example of this is the material asbestos, which is perfectly safe in its primary state (a type of solid rock), but is a significant health hazard when is processed in such a way as to produce the carcinogenic nanometer-scale fibrous particles that become airborne (aerosol) and are therefore readily absorbed in the lungs [3].

13.2 Methods Used for the Characterization of Nanoparticles

There are several techniques used to understand these characterization parameters in nanoparticles: electron microscopy including TEM and SEM, atomic force microscopy (AFM), *Nanoparticle tracking analysis (NTA) for tracking of the Brownian motion*, dynamic light scattering (DLS), X-ray photoelectron spectroscopy (XPS), powder X-ray diffraction (XRD), Fourier transform infrared spectroscopy (FTIR), matrix-assisted laser desorption/ionization time-of-flight mass spectrometry (MALDI-TOF), ultraviolet-visible spectroscopy, dual polarization interferometry, nuclear magnetic resonance (NMR).

13.2.1 Transmission Electron Microscope (TEM) and Scanning Electron Microscopy (SEM)

The electron microscope is a microscope that uses a beam of electrons to create an image of the specimen. It is capable of much higher magnifications and has a greater resolving power than a light microscope, allowing it to see much smaller objects in finer detail. They are large, expensive pieces of equipment, generally standing alone in a small, specially designed room and requiring trained personnel to operate them.

The original form of electron microscopy, *Transmission electron microscopy* (*TEM*) involves a high voltage electron beam emitted by a cathode and formed by magnetic lenses. The electron beam that has been partially transmitted through the very thin (and semitransparent for electrons) specimen carries information about the structure of the specimen. The spatial variation in this information (the "image") is then magnified by a series of magnetic lenses until it is recorded by hitting a fluorescent screen, photographic plate, or light sensitive sensor such as a CCD (charge-coupled device) camera. The image detected by the CCD may be displayed in real time on a monitor or computer. Transmission electron microscopes produce two-dimensional, black and white images.

Resolution of the TEM is also limited by spherical and chromatic aberration, but a new generation of aberration correctors has been able to overcome or limit these aberrations. Software correction of spherical aberration has allowed the production of images with sufficient resolution to show carbon atoms in diamond separated by only 0.089 nm and atoms in silicon at 0.078 nm at magnifications of 50 million times. The ability to determine the positions of atoms within materials has made the TEM an indispensable tool for nanotechnologies research and development in many fields, including heterogeneous catalysis and the development of semiconductor devices for electronics and photonics. In the life sciences, it is still mainly the specimen preparation which limits the resolution of what we can see in the electron microscope, rather than the microscope itself.

The *Scanning electron microscope* (*SEM*) is a powerful and frequently used instrument, in both academia and industry, to study, for example, surface topography, composition, crystallography, and properties on a local scale. The spatial resolution is better than that of the optical microscope although not quite as good as for the transmission electron microscope (TEM). The SEM has an extremely large depth of focus and is therefore well suited for topographic imaging. Besides surface topographic studies the SEM can also be used for determining the chemical composition of a material, its fluorescent properties, and the formation of magnetic domains and so on. The scanning electron microscope (SEM) uses a focused beam of high-energy electrons to generate a variety of signals at the surface of solid specimens. The signals that derive from electron–sample interactions reveal information about the sample including external morphology (texture), chemical composition, and crystalline structure and orientation of materials making up the sample. In most applications, data are collected over a selected area of the surface of the sample, and a 2-dimensional image is generated that displays spatial variations in these properties.

Areas ranging from approximately 1 cm to 5 μm in width can be imaged in a scanning mode using conventional SEM techniques. The SEM is also capable of performing analyses of selected point locations on the sample; this approach is especially useful in qualitatively or semiquantitatively determining chemical compositions, crystalline structure, and crystal orientations [4].

13.2.2 Atomic Force Microscopy (AFM)

Atomic force microscopy (AFM) is a high-resolution surface imaging technique which operates by scanning a sharp probe over the surface of a sample, while measuring the forces experienced by the probe [5]. In the most commonly used imaging mode, the probe and sample are brought in contact and the probe is scanned over the surface while the interaction force is kept constant. Alternatively, the deflection of the cantilever can be measured while the probe scans at constant height. This provides an image of the surface topography with (sub) nanometer scale resolution. A major advantage of the AFM over classical microscopy techniques is that it can simultaneously provide information on local physical properties, such as mechanical properties and interaction forces. In particular, force–distance curves, made by recording the deflection of the cantilever while the sample is moved up and down, allow one to measure directly surface forces in aqueous environments, such as van der Waals and electrostatic forces, solvation forces, steric forces, and intermolecular forces between complementary molecules.

AFM can be applied to a wide variety of samples (conductors, insulators) and may be operated in various environments (vacuum, air, liquids) which make it possible to examine biological systems (biomolecules, cells) under physiological conditions. In this laboratory, AFM is used to probe the nanoscale organization of organic surfaces, including polymers, lipid films, adsorbed protein layers, and living cells. Besides topographic imaging, a special emphasis is put on the quantitative measurement of molecular interactions using functionalized probes.

13.2.3 Nanoparticle Tracking Analysis (NTA) for Tracking of the Brownian Motion

Nanoparticle Tracking Analysis (NTA) provides direct and real time visualization, sizing, and counting of particulate materials between 10 nm and 1 μm in liquid suspension. The technique works on a particle-by-particle basis, relating the degree of movement under Brownian motion to the sphere equivalent hydrodynamic diameter particle size, allowing high resolution particle size distributions to be obtained within minutes. Here we describe application of NTA to the analysis of 100 nm polystyrene nanospheres in liquid suspension. The technique can be adapted for use with nearly all particulate materials with sizes between 10 nm and 1 μm.

Nanoparticle Tracking Analysis (NTA) has been introduced as an alternative technique for dispersed submicron particles. Hereby, the particle size is derived from the average displacement between successive photos due to Brownian motion. As compared to DLS, NTA has the advantage that it has a higher resolution for multimodal samples. In addition, it provides direct visual information from which aggregation phenomena are visually observable [6].

13.2.4 Dynamic Light Scattering (DLS)

Dynamic light scattering is a new method for investigating macromolecular systems. The importance of the technique lies in its noninvasive character. It can be employed on extremely small fluid volumes; the instrumentation is relatively inexpensive and allows the rapid determination of diffusion coefficients as well as providing information on relaxation time distributions for the macromolecular components of complex systems.

DLS technique measures Brownian motion in the liquids and solutions. The Brownian motion of molecules becomes more rapid when the size of particles scattering light is smaller. DLS was discovered shortly after gas lasers were applied as light sources in optical scattering investigations. Lasers are high intensity sharp monochromatic light sources; as a consequence the spectral broadening of the low intensity Rayleigh scattered radiation is relatively easy to detect.

DLS is most commonly used to analyze nanoparticles. Examples include determining nanogold size, protein size, latex size, and colloid size. In general, the technique is best used for submicron particles and can be used to measure particle with sizes less than a nanometer. In the domain of sizes ranging from nanometers to microns, the distinction between a molecule (such as a protein or macromolecule) and a particle (such as nanogold) becomes blurred, even if the measurements are performed in liquid phase (such as in an emulsion). Dynamic light scattering can also be used as a probe of complex fluids such as concentrated solutions. However, this application is much less common than particle sizing.

13.2.5 X-Ray Photoelectron Spectroscopy (XPS)

A sample, introduced in an ultrahigh vacuum chamber, is bombarded with an X-ray beam. The kinetic energy (in electron volts, eV) of emitted electrons of all elements (except H and He) present at the surface (analyzed depth between 1 and 10 nm) is measured with a precision of about 0.2 eV. Shape and position of peaks depend on the chemical state of the element (the so-called chemical shift effect). Area of peaks used in combination with sensitivity factors allow to calculate atom fractions with a detection limit of a few tenths of atom percent. A detailed analysis of certain well-resolved peaks allows quantifying functionalities present at the surface. On most recent systems the minimum spatial resolution is of about 15 μm and 5 μm for XPS analysis and XPS imaging, respectively. In most cases XPS can be considered as a nondestructive technique.

X-ray photoelectron spectroscopy (XPS) is generally regarded as an important and key technique for the surface characterization and analysis of biomedical polymers. This technique, also called ESCA (Electron Spectroscopy for Chemical Analysis), provides a total elemental analysis, except for hydrogen and helium, of the top 10–200 Å (depending on the sample and instrumental conditions) of any solid surface which is vacuum stable or can be made vacuum stable by cooling, chemical bonding information is also provided. Of all the presently available instrumental techniques for surface analysis, XPS is generally regarded as being the most quantitative, the most readily interpretable, and the most informative with regard to chemical information. For these reasons it has been highly recommended and used by researchers for the analysis of polymers.

Surface analysis by XPS involves irradiating a solid in vacuum with mono energetic soft X-rays and analyzing the emitted electrons by energy. The spectrum is obtained as a plot of the number of detected electrons per energy interval versus their kinetic energy. Each element has a unique spectrum. The spectrum from a mixture of elements is approximately the sum of the peaks of the individual constituents. Because the mean free path of electrons in solids is very small, the detected electrons originate from only the top few atomic layers, making XPS a unique surface-sensitive technique for chemical analysis. Quantitative data can be obtained from peak heights or peak areas, and identification of chemical states often can be made from exact measurement of peak positions and separations, as well as from certain spectral features [7].

XPS can be used to characterize the surface of all types of materials, quite exclusively solids (powders or bulk specimen) as it is the case in our group: biomaterials, catalysts, ceramics, fibers, glass, metals, minerals, polymers. The laboratories sharing the XPS facilities have an uncommon expertise in using the method for analysis of microbial cells and adsorbed phases on one hand and of catalysts and dispersed materials on the other hand.

13.2.6 Powder X-Ray Diffraction (XRD)

X-ray powder diffraction (XRD) is a rapid analytical technique primarily used for phase identification of a crystalline material and can provide information on unit cell dimensions. The analyzed material is finely ground, homogenized, and average bulk composition is determined.

Max von Laue, in 1912, discovered that crystalline substances act as three-dimensional diffraction gratings for X-ray wavelengths similar to the spacing of planes in a crystal lattice. X-ray diffraction is now a common technique for the study of crystal structures and atomic spacing.

X-ray diffraction is based on constructive interference of monochromatic X-rays and a crystalline sample. These X-rays are generated by a cathode ray tube, filtered to produce monochromatic radiation, collimated to concentrate, and directed toward the sample. The interaction of the incident rays with the sample produces constructive interference (and a diffracted ray) when conditions satisfy Bragg's Law ($n\lambda = 2d\sin\theta$). This law relates the wavelength of electromagnetic radiation to the diffraction angle and the lattice spacing in a crystalline sample. These diffracted X-rays are then detected, processed, and counted. By scanning the sample through a range of 2θ angles, all possible diffraction directions of the lattice should be attained due to the random orientation of the powdered material. Conversion of the diffraction peaks to d-spacing allows identification of the mineral because each mineral has a set of unique d-spacing. Typically, this is achieved by comparison of d-spacing with standard reference patterns.

All diffraction methods are based on generation of X-rays in an X-ray tube. These X-rays are directed at the sample, and the diffracted rays are collected. A key component of all diffraction is the angle between the incident and diffracted rays. Powder and single crystal diffraction vary in instrumentation beyond this.

X-ray powder diffraction is most widely used for the identification of unknown crystalline materials (e.g., minerals, inorganic compounds). Determination of unknown solids is critical to studies in geology, environmental science, material science, engineering, and biology.

Other applications include: characterization of crystalline materials, identification of fine-grained minerals such as clays and mixed layer clays that are difficult to determine optically, measurement of sample purity, determination of unit cell dimensions.

With specialized techniques, XRD can be used to:

- Determine crystal structures using Rietveld refinement
- Determine modal amounts of minerals (quantitative analysis)
- Characterize thin films samples by:
  - Determining lattice mismatch between film and substrate and to inferring stress and strain
  - Determining dislocation density and quality of the film by rocking curve measurements

- Measuring super lattices in multilayered epitaxial structures
- Determining the thickness, roughness, and density of the film using glancing incidence X-ray reflectivity measurements

• Make textural measurements, such as the orientation of grains, in a polycrystalline sample.

13.2.7 Fourier Transform Infrared Spectroscopy (FTIR)

FT-IR stands for Fourier Transform Infrared, the preferred method of infrared spectroscopy. In infrared spectroscopy, IR radiation is passed through a sample. Some of the infrared radiation is absorbed by the sample and some of it is passed through (transmitted). The resulting spectrum represents the molecular absorption and transmission, creating a molecular fingerprint of the sample. Like a fingerprint no two unique molecular structures produce the same infrared spectrum. This makes infrared spectroscopy useful for several types of analysis.

FTIR spectrometers (Fourier Transform Infrared Spectrometer) are widely used in organic synthesis, polymer science, petrochemical engineering, pharmaceutical industry, and food analysis. In addition, since FTIR spectrometers can be hyphenated to chromatography, the mechanism of chemical reactions and the detection of unstable substances can be investigated with such instruments.

Fourier Transform Infrared (FT-IR) spectrometry was developed in order to overcome the limitations encountered with dispersive instruments. The main difficulty was the slow scanning process. A method for measuring all of the infrared frequencies simultaneously, rather than individually, was needed. A solution was developed which employed a very simple optical device called an interferometer. The interferometer produces a unique type of signal which has all of the infrared frequencies "encoded" into it. The signal can be measured very quickly, usually on the order of one second or so. Thus, the time element per sample is reduced to a matter of a few seconds rather than several minutes.

13.2.8 Matrix-Assisted Laser Desorption/Ionization Time-of-Flight Mass Spectrometry (MALDI-TOF)

Matrix-assisted laser desorption/ionization time of-flight mass spectrometry (MALDI-Tof-MS) has advanced from early-stage research toward applications of real clinical relevance only within the last 10 years. MALDI-Tof-MS has become a popular and versatile method to analyze a range of macromolecules from biological origin from cells to tissues. Its ability to desorb high-molecular-weight thermo labile molecules, its high accuracy and sensitivity, combined with its wide mass range (1–300 kDa), make MALDI-Tof-MS a promising method for the clinical chemistry laboratory for the identification of biomolecules in complex samples, including peptides, proteins, oligosaccharides, and oligonucleotides. The first reports demonstrating successful MALDITof-MS biochemical analysis were published in the late 1980s from the labs of Tanaka et al. [8] and Karas and Hillenkamp [9].

13.2.9 Ultraviolet-Visible Spectroscopy (UV/VIS)

Ultraviolet-visible spectroscopy or ultraviolet-visible spectrophotometry (UV-Vis or UV/Vis) refers to absorption spectroscopy in the ultraviolet–visible spectral region. This means it uses light in the visible and adjacent (near-UV and near-infrared (NIR)) ranges. The absorption in the visible range directly affects the perceived color of the chemicals involved. In this region of the electromagnetic spectrum, molecules undergo electronic transitions. This technique is complementary to fluorescence spectroscopy, in that fluorescence deals with transitions from the excited state to the ground state, while absorption measures transitions from the ground state to the excited state [10].

Molecules containing π-electrons or nonbonding electrons (n-electrons) can absorb the energy in the form of ultraviolet or visible light to excite these electrons to higher antibonding molecular orbitals [11]. The more easily excited the electrons (i.e., lower energy gap between the HOMO and the LUMO), the longer the wavelength of light it can absorb.

UV/Vis spectroscopy is routinely used in analytical chemistry for the quantitative determination of different analyses, such as transition metal ions, highly conjugated organic compounds, and biological macromolecules. Spectroscopic analysis is commonly carried out in solutions but solids and gases may also be studied.

- Solutions of transition metal ions can be colored (i.e., absorb visible light) because d electrons within the metal atoms can be excited from one electronic state to another. The color of metal ion solutions is strongly affected by the presence of other species, such as certain anions or ligands. For instance, the color of a dilute solution of copper sulfate is a very light blue; adding ammonia intensifies the color and changes the wavelength of maximum absorption (λ_{max}).
- Organic compounds, especially those with a high degree of conjugation, also absorb light in the UV or visible regions of the electromagnetic spectrum. The solvents for these determinations are often water for water-soluble compounds, or ethanol for organic-soluble compounds. (Organic solvents may have significant UV absorption; not all solvents are suitable for use in UV spectroscopy. Ethanol absorbs very weakly at most wavelengths.) Solvent polarity and pH can affect the absorption spectrum of an organic compound. Tyrosine, for example, increases in absorption maxima and molar extinction coefficient when pH increases from 6 to 13 or when solvent polarity decreases.

- While charge transfer complexes also give rise to colors, the colors are often too intense to be used for quantitative measurement [11].

13.2.10 Dual Polarization Interferometry (DPI)

Dual polarization interferometry (DPI) is a highly versatile, powerful analytical technique for biophysical characterization of proteins and other biomolecules. It extends the typical dynamic measurements of conventional biosensors by including an additional quantitative, submolecular, conformational measurement. DPI delivers a unique perspective on biochemistry, linking conformational changes to biochemical activity at a resolution normally associated with "big physics."

DPI is an interferometry technique to derive dynamic information concerning the thickness, density, and mass of a molecular layer. As well as mass-based affinity and kinetic determinations possible with other label-free optical and acoustic biosensor technologies, DPI additionally provides real-time data on the orientation of the surface immobilized layer and any conformational changes involved upon binding.

13.2.11 Nuclear Magnetic Resonance (NMR)

Nuclear magnetic resonance spectroscopy (NMR) was first developed in 1946 by research groups at Stanford and M.I.T., in the USA. The radar technology developed during World War II made many of the electronic aspects of the NMR spectrometer possible. With the newly developed hardware physicists and chemists began to apply the technology to chemistry and physics problems. Over the next 50 years NMR developed into the premier organic spectroscopy available to chemists to determine the detailed chemical structure of the chemicals they were synthesizing. Another well-known product of NMR technology has been the Magnetic Resonance Imager (MRI), which is utilized extensively in the medical radiology field to obtain image slices of soft tissues in the human body. In recent years, NMR has moved out of the research laboratory and into the on-line process analyzer market. This has been made possible by the production of stable permanent magnet technologies that allow high-resolution 1H NMR spectra to be obtained in a process environment.

NMR is a spectroscopic technique which relies on the magnetic properties of the atomic nucleus. When placed in a strong magnetic field, certain nuclei resonate at a characteristic frequency in the radio frequency range of the electromagnetic spectrum. Slight variations in this resonant frequency give us detailed information about the molecular structure in which the atom resides.

The NMR phenomenon is based on the fact that nuclei of atoms have magnetic properties that can be utilized to yield chemical information. Quantum mechanical subatomic particles (protons, neutrons, and electrons) have spin. In some atoms (e.g., 12C, 16O, 32S) these spins are paired and cancel each other out so that the nucleus of the atom has no overall spin. However, in many atoms (1H, 13C, 31P, 15N, 19F) the nucleus does possess an overall spin [12].

The nuclear magnetic resonance phenomenon can be described in a nutshell as follows. If a sample is placed in a magnetic field and is subjected to radiofrequency (RF) radiation (energy) at the appropriate frequency, nuclei in the sample can absorb the energy. The frequency of the radiation necessary for absorption of energy depends on three things. First, it is characteristic of the type of nucleus (e.g., 1H or 13C). Second, the frequency depends on chemical environment of the nucleus. For example, the methyl and hydroxyl protons of methanol absorb at different frequencies and amide protons of two different tryptophan residues in a native protein absorb at different frequencies since they are in different chemical environments. The NMR frequency also depends on spatial location in the magnetic field if that field is not everywhere uniform. This last variable provides the basis for magnetic resonance imaging (MRI), for self-diffusion coefficient measurements, and for coherence selection—topics which will not be discussed further in this introductory chapter. For diffusion coefficient measurements and for imaging, the magnetic field is made to vary linearly over the sample. However, for most spectroscopic purposes we generally wish the magnetic field to be as homogeneous as possible over the sample. The homogeneity requirements for NMR spectroscopy are rather stringent; the magnetic field should vary no more than 10 parts per billion (ppb) over the sample volume. After absorption of energy by the nuclei, the length of time and the manner in which the nuclei dissipate that energy can also be used to reveal information regarding a variety of dynamic processes [13].

13.3 Nanoparticle Tracking Analysis (NTA) with NanoSight

The NTA method measures the diffusion coefficients of individual particles and builds the distribution one particle at a time. This compares favorably with an ensemble measurement of the combined light scattering intensity of a population of particles. Consequently, rather than presenting an ideal curve driven by a range of assumptions, the result from NTA is a true high resolution.

This provides a series of key advantages of the NTA method: overcomes intensity-biased results seen in DLS/PCS; size is calculated particle by particle; direct measurement without modeling or assumptions; an absolute method not requiring calibration; provides high-resolution measurement of distribution; requires no information about collection angle, wavelength, or solvent refractive index; minimal sample preparation, with automation capability and immune to interference from dirt and aggregates [14].

NanoSight's Nanoparticle Tracking Analysis (NTA) is a technique for characterizing nanoparticles in their native suspension. This technique provides multiparameter characterization of nanoparticles (down to 10–2,000 nm diameter) achieved by video analysis of a laser illuminated sample through conventional microscopy. Analysis of the Brownian motion of individual particles in the defined field of view allows their size to be determined and their concentration. The Brownian motion of particles (which appears as point scatters) is tracked separately but simultaneously using a CCD camera, from which a high-resolution plot of the particle-size distribution (and profile changes in time during, aggregation or dissolution) is obtained. Sample pre-treatment is carried out only by dilution with an appropriate solvent to acceptable concentrations (between 108 and 109 particles/ml, depending on the type of sample). All types of particles can be measured provided they are sufficiently light to be spread visible. The technique offers significant advantages over existing light scattering techniques for the characterization of poly dispersed populations. Each particle is individually but simultaneously analyzed for: size, fluorescence, high resolution particle size distribution profile, light scattering properties (refractive index), zeta potential through the application of electric fields, and concentration (particles per milliliter) of any given size class of nanomaterial. NTA is widely used as a preliminary characterization technique prior to exposure. This has been applied to a range of materials including metals, metal oxides, and polymers.

13.3.1 Operating Principle

Nanoparticle Tracking Analysis (NTA) is a method that has gained popularity for the study of nanoparticle dispersions. The NTA technique is centered on a sample-analysis module in which a laser diode is configured to pass light through a 250 μl liquid sample containing the nanoparticles. Particles in the beam are visualized by a conventional optical microscope aligned normally to the beam axis (Fig. 13.1). This setup collects light scattered from every particle in the field of view as they move rapidly under Brownian motion.

The nanoparticle-tracking-analysis (NTA) was development technology to visualize and analyze—in terms of both size and scattered intensity—nanoscale particles in liquid with little preparation and at low cost. Analysis of a wide range of particle types in a wide variety of solvents. The only user input required are the sample temperature and solvent viscosity.

A finely focused, 635 nm (or for fluorescent studies 404 nm or 523 nm) laser beam is passed through a prism-edged optical flat (Fig. 13.2), the refractive index of which is such that the beam refracts at the interface between the flat and a liquid layer placed above it. Due to the refraction, the beam compresses to a low profile, intense illumination region in which nanoparticles present in the liquid film can be easily visualized via a long-working distance, 20× magnification microscope objective fitted to an otherwise conventional microscope (Fig. 13.3) [15].

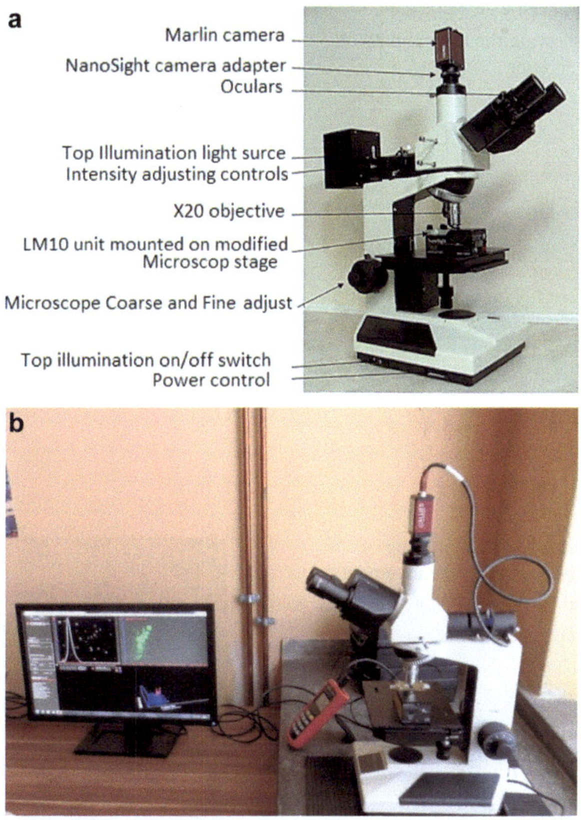

Fig. 13.1 Microscope set (**a**), and LM10 system installed at Physics Faculty, West University of Timisoara, Romania (**b**)

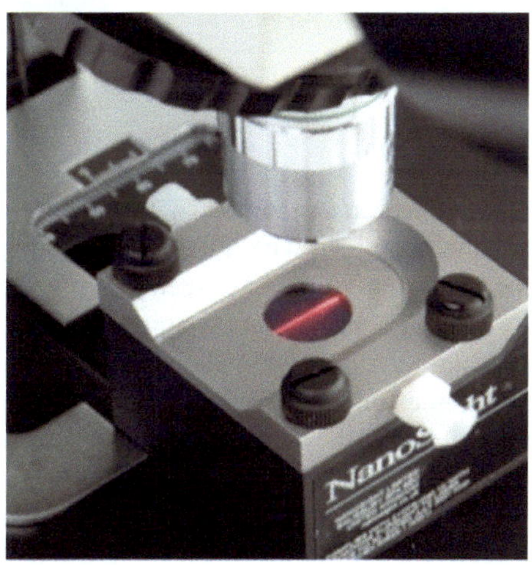

Fig. 13.2 Laser module showing the beam passing through a sample

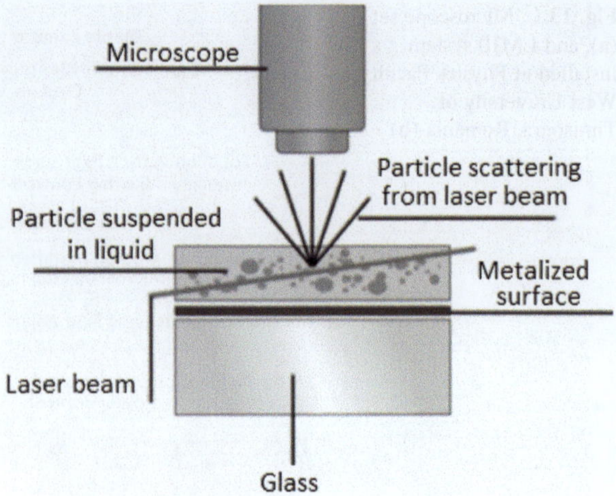

Fig. 13.3 Schematic showing the optical path of the laser

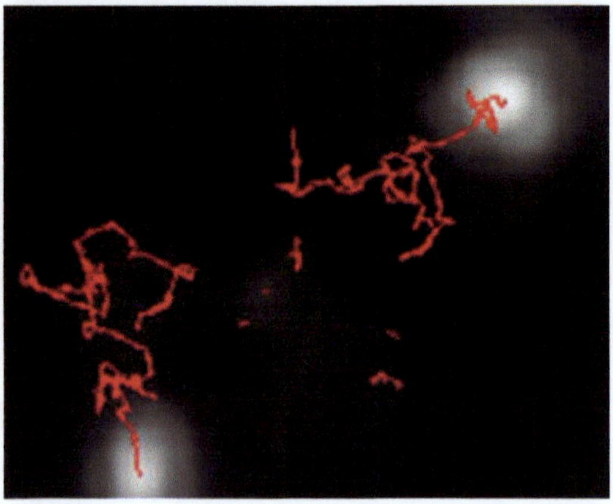

Fig. 13.4 Monitoring of the nanoparticle movements in NTA equipment

Mounted on a mount, a CCD camera, operating at 30 frames per second, is used to capture a video with a field of view approximately 100 μm × 80 μm. Particles in the scattering volume are seen moving rapidly under Brownian motion.

The NTA programmer simultaneously identifies and tracks the center of each particle on a frame-by-frame basis throughout the length of the video (typically 900 frames or 30 s). Figure 13.4 shows an enlarged image of two such particles and the trajectory they have taken over several frames as tracked by the NTA image analysis programmer.

The average distance each particle moves in x and y in the image is automatically calculated. From this value, the particle diffusion coefficient, D, can be obtained and, knowing the sample temperature T, and solvent viscosity η the particle hydrodynamic diameter d identified. That 3 dimensional Brownian movement is tracked only in 2 dimensions (x and y) is accommodated by use of the following variation of the Stokes–Einstein equation (Eq. 13.1) [16], where K_B is Boltzmann's constant, T the temperature, and η the viscosity.

$$\frac{\overline{(x,y)}^2}{4} = D = \frac{TK_B}{3\pi\eta d} \tag{13.1}$$

The range of particle sizes which can be analyzed by NTA depends on the particle type. The lower size limit is defined by the particle size and particle refractive index. For very high particles such as colloidal gold, accurate determination of size can be achieved down to 10 nm diameter. For lower refractive index particles, such as those of biological origin, the smallest detectable size might only be between 25 and 35 nm. This minimum size limit allows, however, the analysis of most types of virus [13].

In NTA method, there are some factors and parameters which can seriously affect the accuracy of final particle size result, for example, sample concentration, camera parameters, and analysis parameters setting. Therefore setting and selecting those parameters are much important for obtaining accurate and reliable results.

13.3.2 Parameters Measurements

Capabilities of NTA measurements consist of direct and live view of particles in suspension, particle-by-particle, high-resolution particle sizing, especially applicable to polydisperse systems, measurement of concentration and particle count, scattering intensity distribution, evidence of nonsphericity and of aspect ratio, sizing down to 10 nm, material dependent, number vs. intensity vs. size is provided for each particle size class, particle-by-particle measurement of electrophoretic mobility, and subsequent calculation of zeta potential.

The direct observation of particle motion and scattering behavior provides information beyond particle size. These real-time observations validate the reported particle size distributions and provide instant insight into polydispersity and the state of aggregation. Measurements take just minutes, allowing being quantified time-based changes and aggregation kinetics.

13.3.2.1 Size vs. Intensity

The benefit of NTA which are able to simultaneously measure two independent parameters such as particle scattering intensity and particle diameter (from dynamic behavior) is valuable in resolving mixtures of particle. The small differences in particle size within a population can be resolved with far higher accuracy than would be achieved by other ensemble light scattering techniques. Accordingly, by gating on size and light scatter it is possible to obtain reproducible counts and

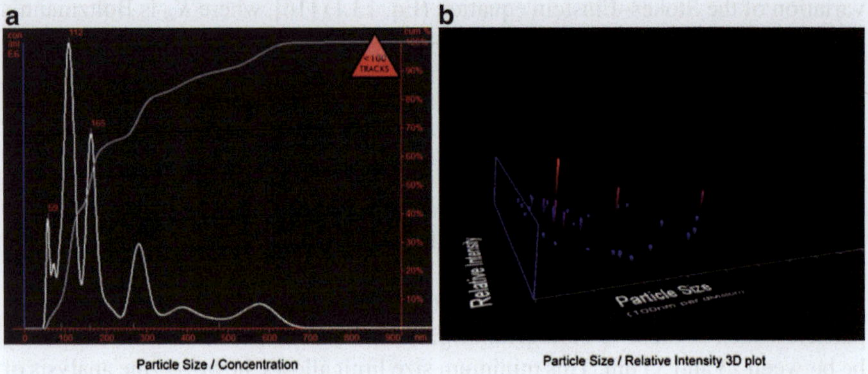

Fig. 13.5 (**a**) Particle size distribution profile of mixture; (**b**) three-dimensional plot of particle size vs. scattering intensity of each particle. The particle concentration is on the vertical axis

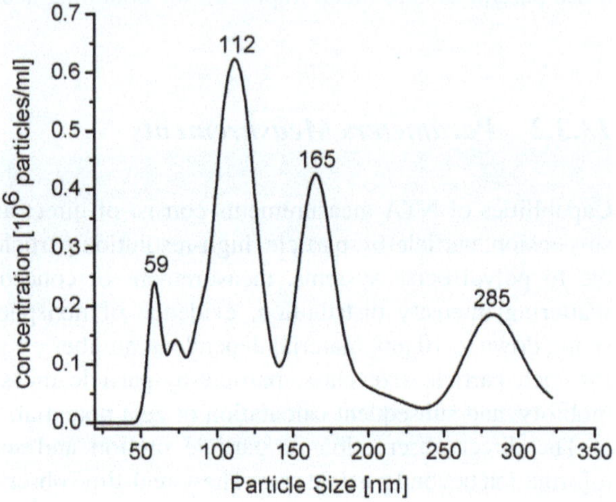

Fig. 13.6 A particle size distribution profile of a mixture of particle, obtained with the visualization system of nanoparticles, Nano Sight LM10

particle size for individual populations (Fig. 13.5). For particles of similar composition, particle scattering intensity increases with particle size.

For mixtures of particles of different composition the ability to plot, for each particle, scattered intensity against particle diameter allows information to be gained about the composition of the mixture.

As example, Fig. 13.6 illustrates the size/concentration distribution of particle for a sample, followed by the analysis report generated by the LM10 software. The distribution diagram indicates that there were four significant groups of nanoparticles in the sample.

13.3.2.2 Size vs. Fluorescence

The ability of NTA to simultaneously measure particle size and particle scattering intensity and with fluorescence allows heterogeneous particle mixtures to be resolved. Particle concentration can be estimated directly, the particle size distribution profile obtained by NTA being a direct number/frequency distribution. The method allows fluorescence measurement for specific particle types to be characterized (sized, counted) in complex mixtures through, for instance, the use of fluorescently labeled antibodies and other probes.

The accuracy of results obtained by the NANOSIGHT NTA Nanoparticle Tracking system is dependent on a number of factors, primarily particle concentration analyzed, the length of time over which the sample is analyzed, and the size of the particles present. While the sample chamber is approximately 250 μl in volume, the section of the laser beam visualized (the scattering volume) is very small (a cylinder approx. 70 μm diameter × 80 μm length). Accordingly, for a statistically significant number of particles to be present on the beam, sample concentrations should have values between 105 and 1010 particles/ml. Higher concentrations (e.g., 1010/ml) with a measurement accuracy of a shorter analysis (see Fig. 13.7). Similarly, observing a sample for 10 s (300 sequential images at 30 fps) will result in better statistical accuracy than an analysis of 1 s. Table 13.1 shows the effects of increasing particle concentration and analysis time for a suspension of ash from the incinerator.

13.3.3 Sample Preparation and Dispersion

The NanoSight device as a particle sizing system based on the scattering of light measures all particles visible in the sample and each separate light scattering center is seen as an individual particle, whether or not it is made up of a number of small particles, i.e., is an aggregate and agglomerate of primary particles. Should the user be interested in the detection and analysis of such aggregates, sample preparation is restricted simply to effecting a suitable dilution for analysis.

Should the number of aggregates will be low (compared to the number of primary particles within the sample population) the unique capability of the NanoSight technique to simultaneously analyze each and every particle on an individual basis means the presence of low numbers of aggregates will not interfere with the analysis of the bulk of primary particles [17].

(a) Suspending solvent. The presence of aggregates (and agglomerates or flocks) is undesirable in so far as they are present at such a high number that they interfere with or dominate the analysis and result. Aggregates can be dispersed in a number of ways using physical processes such as ultrasonication or the application of high shear forces. The standard LM viewing unit is configured to operate optimally with water-based samples. While the device can be used

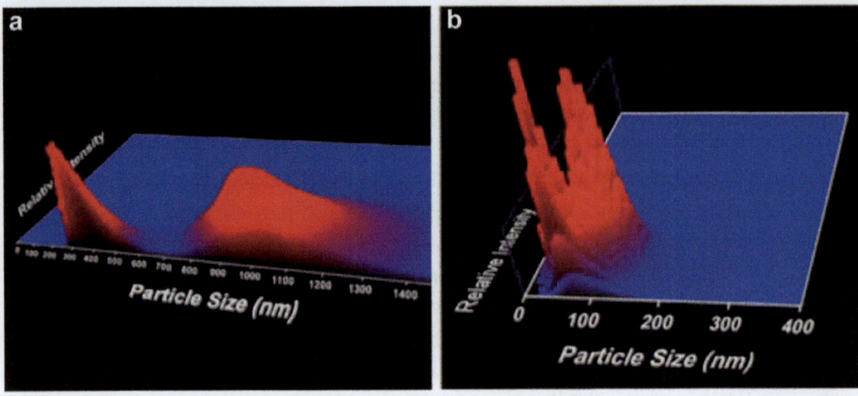

Fig. 13.7 (**a**) Measurement of size and number by light scattering was made using the standard red laser. (**b**) Using a second laser emitting at a shorter wavelength only light emitted from the fluorescent beads was measured

Table 13.1 Particle size estimation as a function of particle concentration and analysis time

Bin Centre (nm)	Concentration (E6 particles/ml)	Percentile Undersize (%)	Bin Centre (nm)	Concentration (E6 particles/ml)	Percentile Undersize (%)
10	0.000	0.000	710	0.005	99.999
30	0.000	0.000	730	0.001	100.000
50	1.334	2.189	750	0.000	100.000
70	2.910	6.963	770	0.000	100.000
90	4.252	13.937	790	0.000	100.000
110	11.261	32.410	810	0.000	100.000
130	5.701	41.761	830	0.000	100.000
150	3.963	48.262	850	0.000	100.000
170	7.276	60.197	870	0.000	100.000
190	2.206	63.816	890	0.000	100.000
210	0.989	65.438	910	0.000	100.000
230	0.592	66.409	930	0.000	100.000
250	1.008	68.062	950	0.000	100.000
270	2.884	72.793	970	0.000	100.000
290	3.366	78.314	990	0.000	100.000
310	1.676	81.062	1000-2000	0.000	100.000
330	0.697	82.206			
350	0.642	83.259			
370	0.822	84.607			
390	0.921	86.118			
410	0.816	87.457			
430	0.633	88.495			
450	0.499	89.314			
470	0.424	90.009			
490	0.408	90.679			
510	0.490	91.483			
530	0.670	92.582			
550	0.888	94.038			
570	1.036	95.738			
590	1.005	97.387			
610	0.781	98.668			
630	0.476	99.449			
650	0.225	99.818			
670	0.082	99.952			
690	0.023	99.990			

Results	
Mean:	219 nm
Mode:	112 nm
SD:	149 nm
D10:	91 nm
D50:	162 nm
D90:	479 nm
User Lines:	0 nm, 0 nm
Concentration:	0.61 E8 particles/ml
Completed Tracks:	34

Measurement Conditions	
Temperature:	25.10 °C
Viscosity:	0.89 cP
Frames Per Second:	30.00
Measurement Time:	0 of 10 s
Drift Velocity:	857 nm/s
Camera Shutter:	8 ms

Analysis Conditions	
Blur:	Auto
Detection Threshold:	10 Multi
Min Track Length:	Auto
Min Expected Size:	Auto - failed

to visualize particles in any solvent type, the suspending medium must be sufficiently nonabsorbing, nonindex matching and gas free.

(b) Particle Size Range Applicable. The NanoSight system allows particles as small as 10–20 nm and particles as large as a micron to be seen to be visualized (depending on particle and solvent type). The large particles (>1 μm) will scatter significant amounts of light and may mask the presence of smaller particles if present at too high a concentration. The very large particle aggregates (>10 μm) may affect image quality or might block the sample inlet/outlet ports and should be removed by filtration or centrifugation before the sample is analyzed.

(c) Particle concentration. For particles to be resolved on an individual basis, it is necessary for the sample to be diluted to a particle number concentration of $<10^{10}$ per ml though this limit will vary with sample type. It is best to adjust sample concentration until a clear image is obtained of a population of at least 100 particles in the scattering volume. The laser beam is focused, on manufacture, to generate a beam waist of approx. 100 μm widths of which a length of beam of between 100 μm is observed.

(d) Dilution. The sample must be diluted to a number concentration of between 10^6 and 10^{10} particles per milliliter (depending on particle type) and not contain particles larger than 10 μm diameter. Such particles will degrade image quality and may sediment in the sample chamber necessitating frequent cleaning of the optical surfaces. Should the sample require dispersion before analysis, gentle agitation can be used, although the introduction of bubbles into the sample will degrade image quality (bubbles are very effective light scattering centers). Small quantities of dispersant, or a suitable detergent, may be employed, but formation of foam should be avoided.

(e) Physical removal of large particles. Physical removal of large particles can be performed simply by filtration or centrifugation of the sample or allowing the sample to settle over a suitable time period.

(f) Dispersion and agglomeration. Choosing the correct dispersing agent is central to the formation of a stable dispersion, and the identification and selection of a suitable dispersant requires a significant understanding of the sample and its interaction with its suspending solvent. The agent should not interfere with the measurement itself, it should be able to contact the primary particle without causing it to dissolve and it should prevent aggregation of the primary particle. In the absence of an established dispersion protocol for any given sample type, empirical methods are the only way by which optimum dispersion protocols can be arrived at.

13.3.4 LM10 Usage

Use NanoSight LM is done by respecting several stages.

(a) *Assembling the LM sample unit*
The LM sample unit can be assembled as follows: switch off unit and remove power lead, clean all necessary parts, cleaned glass element into unit, ensure

the "O"-ring is bedded correctly in its channel in the top plate of the device, lay top plate on glass element and adjust gently until plate locates, drop in four screws and tighten. The LM10 unit is ready to inject a sample.

(b) *Injection into the LM unit*
Load the sample into the sample chamber using a syringe (without needle). The Luer fittings are designed to accept standard syringe bodies of all sizes. Note that high pressures can inadvertently be generated when narrow bore (e.g., 1–2 ml) syringe bodies are used. To avoid generating pressures which might result in the sample bypassing the seals in the sample chamber or damaging the window, care must be taken to introduce the sample slowly.

(c) *Adjusting the microscope*
Once the sample is loaded and checked as above, place the LM unit onto the microscope stage and adjust the position and height of the microscope objective to be used to obtain a clear image of particles present within the beam. The laser beam exhibits some divergence as it passes through the chamber. Accordingly, beam width will vary depending on which part of the beam is located.

(d) *View or capture images*
Once an image can be seen either by viewing via the oculars or by the camera, a movie can be recorded using suitable camera settings.

(e) *Loading a new sample*
To remove a sample, extract using a syringe and rinse the chamber using fresh particle-free water or solvent prior to loading with a different sample. Between samples which differ significantly in terms of solvent, particle type or particle loading, it is advisable to disassemble the top plate and rinse and dry thoroughly the optical element and top plate.

(f) *Cleaning*
Depending on sample type and length of use, it may prove necessary to clean the chamber and optical surface of the LM sample unit.

(g) *Storage*
Between uses of the NanoSight LM unit, or for longer term storage, the unit must be cleaned as described above and stored with no fluids in the chamber.

13.4 Applications

There are many applications where NTA is being used routinely. This is evidenced by the publication of many papers now available in the literature and systems installed in industrial and academic institutions worldwide science. In this section, examples of applications will be reported with references to where further examples may be found. In this section are presented some applications of NTA relevant to nanoparticle characterization (assessment of NTA, nanoparticle toxicology, virology and vaccine production, protein aggregation, nanobubbles, orthopedic implants, micro biota, virology).

13.4.1 Assessment of NTA

Vagner et al. [18] present a method for improving polydisperse particle size distributions based on the logarithmic transform of the estimated diameters to reduce the heteroscedasticity, which is partly due to the constant coefficient of variation of the value of diameter. Calculated cluster proportions were validated and verified by Monte-Carlo simulations of polydisperse suspensions, NTA measurements, and SEM images.

In Nanoparticle Tracking Analysis for the Multiparameter Characterization and Counting of Nanoparticle Suspensions Wright [19] describes application of NTA to the analysis of 100 nm polystyrene nanospheres in liquid suspension. The technique can be adapted for use with nearly all particulate materials with sizes between 10 nm and 1 μm.

13.4.2 Nanotoxicology

Montes-Burgos et al. [20] investigated the use of NTA (with Nanosight software) as a tool for the characterization of particle size and coating properties preceding a nanotoxicology test. Long videos (166 s) were used for measuring diluted samples of standard Au nanoparticles with 60-nm nominal size and mixtures of gold with human plasma in stable conditions (without aggregation). A correction for the broadness of the size distribution of the particles was also presented.

In Nanoparticle reactivity toward dithiothreitol, Sauvain et al. [21] studied the ability of particles to catalyze the transfer of electrons from dithiothreitol to oxygen, in an approach to measure the oxidative potential of particles. The agglomeration-size distribution was determined for carbon-based NPs (carbon black and diesel standards), titanium-dioxide (two different types), and silica NPs using NTA.

The influence of proteins on the dispersibility and aggregation kinetics of PVP-coated silver nanoparticles (AgNPs) and citrate-stabilized silver nanoparticles was studied using a self-constructed device and software that monitored the individual movement of single particles in biological medium [22]. The many individual samples were analyzed for determination of size and fitted into a distribution histogram.

Patrick Hole et al. [23] in his study looks at the reproducibility using technique Nanoparticle Tracking Analysis (NTA) on a range of particle sizes and materials in several different media. In his study are presented analytical and experimental facilities in nanotechnology, medicine and natural sciences with the goal of developing and implementing best practice and quality in all aspects of nanosafety assessment.

13.4.3 Ecotoxicology Applications

Farkas et al. [24] utilized NTA with two CCD cameras to determine the particle-size distribution of Ag nanoparticles and Au nanoparticles dispersed in different growth media for studying the ecotoxicological effects of these particles on rainbow-trout hepatocytes.

Van Hoecke et al. [25] utilized NTA to determine the mode and the mean size of cerium-dioxide nanoparticles in a study to determine for alga Pseudokirchneriella subcapitata ecotoxicity toward green, two crustaceans (*Daphnia magna* and *Thamnocephalus platyurus*), and embryonic development of *Danio rerio*.

13.4.4 Microbiota

Synthesis of silver nanoparticles (Ag NPs) by the leaf extract of *Murraya koenigii* (Indian curry leaf tree) is reported in the study of Bondea et al. [26]. The Ag nanoparticles thus obtained demonstrated remarkable antibacterial activity against three human pathogenic bacteria when used in combination with commercially available antibiotics.

Besinis et al. [27] investigated the toxicity of silver (Ag), titanium dioxide, and silica nanoparticles (NPs) against the oral pathogenic species of *Streptococcus* mutants, compared to the routine disinfectant, chlorhexidine. The bacteria were assessed using the minimum inhibitory concentration assay for growth, fluorescent staining for live/dead cells, and measurements of lactate. Overall, Ag nanoparticles were the best disinfectant and performed better than chlorhexidine.

13.4.5 Virology

In the study [28], Kramberger et al. used NanoSight LM10 with Nanoparticle tracking analysis (NTA) software in order to evaluate the quantification of latex particles, adenovirus 5, and influenza virus. NTA analysis was used to quantify chromatographic fractions of adenovirus and influenza virus after purification on a CIM monolithic column. The results demonstrated that nanoparticle tracking analysis is a method for fast estimation of virus concentration in different samples and it can provide a better insight into the sample status.

Conclusions

NTA proved to be very suitable for analyzing nanoparticles. This technique is suitable for analyzing protein aggregates and the sample preparation does not

(continued)

influence the aggregate distribution in the measurement cell. It also gives the possibility of live monitoring heat-induced aggregation, providing information about the aggregation kinetics. In NTA measurement method, sample concentration and parameters (shutter, gain, duration, and threshold) settings in video capture and video analysis are important factors, which might have a significant effect on the accuracy and reliability of final results.

Acknowledgements This work was supported by a grant of the Romanian National Authority for Scientific Research, CNCS—UEFISCDI, project number PN-II-ID-PCE-2011-3-0762.

References

1. Owen B, Hole P, Vincent P, Siupa A, Carr B. Nanosuspension characterization: application of NTA to research from drug delivery to exosomes research. NanoSight Ltd., Minton Park, Amesbury, SP4 7RT, UK. Ben.Owen@nanosight.com
2. Buzea C, Pacheco Blandino II, Robbie K (2007) Nanomaterials and nanoparticles: sources and toxicity. Biointerphases 2(4):MR17–MR172
3. L'opez-Serrano A, Munoz Olivas R, Sanz Landaluze J, C'amara C (2014) Nanoparticles: a global vision. Characterization, separation, and quantification methods. Potential environmental and health impact. Anal Methods 6:38
4. http://serc.carleton.edu/research_education/geochemsheets/techniques/SEM.html
5. Binnig G, Quate CF, Gerber C (1986) Atomic force microscope. Phys Rev Lett 56:930–933
6. Saveyn H, De Baets B, Hole P, Smith J, Van der Meeren P (2010) Accurate particle size distribution determination by Nanoparticle Tracking Analysis based on 2-D Brownian dynamics simulation. J Colloid Interface Sci 352(2):593–600
7. Moulder JF, Stickle WF, Sobol PE, Bomben KD (1995) X-ray photoelectron spectroscopy. Physical Electronics USA, Inc., Minnesota
8. Tanaka K, Waki H, Ido Y, Akita S, Yoshida Y, Yoshida T (1988) Protein and polymer analyses up to m/z 100,000 by laser ionization time-of-flight mass spectrometry. Rapid Commun Mass Spectrom 2:151–153
9. Karas M, Hillenkamp F (1988) Laser desorption ionization of proteins with molecular masses exceeding 10,000 daltons. Anal Chem 60:299–301
10. Skoog DA (2007) Principles of instrumental analysis, 6th edn. Brooks/Cole, Singapore, pp 169–173
11. http://pharmaxchange.info/press/2011/12/ultraviolet-visible-uv-vis-spectroscopy-principle/
12. John C. Edwards Ph.D., Principles of NMR, Process NMR Associates LLC, 87A Sand Pit Rd, Danbury CT 06810
13. James TL. Fundamentals of NMR. Department of Pharmaceutical Chemistry, University of California San Francisco, CA 94143-0446 USA
14. Malloy A (2011) Count, size and visualize nanoparticles. Materials Today 14:4
15. Brzana A, Carr B, Wright M, Malloy A, Hole P, Smith J. The real time and direct multiparameter analysis of individual nanoparticles in solution. NanoSight Ltd., Minton Park, Amesbury, Salisbury, SP4 7RT, UK, EU
16. Applications of nanoparticle tracking analysis (NTA) in nanoparticle research. NanoSight_Application_Review_NTA_April_2009_M201B.pdf (accessed 26/09/2009), part of NanoSight. http://www.nanosight.co.uk. Accessed 26 Sept 2009

17. Carr B, Malloy A. NanoParticle tracking analysis—the NANOSIGHT system, 2007 NanoSight Ltd. 2 Centre One, Lysander Way, Old Sarum Business Park Salisbury SP4 6BU UK. www.nanosight.co.uk
18. Wagner T, Wiemann M, Schmitz I, Lipinski H-G (2013) A cluster-based method for improving analysis of polydisperse particle size distributions obtained by nanoparticle tracking. J Nanopart Article ID 936150, 9 pages
19. Wright M (2012) Nanoparticle tracking analysis for the multiparameter characterization and counting of nanoparticle suspensions. Nanoparticles in biology and medicine. Methods in molecular biology. Clifton, NJ, series no. 906, vol 6, pp 511–524
20. Montes-Burgos I, Walczyk D, Hole P, Smith J, Lynch I Dawson K (2010) Characterisation of nanoparticle size and state prior to nanotoxicological studies. J Nanopart Res 12(1), January
21. Sauvain JJ, Deslarzes S, Riediker M (2008) Nanoparticle reactivity toward dithiothreitol. Nanotoxicology 2:121–129
22. Kittler S, Greulich C, Gebauer JS, Diendorf J, Treuel L, Ruiz L, Gonzalez-Calbet JM, Vallet-Regi M, Zellner R, Koller M, Epple M (2010) The influence of proteins on the dispersability and cell-biological activity of silver nanoparticles. J Mater Chem 20:512
23. Hole P, Sillence K, Hannell C, Manus Maguire C, Roesslein M, Suarez G, Capracotta S, Magdolenova Z, Horev-Azaria L, Dybowska A, Cooke L, Haase A, Contal S, Manø S, Vennemann A, Sauvain J-J, Crosbie Staunton K, Anguissola S, Luch A, Dusinska M, Korenstein R, Gutleb AC, Wiemann M, Prina-Mello A, Riediker M, Wick P (2013) Interlaboratory comparison of size measurements on nanoparticles using nanoparticle tracking analysis (NTA). J Nanopart Res 15:2101
24. Farkas J, Christian P, Urrea JAG, Roos N, Hassellov M, Tollefsen KE, Thomas KV (2010) Effects of silver and gold nanoparticles on rainbow trout (Oncorhynchus mykiss). Aquat Toxicol 96:44
25. Van Hoecke K, Quik JTK, Mankiewicz-Boczek J, De Schamphelaere KAC, Elsaesser A, Van der Meeren P, Barnes C, Howard CV, Van De Meent D, Rydzynski K, Dawson KA, Salvati A, Lesniak A, Lynch I, Silversmit G, De Samber B, Vincze L, Janssen CR (2009) Fate and effects of CeO2 nanoparticles in aquatic ecotoxicity tests. Environ Sci Technol 43:4537
26. Bondea SR, Rathoda DP, Inglea AP, Adea RB, Gadea AK, Rai MK (2012) Murraya koenigii-mediated synthesis of silver nanoparticles and its activity against three human pathogenic bacteria. Nanosci Methods 1:25–36
27. Besinis A, De Peralta T, Handy RD (2014) The antibacterial effects of Ag, TiO2 and SiO2 nanoparticles compared to the dental disinfectant chlorhexidine on Streptococcus mutants using a suite of bioassays. Nanotoxicology 8(1):1–16
28. Kramberger P, Ciringer M, Strancar A, Peterka M (2012) Evaluation of nanoparticle tracking analysis for total virus particle determination. Virol J 9:265

Part IV
Methods for Sorting, Separating and Manipulating Nanoparticles

Chapter 14
Dielectrophoresis Used for Nanoparticle Manipulation in Microfluidic Devices

Mihai Lungu, Madalin Bunoiu, and Adrian Neculae

Abstract Nano-sized particles have received considerable interest in the past two decades. The filtration of nanoparticles is becoming an important issue as they are produced in large numbers from material synthesis or combustion emission, and their effect on human health is relatively high. Dielectrophoresis (DEP), phenomenon that induces spatial movement of particles placed in nonuniform electric field, depending on the dielectric properties of the particles and the surrounding medium, the geometry of the electrodes, and the amplitude and frequency of the applied signal, proved to be the most adequate tool in order to manipulate particles at submicron scale. First, this work presents an overview of the various applications of the dielectrophoresis. Next, the theoretical description of the main forces implied in the spatial control of submicron particles is given. Finally, a mathematical model describing the filtration of nanoparticles suspended in flue gas by a combination of dielectrophoretic and electrohydrodynamic forces, and a set of numerical results obtained by simulations performed in the frame of this model are presented. The dielectrophoretic force and the nanoparticles concentration profile in a DEP-based separation micro system consisting of a micro channel are numerically investigated using the COMSOL Multiphysics finite element code. The performances of the filtration device are analyzed in terms of a specific quantity related to the separation process, called *Filtration rate*. The simulations provide the optimal set of values for the control parameters of the separation process in order to obtain a desired performance, and represent a useful tool in designing of microfluidic devices for separating nanoparticles from flue gas.

M. Lungu (✉) • M. Bunoiu • A. Neculae
Department of Physics, West University of Timisoara, Blv. V. Parvan No. 4, 300223 Timisoara, Romania
e-mail: lmihaister@gmail.com

M. Lungu et al. (eds.), *Nanoparticles' Promises and Risks*,
DOI 10.1007/978-3-319-11728-7_14

14.1 Introduction and Overview

Nanoparticles have received considerable interest in the past two decades. Their toxicity for human health is relatively high because they can readily enter the human body through inhalation and have a large specific surface area. The presence in the environment of nanoparticle with size ranging from 50 to 200 nm has a profound impact on human health. Once inhaled, due to their tendency to remain trapped in the inner respiratory ways, the nanoparticles infiltrate into the blood and cannot be eliminated, because the macrophage cells cannot identify them. Inhaled nanoparticles may generate free radicals, affect the DNA, and alter the genes, being responsible for mutagens and carcinogenic effects or causing a variety of lung-disease typologies [1, 2].

Filtration of submicron particles suspended in flue gas is an important technological challenge, as they are produced in large numbers from material synthesis and combustion emission; in urban environment, the burning processes including incinerators of waste or diesel emissions are responsible for the emission of a significant amount of nanoparticles [2, 3]. Along with carbon, dioxide and water, the gaseous products resulted from combustion processes contain at least traces of finely divided carbon (VOC), carbon monoxide, oxides of nitrogen, phosphorus and sulfur, halogen acids, metal oxides and heavy metal vapors. By the de novo synthesis, VOCs and carbon monoxide in the presence of halogenated acids pass into the halogen derivatives of dioxins and furans, which are included as such in the form of nanoparticles or adsorbed on finely divided carbon evacuated through the flue gas stream, along with the other mentioned components. All chemical compounds contained in flue gas have adverse effect on the atmosphere, biodiversity, and human body. Spreading both as nanoparticles or adsorbed on finely divided carbon, they maintain long as very fine suspension in the atmosphere. These are the most direct ways for pulmonary and blood assimilation, and cells establishment of these emissions [1, 4, 5].

From a public health standpoint, the size of a particle is as important as its composition, recent research showing that although raw materials may not be dangerous, they can become toxic under the form of nanoparticles [2, 3]. Although the nanoparticles have smaller masses than microparticles, their number is at least four orders of magnitude higher than the number of all other particles found in the flue gas. The pollutant emitting sources are generally equipped with various filters for particles, but they work only on micrometric particles, while the nanometric particles almost entirely escape in the environment [1, 3].

In many scientific and technical areas, a considerable interest is shown to the separation of nanoparticles in accordance with their physical or chemical characteristics, and several methods of nanoparticle manipulation were explored. Due to the quantum confinement effect, the nanoparticles show distinct physical and chemical properties depending on the sizes and shapes, particularly when the sizes are close to or smaller than the dimensions of exciton of the corresponding bulk materials. However, their wide distributions of sizes always give rise to the major limitations for precise investigation of their unique physical and chemical characteristics [6, 7].

The traditional methods for nanoparticle manipulation (retention and separation) have not been successful. In a typical particle-capture device, only a small fraction is collected and the method is successful only when the nanoparticles are attached to larger ones. Mechanical methods or other devices based on controlling the particles towards a desired location are becoming of less value, while chemical methods of sorting particles are usually slow and may contaminate the particles under manipulation. Mechanical devices used for filtering the postcombustion gases resulted in waste incinerators (cyclones, bag filters, sedimentation chambers) are less effective at this scale because of the nanoparticles' low weight while Corona electrostatic filters have high micrometric particle retention efficiencies (93–99 %), yet most nanoparticles still remain undetected [1, 2]. Optical methods sometimes used in trapping nanoparticles have the significant disadvantage that they produce heating of the fluid in which the particles are suspended. They also require the use of specific optical equipment, which is often large and difficult to integrate on a micro- or nano-analysis device. Methods such as transmission electron microscopy (TEM) and size-exclusion chromatography were introduced so far for identifying and separating nanoparticles. These methods, however, have some inherent problems in the detection processes involving the degradation of sample, irreversible adsorption, etc. [3, 4].

The methods utilizing electric fields are emerging as most promising techniques for nanoparticle manipulation that involves within microfluidic systems many processes including patterning, focusing, sorting, trapping, handling, and separation, where electrical forces can act both on particles and on the suspending fluid. The most promising technique for nanoparticle trapping—defined as electric field-induced particle immobilization at certain regions in a device—and controlled spatial separation is a method based on dielectrophoresis (DEP), phenomenon that emerges upon application of an electric field (DC, or AC) to a suspension of particles in a fluid medium [6, 8, 9]. The application of an electric field across a suspension of colloidal particles leads to their polarization. The DEP force arises when the particles′ induced dipoles interact with a nonuniform electric field leading to particle movement. Hence, the particle undergoes a DEP motion under the resulting translational force, when the DEP force overcomes other competing forces such as the buoyancy, thermal force, hydrodynamic force, and Brownian motion. This movement was termed "dielectrophoresis" by Pohl [10, 11], a combination of the word for force, "phoresis" in Greek, and the word "dielectric." This force does not require the particle to be charged.

The strength of the DEP force depends strongly on the medium and particles' electrical properties, on the particles' shape and size, as well as on the frequency of the electric field. Consequently, fields of a particular frequency can manipulate particles with great selectivity. Since the relative dielectric polarization of the nanoparticles depends on the driving frequency of the applied electric field, an alternating (AC) electric field is usually applied to generate DEP forces of different magnitudes and directions. The force depends on the magnitude of the field and the gradient, together with the particle volume. The direction of the force depends on the Clausius Mossotti factor, which is a measure of the polarizability of the particle in the suspending medium and importantly varies with the frequency of the applied potential (as will be shown in the theoretical part). The particles are either attracted (positive DEP) to the region of maximum field intensity or repelled (negative DEP) from it, depending on the effective particle polarizability relative to the media. The particle "chaining" force is described as a result of the attraction of the induced dipoles within the particles. The chaining force acting between particles of the same type is always positive and attractive. AC electric fields can be used to manipulate many types of colloidal particles in different media by simply adjusting AC electric field parameters (magnitude, frequency, wave shape, wave symmetry, and phase) [6–8].

Due to its ability to manipulate particles based solely on their dielectric properties and size, DEP is used for a wide variety of applications. DEP methods can be used in many forms (electrorotation, traveling wave DEP, negative and positive DEP) to manipulate and more generally, control the position, orientation, and velocity of micro- and nanometer scale particles, including carbon nanotubes and biological particles such as viruses, DNA, bacteria, and cells of various kinds [7, 12].

DEP-based separation methods usually employ fluid flow as a force competing with DEP in order to achieve the separation of two or more populations of particles. By using this method it is possible to sort particles according to size (because the DEP force depends on the volume of the particle, while drag depends on the radius of the particle) and dielectric properties. The DEP force depends (amongst other things) on the particle volume, therefore high electric field gradients are required to move nanometric particles. Because of electrohydrodynamic (EHD) effects, the high intensity electric fields give rise to fluid displacement and may cause particle movement through the viscous drag force. On the other hand, employing of nonuniform, AC electric fields avoids the problem of electrolysis that is caused when using DC electrophoresis for similar tasks.

The fundamental aspects of nanoparticle dynamics in fluid flows are also of great interest for numerous applications in medicine, biotechnology, and pharmaceutical research, therefore the manipulation of colloidal particles and fluids in microsystems by using electrical forces has many existing and potential applications. The ability to manipulate particles is of particular importance for micro- and nanofluidics, where the particles can be physical or biological objects manipulated with different methods (electrokinetic or magnetic forces, ultrasound, or optical tweezers) [7, 11, 13].

Particle dynamics in fluid flows has been the focus of attention for a long time, in various contexts and at different scales, therefore the manipulation of nanoparticles in microsystems by using DEP forces has many existing and potential applications, presenting the advantages of voltage-based control and dominance over other forces. For most applications involving micrometer and submicrometer particles, the forces that tend to dominate in microdevices are viscous forces and electrical forces; in the range above a few millimeters, the electrical forces are rather ineffective, but in the micrometer (and submicrometer) scale, the electrical forces dominate [9].

As a recently introduced technique, microfluidics involves the manipulation of fluids and particles using dielectrophoresis within a microscale chip. Physical parameters that characterize microfluidics include Reynolds number, diffusion, fluidic resistance, surface area to volume ratio, and surface tension [7]. Within microfluidic systems, the manipulation of particles is of fundamental importance and the characterization of particles within microfluidic systems provides a quantitative and analytical approach to interrogate the physicochemical properties of particles, such as size, permittivity, and conductivity. In terms of nanoparticle handling, a number of different methods can be combined to produce a complex technological platform [6–8].

Nanoparticle manipulation and separation have been achieved using a wide range of different electrode configurations, the three classical electrode configurations that have been used for DEP manipulation being castellated electrodes, polynomial electrodes, and interdigitated or parallel finger paired electrode array, schematically presented in Fig. 14.1 [6, 9, 11]:

For example, nanometer sized particles can be both stably trapped in nanofabricated electrode arrays and, more importantly, a heterogeneous mixture can be separated into two populations using dielectrophoresis. By varying electrode shape and applied voltage, DEP forces have been used for particle trapping, arraying, levitation, translation, fractionation, filtration, orientation, and characterization based on dielectric properties. A parallel, planar, interdigitated electrode array is depicted in Fig. 14.2 [8, 14, 15]. When a potential is applied across alternating electrodes, this array establishes a nonuniform electric field dependent on dimensional and material properties. Particles suspended in a fluid experience a spatially varying DEP force that governs particle kinetics, being possible to separate heterogeneous mixtures comprising a wide range of nanoparticles into constituent particle types.

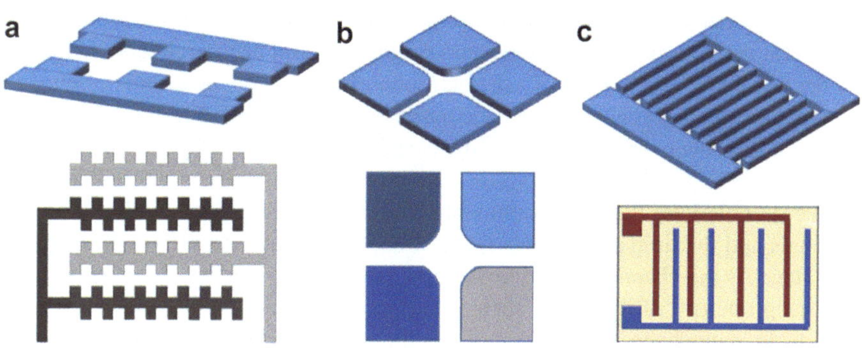

Fig. 14.1 Schematic of the DEP patterning chamber: classical interdigitated castellated microelectrode pattern (**a**), polynomial electrodes (**b**), and parallel finger paired electrodes (**c**)

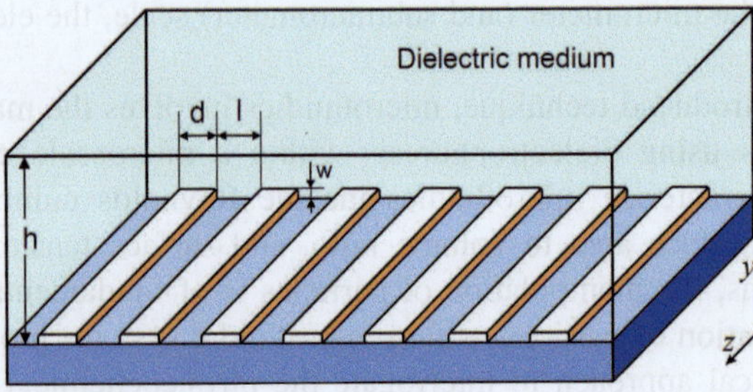

Fig. 14.2 Schematic of the DEP patterning chamber with interdigitated bar electrode array at the bottom surface (nonconducting substrate). The geometric parameters of the device include chamber height (h), electrode spacing (l), electrode width (d), and electrode height (w). The electrodes are connected in different phase sequences to apply the DEP effect

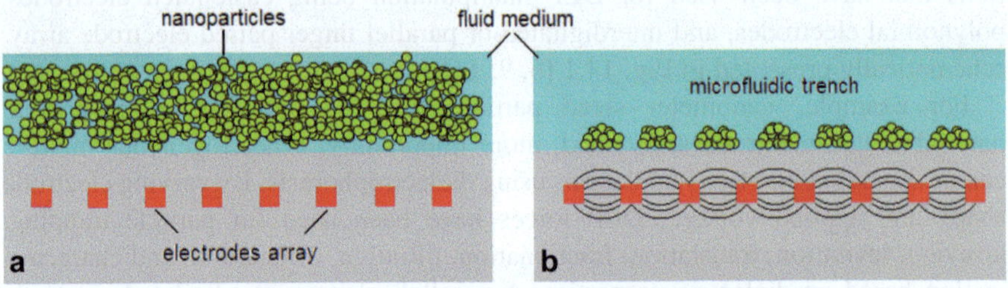

Fig. 14.3 Random distribution of particles contained within the fluidic trench of a lab-on-a-chip system (**a**), Particles arranged by dielectrophoresis into a periodic grating pattern (**b**)

An increasingly important application of DEP is the selective separation of nanoparticles in lab-on-a-chip systems, where all of the components are built in a single device [16]. One of the common strategies for dielectrophoretic separation in lab-on-chip systems is flow separation: flow is used to carry two kinds of different particles in a microchannel. One kind of particles is trapped on the electrode arrays located at the bottom of the channel, and the other would be brought out of channel by the flow [17]. The scheme of a dielectrophoresis-based lab on a chip designed to manipulate and detect nanometer scale particles is presented in Fig. 14.3, where the central components are the microfluidic trench and the dielectrophoresis electrode array.

We do note here that one prominent difference between fluid motion in micro channels and that in nanochannels is the strong fluid-wall interactions observed in the latter. As the channel size decreases, the surface-to-volume ratio increases and various properties of the walls, such as surface roughness, greatly affect the fluid motion in nanochannels. In the case of the nanoparticle flow above a surface covered with electrodes, some parts of the surface are more accessible to diffusing particles than others [18]. Hence, the current density across the surface may strongly vary from one place to another, and the boundaries or interfaces can be microscopically wavy-rough. If one considers uncharged electrodes (their volume charge density is zero), or the amplitude of surface roughness in micro channel is with some orders of magnitude lower than the height of micro channel, the roughness effect inside the micro channel geometry can be neglected.

Another specific effect that can occur at the microscale level between the particles and the electrodes is the Casimir–Lifshitz effect [8, 19]. Specifically, this effect states that one can observe an attraction between bulk material bodies, due to the modification of the zero-point energy associated with the electromagnetic modes in the space between them. The Casimir–Lifshitz can be strongly influenced by the surface roughness of the particles therefore when neglect the roughness effect, the Casimir effect can be neglected too.

Due to the advances in microfabrication techniques that allowed progressively smaller microstructures to be constructed, DEP can be used for the manipulation of nanobioparticles, such as macromolecules, viruses, and spores. DEP of viral particles is becoming an important technique for the separation, concentration, and identification of viruses; in addition, it allows for fast detection and concentration of microorganisms in a single step. Microdevices based on DEP could be employed as online detectors of the viral particles for clinical, analytical, and environmental applications [20].

The manipulation, concentration, and separation of DNA molecules are important in different fields such as genetics, microbiology, medicine, and biochemistry; therefore, methods that allow to concentrate, trap, and separate DNA have important analytical applications [21–23]. Traditional methods for sample concentration as filtration, liquid–liquid extraction, centrifugation, and adsorption are useful in some situations, but processing times are long and they are not suitable for single molecule manipulation. Because DNA is a uniformly charged molecule, with the mass/charge ratio constant at all lengths, means that in a DEP separation, all the molecules and fragments of DNA would move at the same speed, resulting in no size fractionation, DEP being an ideal technique for trapping, concentration, and stretching of DNA molecules achieved in microdevices [20].

On the other hand, DEP seems highly suited for protein separation methods as it has the potential to provide a concentration tool and to improve current separation approaches especially in combination with other techniques [24]. The analysis of proteins often requires powerful separation, fractionation, and preconcentration techniques, in many cases achieved only through the combination of different techniques. Sensitive protein detection as well as purification is also an important aspect in diagnosis as well as techniques to identify and manipulate proteins with label free methods. Although the studies with DEP manipulation of proteins are not as numerous as the studies focused on DNA, many research groups have obtained important results in this field, since DEP is a nondestructive technique without modification of proteins' biological functionality [20–23]. In this sense, protein dielectrophoresis has the potential to play an important role in manipulation, fractionation, preconcentration, and separation method in bioanalysis or as a manipulation tool for nanotechnological applications. As a novel technique used to the protein manipulation, dielectrophoresis has therefore a great potential for protein concentration, and it could be employed as a separation and concentration technique for the pharmaceutical industry and to improve protein analysis techniques. Potential applications of DEP for protein manipulation in the pharmaceutical, analytical, and genetic fields are a large potential, more applications employing protein DEP being demonstrated in the last 20 years [24].

Furthermore, it has also been demonstrated that it is possible to use DEP to separate human breast cancer cells from blood, again because the dielectric properties of these differ significantly [25]. The advancement of these methods will enable levels of control of biological systems that have not been achieved up to now by any other method [23, 26].

Recently, several attempts to use dielectrophoresis as a novel separation method for nanoparticles with different chemical groups on their various surfaces containing inorganic and polymer materials (Janus particles) were done [27]. "Janus" particles (JP), named after the Roman mythology god (who possessed two faces), are anisotropic dissymmetric particles, with two different physical/chemical properties, including biphasic material, biocompartmental, dumbbell-like, snowman-like, acorn-like, or half-raspberry-like particles (Fig. 14.4), and offer new possibilities in microfluidic research fields [28, 29]. They show particular properties and their use is of great interest in biology or medicine diagnosis, an active research field being the use of colloidal particles to sense and manipulate cells that present specific proteins on their surface [30].

Janus particles proved to be capable to sense protein–cell interactions or to organize cell on a substrate antigen–antibody interactions by adsorbing antigenes on their polystyrene side, a fine control of their handling being demonstrated by using DEP. The DEP manipulation of Janus particle coupled with a selective grafting of proteins on one of their side could lead to a powerful tool regarding protein–protein or protein–cell interactions studies. The grafting of proteins on a single side of a JP is use to preactivate this side only with a molecule that will bind the protein [24]. On the other hand, Janus particles demonstrate other multitude of interesting effects in external electric fields, such as the assembly of targeted

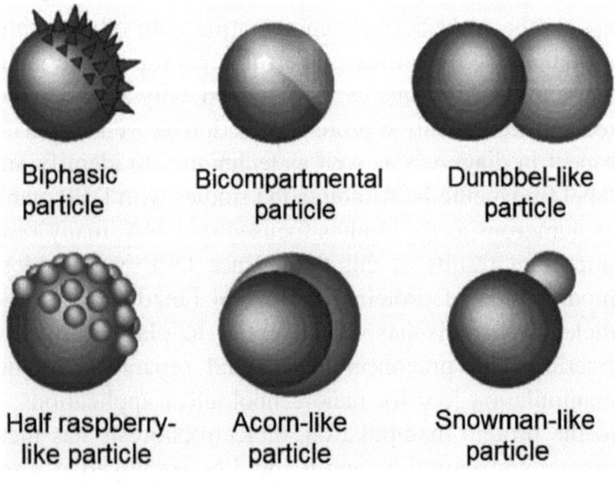

Fig. 14.4 Schematic representation of Janus and comparable particles

structures, where few experimental studies report how such metallodielectric JP assemble respond to external AC fields under DEP [31]. The metallodielectric particles assemble into new types of chain structures, where the metallized halves of neighboring particles align into lanes along the direction of the electric field, while the dielectric halves face in alternating direction. These particles could form clusters rather than strings, the staggered chains may also assemble in various orientations to form different types of two-dimensional metallodielectric crystals. The JP could display electrohydrodynamic mobility in the direction perpendicular to the electric field, form crystals of unique symmetries by confining staggered chains of Janus particles in a small area, or form more complex three-dimensional (3D) bundle structures. These structures disassembled once the electric field turns off, proving the reversibility of this process. The assembly of Janus metallodielectric particles may find applications in liquid-borne microcircuits and materials with directional electric and heat transfer The metal-coated hemispheres of the particles have a main role in the formation of different structures and in the electrohydrodynamic mobility of the particles once the electric field intensity within the experimental cell becomes strong enough to overcome Brownian motion. The knowledge of Janus particles can be applied to other types of anisotropic particles, which may form different types of novel structures and potentially lead to the fabrication of new materials [27].

Another significant application of DEP is for sorting carbon nanotubes [6, 32], which is an essential step in the fabrication of carbon-nanotube based sensors, because different nanotube types have different dielectric properties and dimensions. DEP is also used for assembling carbon nanotube networks on electrodes. This will enable the construction of novel sensors based on carbon nanotube response properties. Sensors for temperature and fluid flow have also been manufactured using DEP to deposit carbon nanotubes to electrodes [33, 34].

The aim of the present study is to analyze dielectrophoresis, both theoretically and numerically, in order to reveal its potential for novel applications in the field of the manipulation of nanoparticles, focusing on the enhancing of the performance of filtering devices and the reduction of nanoparticles emissions in the air through the optimization of the residual gas filtering conditions. After reviewing the basic equations used for modeling the dielectrophoretic process, this paper presents some aspects regarding the numerical investigation of the behavior of a suspension of submicron particles under the action of DEP force in a system consisting of a micro-channel controlled with an interdigitated electrode array. In particular, we want to identify a suitable way to predict accurately the particle entrapment by using numerical solutions of the DEP-flow equations for the fluid and find the optimal values of the control parameters for separation process of nanoparticles from flue gas, in order to be useful in designing of appropriate microfluidic devices.

14.2 Theoretical Considerations

14.2.1 *Dielectrophoresis*

The major electrical forces acting on small particles suspended in a fluid are due to electrophoresis and dielectrophoresis phenomena. Electrophoresis occurs due to the action of the electric field on the fixed, net charge of the particle, while dielectrophoresis only occurs when there are induced charges, and only results in motion in a nonuniform field (this can be a DC or an AC field) (Fig. 14.5).

Pohl [10] used this term to describe the force exerted on uncharged dielectric particles by their polarizability, having the following main characteristics:

- Particles experience a DEP force only when the electric field is nonuniform.
- The DEP force does not depend on the polarity of the electric field and is observed with AC as well as DC excitation.
- Particles are attracted to regions of stronger electric field when their permittivity ε_p exceeds that of suspension medium ε_m, i.e., when $\varepsilon_p > \varepsilon_m$
- Particles are repelled from regions of stronger electric field when $\varepsilon_p < \varepsilon_m$

The utilization of the difference between dielectrophoretic forces exerted on different particles in nonuniform electric fields is known as DEP separation. The exploitation of DEP forces has been classified into two groups: DEP migration and DEP retention. DEP migration uses DEP forces that exert opposite signs of force on different particle types to attract some of the particles and repel others [35]. DEP retention uses the balance between DEP and fluid-flow forces. Particles experiencing repulsive and weak attractive DEP forces are eluted by fluid flow, whereas

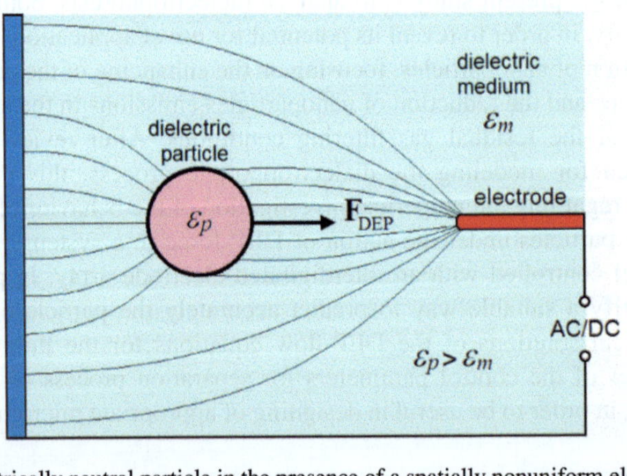

Fig. 14.5 Electrically neutral particle in the presence of a spatially nonuniform electric field. The dipole moment induced within the particle results in a translational force and the dielectric spherical particle undergoes a DEP motion. For AC bias voltage, the potential will be interchanged periodically on the electrodes and the resulting force on the particle is still acting on the same direction due to uneven electric field distribution

particles experiencing strong attractive DEP forces are trapped at electrode edges against flow drag [6].

To determine the electric field and then the dielectrophoretic forces, the electric potential is solved for a defined space and set of boundary conditions that represent the electrode array.

Using the phasor notation, an arbitrary electric potential oscillating at frequency ω can be defined as [8, 36]:

$$V(\mathbf{x}, t) = \mathrm{Re}\left\{\widetilde{V}(\mathbf{x})e^{j\omega t}\right\}, \tag{14.1}$$

where $j = (-1)^{1/2}$, $\mathbf{x}$ is the spatial coordinate, t is the time, Re{ } indicates the real part of the complex quantity, and $\widetilde{V} = V_\mathrm{R} + jV_\mathrm{I}$, with V_R and V_I the real and imaginary part of the electric potential, respectively. The corresponding electric field is $\mathbf{E}(\mathbf{x}, t) = \mathrm{Re}\left\{\widetilde{\mathbf{E}}(\mathbf{x})e^{j\omega t}\right\}$, where $\widetilde{\mathbf{E}} = -\nabla\widetilde{V} = -\nabla(V_\mathrm{R} + jV_\mathrm{I})$. For an uncharged homogeneous medium, the electrical potentials satisfy the Laplace equation:

$$\nabla^2 V_\mathrm{R} = 0 \quad \text{and} \quad \nabla^2 V_\mathrm{I} = 0. \tag{14.2}$$

In the dipole approximation, the dielectrophoretic force acting on a dielectric particle in a nonuniform electric field can be written as [36]:

$$\mathbf{F} = (\widetilde{\mathbf{p}} \cdot \nabla)\widetilde{\mathbf{E}}, \tag{14.3}$$

and the time-averaged force on the particle is:

$$\langle \mathbf{F} \rangle = \frac{1}{2}\mathrm{Re}\left\{(\widetilde{\mathbf{p}} \cdot \nabla)\widetilde{\mathbf{E}}^*\right\}, \tag{14.4}$$

where $\widetilde{\mathbf{p}}(\omega)$ is the induced dipole moment of the particle and "*" indicates the complex conjugate. For a homogeneous dielectric sphere of radius a, the induced dipole moment is given by [8]:

$$\widetilde{\mathbf{p}}(\omega) = 4\pi a^3 \varepsilon_\mathrm{m} \widetilde{K}(\omega)\widetilde{\mathbf{E}}, \tag{14.5}$$

where $\widetilde{K}(\omega)$ is the complex Clausius–Mossotti (CM) factor, which, for a spherical particle, can be expressed as [37]:

$$\widetilde{K}(\omega) = \frac{\widetilde{\varepsilon}_\mathrm{p} - \widetilde{\varepsilon}_\mathrm{m}}{\widetilde{\varepsilon}_\mathrm{p} + 2\widetilde{\varepsilon}_\mathrm{m}}, \tag{14.6}$$

with $\widetilde{\varepsilon}_\mathrm{p}$ and $\widetilde{\varepsilon}_\mathrm{m}$ the absolute complex permittivity of the particle and the medium, respectively.

After some mathematical manipulation, it can be shown that the force in Eq. (14.4) comprises two independent contributions (the subscripts R and I denote the real and imaginary parts of the corresponding complex quantities):

$$\langle \mathbf{F} \rangle = 2\pi\varepsilon_{\mathrm{m}} a^3 K_{\mathrm{R}} \nabla |\widetilde{\mathbf{E}}|^2 - 4\pi\varepsilon_{\mathrm{m}} a^3 K_{\mathrm{I}} \nabla \times (\mathbf{E}_{\mathrm{R}} \times \mathbf{E}_{\mathrm{I}}). \tag{14.7}$$

The first term relates to an electric field which is nonuniform in magnitude but does not exhibit any phase variation. Equation (14.7) suggests that dielectrophoresis can be used as an effective method for separating particles, solely according to their dielectric properties and sizes, an important step forward in nanoparticle manipulation. In a dielectric medium, the direction of the DEP force is influenced by the polarizability of the particle, which depends on the permittivities of the particle and the suspending medium. Replacing the complex permittivity, $\widetilde{\varepsilon} = \varepsilon - j\sigma/\omega$, where ε represents the electric permittivity and σ the conductivity of the dielectric, in Eq. (14.6), the CM factor can be expressed as [15, 38]:

$$\widetilde{K}(\omega) = \frac{(\varepsilon_{\mathrm{p}} - \varepsilon_{\mathrm{m}}) + j/\omega \cdot (\sigma_{\mathrm{p}} - \sigma_{\mathrm{m}})}{(\varepsilon_{\mathrm{p}} + 2\varepsilon_{\mathrm{m}}) + j/\omega \cdot (\sigma_{\mathrm{p}} + 2\sigma_{\mathrm{m}})}. \tag{14.8}$$

The variation in this factor is unique for a particular type of particle. According to Eq. (14.8) the CM factor depends on the dielectric properties of the particle and medium, and on the frequency of the applied field; at low frequencies the value and the sign are determined by the electrical conductivities of the particle and the medium and at higher frequencies by the permittivities. The high and low frequency limits for the real part of the CM factor, K_{R}, are:

$$\lim_{\omega\to 0} K_{\mathrm{R}} = \frac{\sigma_{\mathrm{p}} - \sigma_{\mathrm{m}}}{\sigma_{\mathrm{p}} + 2\sigma_{\mathrm{m}}} \tag{14.9}$$

$$\lim_{\omega\to \infty} K_{\mathrm{R}} = \frac{\varepsilon_{\mathrm{p}} - \varepsilon_{\mathrm{m}}}{\varepsilon_{\mathrm{p}} + 2\varepsilon_{\mathrm{m}}} \tag{14.10}$$

Equations (14.9) and (14.10) show that the relative differences in ohmic losses dominate the low frequency behavior of, while dielectric polarization effects are more significant at high frequencies. These equations also show that K_{R} is bounded $(-0.5 < K_{\mathrm{R}} < 1)$ regardless of frequency.

This behavior of K_{R} is crucial for the dielectrophoretic control of particles because this factor determines the direction of the dielectrophoretic force: when the sign of K_{R} is positive, the particle is more polarizable than its surrounding medium and its movement is oriented towards regions with the highest field strength. When K_{R} is negative, particles with polarizability less than that of the medium move towards the region with the lowest field gradient. These phenomena are known as *positive dielectrophoresis* (pDEP) and *negative dielectrophoresis* (nDEP), respectively. As $K_{\mathrm{R}}(\omega)$ has a complex dependence on the properties of the particle (permittivity, conductivity) and the frequency of the applied field, a

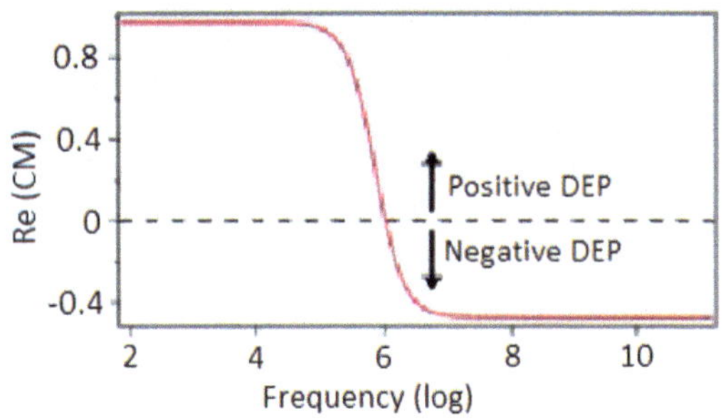

Fig. 14.6 Variation with frequency of the real part of the CM factor for $\sigma_p > \sigma_m$ and $\varepsilon_p < \varepsilon_m$

particle can experience both positive and negative DEP forces based on the various possible combinations of the above variables.

Figure 14.6 shows an example of dielectrophoretic spectrum with $\sigma_p > \sigma_m$ and $\varepsilon_p < \varepsilon_m$.

The second term of Eq. (14.7) is nonvanishing if the electric field has a spatially dependent phase, and the dielectrophoresis resulting from such an electric field phase gradient is known as "*traveling wave dielectrophoresis*" (twDEP). In traveling wave dielectrophoresis, the positive DEP force pushes the particles along the direction of the traveling wave while the negative DEP force pushes them in the opposite direction.

After inserting the expression for the electric potential phasor in Eq. (14.7), the time-averaged DEP and twDEP forces can be expressed as [35, 39]:

$$\langle \mathbf{F}_{\mathrm{DEP}} \rangle = 2\pi\varepsilon_m a^3 K_R \nabla \left(|\nabla V_R|^2 + |\nabla V_I|^2 \right), \tag{14.11}$$

and, respectively:

$$\langle \mathbf{F}_{\mathrm{twDEP}} \rangle = -4\pi\varepsilon_m a^3 K_I (\nabla \times (\nabla V_R \times \nabla V_I)). \tag{14.12}$$

In conclusion, dielectrophoresis can be used as an effective means of separating particles, solely according to their dielectric properties and size. The real part of the Clausius–Mossotti factor gives the DEP force in the direction perpendicular to the electrode array, while the imaginary part gives the twDEP force in the parallel direction. Nevertheless, a disadvantage to use of DEP is that the DEP forces are inherently transient and disappear when the field is removed.

14.2.2 Electrorotation

Electrorotation occurs when a dipole sits in a rotating electric field generated between four electrodes, by applying four 90° phase-shifted AC signals and the interaction between the electric field and this dipole moment leads to a torque on the particle [37, 38]. If the field vector changes direction, the vector of the dipole moment will follow the changing of field vector. If the field vector rotates then the particle will also rotate. The interaction between a rotating field of magnitude **E** and the dipole induced in the particle results in a frequency-dependent torque $\mathbf{T}(\omega)$ on the particle, the time-averaged rotating torque given by [11]:

$$\langle \mathbf{T}(\omega) \rangle = \frac{1}{2} \mathrm{Re}\left[\widetilde{\mathbf{p}} \times \widetilde{\mathbf{E}}^{*} \right], \tag{14.13}$$

that for a spherical particle of radius a becomes [40]

$$\langle \mathbf{T}(\omega) \rangle = -4\pi\varepsilon_{\mathrm{m}} a^{3} K_{\mathrm{I}} \left| \widetilde{\mathbf{E}} \right|^{2}. \tag{14.14}$$

Equation (14.14) shows that the frequency-dependent property of the torque depends on the imaginary part of the CM factor. The particle will rotate with or against the electric field, depending on whether the imaginary part of the CM factor is negative or positive; a balance between the torque and hydrodynamic friction of the particle determines its rotation. In a viscous medium, the particle rotates at a constant angular velocity, and the magnitude of steady-state electrorotation rate (in rad s^{-1}) is [40]:

$$\Omega = \frac{\Gamma}{R}, \tag{14.15}$$

where R is the friction coefficient that depends on the viscosity η of the surrounding medium and on the geometry and surface properties of the particles. For a spherical particle the friction coefficient is $R = 8\pi a^{3}\eta$.

14.2.3 Other Forces Exerted on Nanoscale Particles

Particles actuated by dielectrophoresis move in a fluidic medium; therefore other forces have to be taken into consideration when describing their motion (Fig. 14.7), of which the most considerable are the hydrodynamic forces. The velocity of the particles will be significantly slowed by the resistance of drag forces while buoyancy forces may cause them to naturally float. In addition to these, under certain conditions forces that have been considered insignificant when manipulating micron scale particles (e.g., cells) can become more pronounced at the nanometer scale (e.g., viruses). This includes both electrothermal effects that induce additional drag and random Brownian forces. If the dielectrophoretic force is not strong enough, these forces will cause the particles to move at unreasonably slow speeds, move to unintended locations or, possibly, not move at all. The total force on a polarizable particle in a nonuniform AC field can be written as the sum of a number of independently acting forces (Fig. 14.7) [41]:

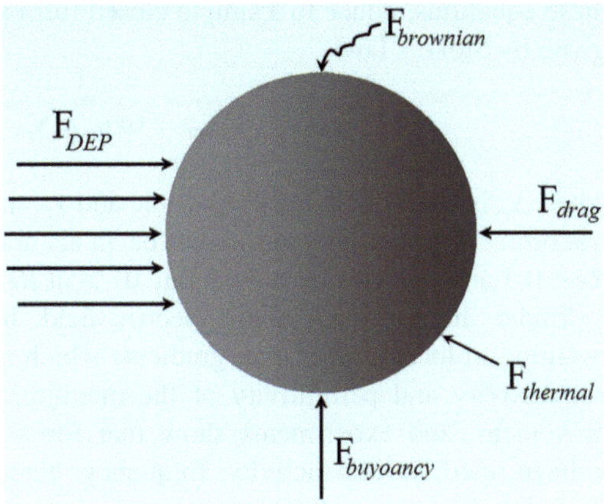

Fig. 14.7 Forces exerted on a particle moving in fluid, under the influence of dielectrophoresis

$$\mathbf{F} = \mathbf{F}_{\text{DEP}} + \mathbf{F}_{\text{drag}} + \mathbf{F}_{\text{buyoancy}} + \mathbf{F}_{\text{thermal}} + \mathbf{F}_{\text{brownian}} \tag{14.16}$$

The forces in this equation include, besides the dielectrophoretic forces, sedimentation, thermal, and buoyancy forces. The random force is due to Brownian motion and for particles with a diameter of the order of 100 nm this force is considerable. The mean free path of the movement is inversely dependent on mass, implying that the decrease in the particle diameter requires significant increases in the applied electrostatic energy (or field strength). For particles of diameter less than 1 μm, thermal effects can dominate, but the DEP force can be sufficient to produce deterministic particle movement.

In the following a brief review of these forces as they relate to dielectrophoresis will be provided.

14.2.3.1 Hydrodynamic Forces

In fluid mechanics, the Reynolds number is the ratio used to measure the relative importance of inertial forces to viscous forces and is defined as [16]:

$$Re = \frac{\rho_{\text{m}} L \text{v}_{\text{m}}}{\eta_{\text{m}}} \tag{14.17}$$

where ρ_{m} is the fluid density, v_{m} is the mean fluid velocity, L is the characteristic length of the system, and η_{m} is the dynamic fluid viscosity. Nanometer scale particles have very small Reynolds numbers; therefore they experience laminar Stokes flows in which inertia is negligible.

The motion of fluids is described by a set of partial differential equations known as the Navier–Stokes equations [41]. For the case of a sphere undergoing a small Reynolds number undergoing laminar flow in an incompressible, Newtonian fluid, these equations reduce to a simple closed form and the drag force in Eq. (14.16) is given by Stoke's law:

$$\mathbf{F}_{\text{drag}} = 6\pi\eta_{\text{m}}a(\mathbf{v}_{\text{m}} - \mathbf{v}_{\text{p}}) \tag{14.18}$$

where $\mathbf{v}_{\text{p}}$ is the velocity of the particle and $\mathbf{v}_{\text{m}}$ the velocity of medium. Stoke's law has been experimentally verified to be an accurate estimate of the drag force when $Re < 0.5$ and deviates by only about 10 % at $Re = 1$ [16].

Under the influence of an electric field, heat is generated in the medium, resulting in local temperature gradients which in turn give rise to gradients in the conductivity and permittivity of the medium. These gradients can induce fluid movement, and experiments show that for a given set of parameters (applied voltage, medium conductivity, frequency, electrode geometry, etc.) the resulting fluid flow has a reproducible pattern. As a consequence the drag force exerted by the moving fluid must be considered in the total force **F** expression.

The other hydrodynamic force exerted on particles manipulated by dielectrophoresis is buoyancy [42]:

$$\mathbf{F}_{\text{buoyancy}} = V_{\text{p}}(\rho_{\text{p}} - \rho_{\text{m}})\mathbf{g} \tag{14.19}$$

where **g** is the acceleration due to gravity and the ρ_{p} and V_{p} represent the density and volume of the particle. Since the volume of a nanoparticle is small, the magnitude of the buoyancy force is also small. However, the densities may be such that will have to overcome particles' natural tendency to float or sediment over time.

14.2.3.2 Electrothermal Forces

The high intensity electric fields often needed to manipulate particles have been observed to produce Joule heating inside the fluidic medium [43], especially when dielectrophoresis is occurring at AC frequencies in the MHz range. This ohmic heating causes a temperature gradient that in turn results in spatial conductivity and permittivity gradients within the suspending medium. The variation of electrical properties within the medium results in Coulombic and dielectric body forces that will induce extra fluid flow. The time-averaged body force on the fluid is given by [44]:

$$\langle \mathbf{F}_{\text{thermal}} \rangle = Re\left\{ \frac{\sigma_{\text{m}}\varepsilon_{\text{m}}(\alpha - \beta)}{\sigma_{\text{m}} + j\omega\varepsilon_{\text{m}}} \left(\nabla T \cdot \widetilde{\mathbf{E}}\right)\widetilde{\mathbf{E}}^{*} - \frac{1}{2}\varepsilon_{\text{m}}\alpha\left|\widetilde{\mathbf{E}}\right|^{2}\nabla T \right\} \tag{14.20}$$

where α and β are the linear and volumetric coefficients of thermal expansion and T is the absolute temperature. The additional drag force from the electrothermally induced flow can be found by solving the Navier–Stokes equations and using Eq. (14.18) as the volume force term.

14.2.3.3 Random Brownian Force

Brownian motion is the random movement of particles suspended in a fluid [45]. Since water molecules move at random, a suspended particle receives a random number of impacts of random strength and direction in any short period of time. Water molecules are about 1 nm in size; therefore particles such as viruses are small enough to feel the effects of these impacts. Due to its random nature, no net movement results from these impacts. It will follow a Gaussian profile with a displacement given by [16]:

$$\Delta x = \sqrt{\frac{k_B T}{3\pi a \eta_m} t} \tag{14.21}$$

where k_B is Boltzmann's constant and t is the period of observation. In order to move an isolated particle in a deterministic manner during this period, the displacement due to the dielectrophoretic force should be greater than Δx. The magnitude of the dielectrophoretic forces that can be generated by a lab on a chip is large enough that random displacement due to Brownian motion won't hinder our ability to manipulate particles.

The effect of Brownian motion was believed to be so large for nanoparticles such that deterministic movement of submicron particles could not be achieved by using DEP. Pohl showed that excessively large electrical field gradients would be required to move a particle of 500 nm meter because the force on a particle due to Brownian motion increases as the particle's volume is reduced [11].

For nanoparticles, several other effects become significant besides dielectrophoresis. For example the usage of high electric field strengths produces fluid flow and heating of the suspending electrolyte. The electric field can interact with the fluid to produce frequency dependent forces such as electro-osmosis and electrothermal force [45]. The resulting flow exerts a drag force on the particles and produces an observable motion. Recently, a new type of force has been observed on microelectrodes due to the electric double layer (EDL) of a particle in an AC electric field [42]. EDL is also believed to enhance the dielectrophoretic effects on submicron particles.

14.2.4 Dielectrophoretic-Field Flow Fractionation (DEP-FFF)

Field flow fractionation (FFF) is a technique designed to separate different types of particles, based on principle that DEP forces are combined with hydrodynamic forces in order to fractionate different types of particles in a liquid flow according to their physical properties (volume, mass, polarizability) [38, 46]. Figure 14.8 shows the principle of DEP-FFF separation.

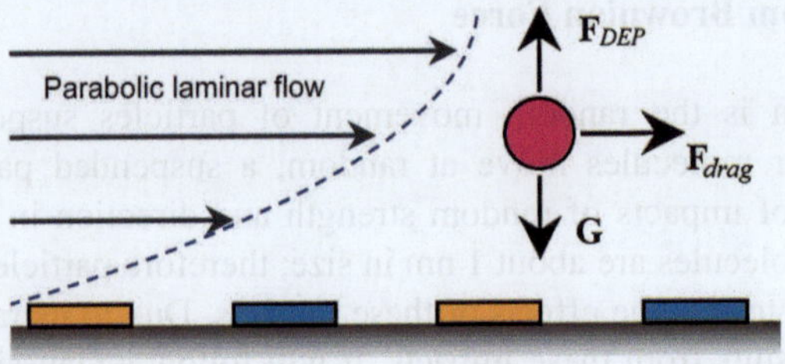

Fig. 14.8 Diagram showing the principle of a DEP FFF system. Particles are separated in the vertical direction by a balance of nDEP and gravitational forces. A parabolic fluid flow is applied and particles are separated

Particles are injected into a carrier flow that passes through the separation chamber, with a DEP force being applied perpendicular to the flow. The induced flow field has different velocities at different heights from the bottom surface. Particles are repelled from the electrodes under nDEP forces. The particles with different dielectrophoretic properties would have different height therefore move with different velocities. If another force field is applied vertically (for example, gravitation), the particles will gather at different places on the lower solid wall. Assuming that the particle is large, Brownian motion can be neglected, and the levitation height is dictated by the equilibrium between gravity and the negative DEP force [11]:

$$\mathbf{G} + \langle \mathbf{F}_{\mathrm{DEP}} \rangle = \mathbf{0} \tag{14.22}$$

For a spherical particle of radius a, suspended in a fluid medium, the gravitational force can be calculated as:

$$\mathbf{G} = \frac{4}{3}\pi a^3 (\rho_\mathrm{p} - \rho_\mathrm{m})\mathbf{g}. \tag{14.23}$$

The levitation height for a particle, where gravity balances DEP force, can be written as [38]:

$$h_{\mathrm{eq}} = h_0 + A \ln\left(\frac{K_\mathrm{R}}{(\rho_\mathrm{p} - \rho_\mathrm{m})g}\right) \tag{14.24}$$

where A and h_0 are constants depending on the electrode geometry and the applied electric signal. This key formula expresses the correlation between the levitation height, the density, and dielectric properties of the particle and of the suspension medium. Particles that experience different DEP forces will be levitated to different heights.

The applied velocity field in the channel exerts different hydrodynamic drags on the nanoparticles that attain different positions away from the chamber wall, which, in turn, exhibit different characteristic concentration profile. Since different types of the particles are transported at different rates, a heterogeneous sample can be separated and fractionated along the channel. The larger particles stay behind and are separated from the smaller particles.

14.3 Results and Discussions

14.3.1 Materials Characterization

In order to obtain relevant input data for the simulations, we investigated three probes, named as A, B, and C, consisting in samples of ash resulted from the combustion of different wastes, collected monthly from filters of Pro Air Clean Timisoara hazardous waste incinerator, within a period of 3 months. The incineration procedure is based on the complete burning of waste at 850–1,000 °C and postcombustion at temperatures between 950 and 1,300 °C of the resulting gas, followed by sudden cooling, a process that ensures the complete destruction of dioxins, furans, and other toxic components resulting from special waste incineration. The plant allows any type of solid waste incineration (excluding radioactive wastes) in the inferior chamber, and liquid waste incineration such as pesticides, solvents, and oils through direct injection in the postcombustion chamber. Although the technological process is automatic and developed for advanced control of emissions in the atmosphere, it cannot filter and control the nanoparticle contained in flue gas [1, 3]. These probes were analyzed from dimensional and dielectric point of view.

14.3.1.1 Dimensional Analysis

For dimensional characterization, we prepared for each of the three probes a mixture of 5 mg ash in 100 ml distilled water at room temperature, and put it to rest for 20 min, in order to decant the microparticles. Then we collected the remaining slurry liquid and analyzed the particle size/concentration distribution using a Nano Sight LM 10 nanoparticle visualization system. This high-performance device determines the size distribution and the number of nanoparticle in polydispersed and heterogeneous systems using nanoparticle-tracking analysis method. Figure 14.9a reveals a sample video frame with suspended nanoparticle, while Fig. 14.9b illustrates the particle size/concentration distribution for probe A. The distribution diagram indicates that it contains four significant groups of nanoparticle, having sizes of 55 nm, 100 nm, 155 nm, and 275 nm, respectively.

Figure 14.10 illustrates the size/concentration distribution of particles for probe B (Fig. 14.10a) and probe C (Fig. 14.10b), after a decantation time of 20 min. The distribution diagrams indicate four significant groups of nanoparticle, with sizes of 44, 67, 109, and 180 nm in probe B, and five significant groups of nanoparticle having sizes of 40, 60, 102, 138, and 175 nm in probe C.

These analyses show that the gas resulting from the combustion of waste contains nanoparticle. Recall that these nanoparticles are only those that probably were attached to larger particles, and stocked during the mechanical filtering process. They are, however, relevant to our study because the results suggest the existence of a much larger quantity of nanoparticle in the combustion waste gases

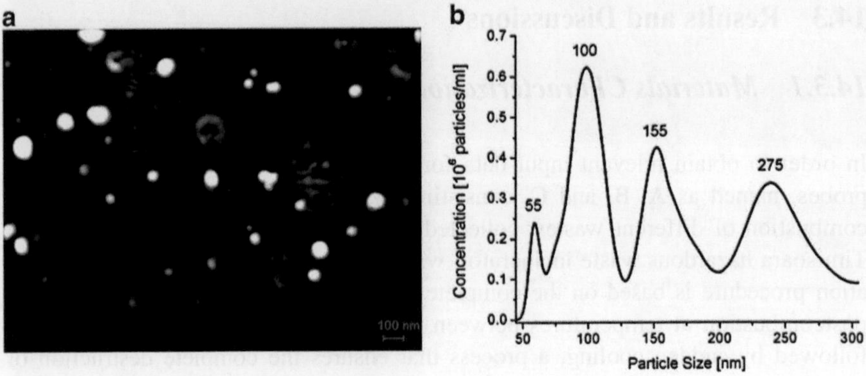

Fig. 14.9 Sample video frame with suspended nanoparticle (**a**), and the particle size/concentration distribution diagram (**b**) for probe A, after 20 min of decantation time

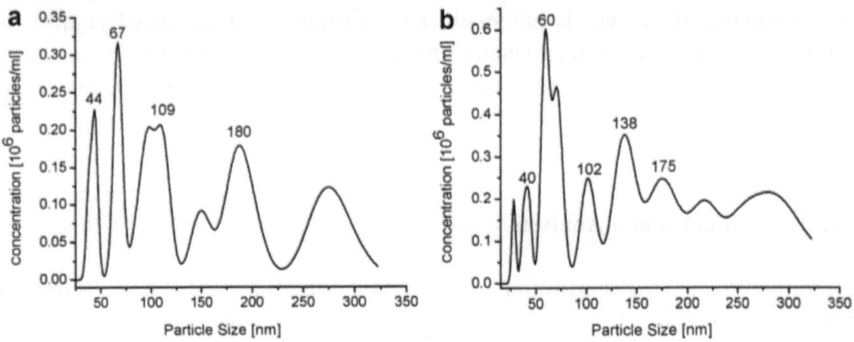

Fig. 14.10 The size/concentration distribution of nanoparticle for probes B and C, obtained with the visualization system of nanoparticle, Nano Sight LM10, after 20 min of decantation time

than originally detected. For each of the three probes one observes higher concentrations for particles with radii of about 50, 100, and 150 nm, values that will be considered relevant in our future analysis.

14.3.1.2 Dielectric Measurements

The frequency dependence of the real part of the CM factor for the mentioned ash probes was determined by dielectric measurements, based on the frequency dependence of the complex dielectric permittivity of the probes realized in the frequency range 25 Hz–2 MHz. The measurements of the real and imaginary components of the complex dielectric permittivity $\widetilde{\varepsilon} = \varepsilon - j\sigma/\omega$, were performed using an RLC-meter Agilent type E4980A, to which a cylindrical capacitor containing the analyzed material was connected [47]. Figure 14.11 shows the measured frequency

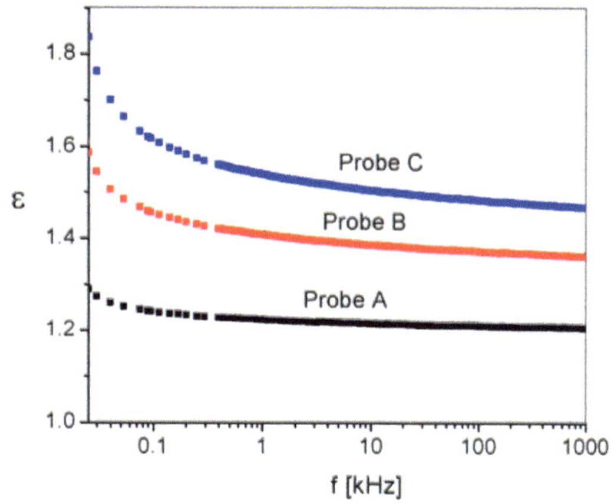

Fig. 14.11 The frequency dependence of the measured real component, ε of the effective complex dielectric permittivity of the probes A, B, and C

dependence for the real component, ε, of the effective complex dielectric permittivity of the probes A, B, and C.

The frequency dependence of the real part of the CM factor for the mentioned powder samples was determined by the dielectric measurements using the Maxwell–Wagner model [47]:

$$K_{\mathrm{R}} = \frac{\varepsilon^2 + \varepsilon - 2}{\varepsilon^2 + 4\varepsilon + 4}, \tag{14.25}$$

and consequently, knowing the frequency dependence of the real component of complex permittivity, ε, we obtained the frequency dependence of the real part of CM factor for the probes A, B, and C, represented in Fig. 14.12:

One observes that the real component of the CM factor acquires highest values at low frequencies and decreases continuously but remaining positive while frequency increases. More precisely, in the investigated frequency range, the behavior is described by three distinct variation regions: first, it decreases slightly for frequencies less than 200 Hz, then falls sharply for frequencies between 200 and 4 kHz, and finally remains nearly constant for higher frequencies. Due to the positive values of the real part of CM factor, it follows that the use of positive DEP force is appropriate for manipulating the ash nanoparticle from the flue gas. For further analysis regarding the filtration process, only devices working at frequencies up to 200 Hz are of interest, because in this domain, the CM factor has the highest value, and consequently one reaches the highest strength of pDEP force acting on nanoparticle in the flue gas. Therefore, for further calculations we will use the value $K_{\mathrm{R}} = 1$.

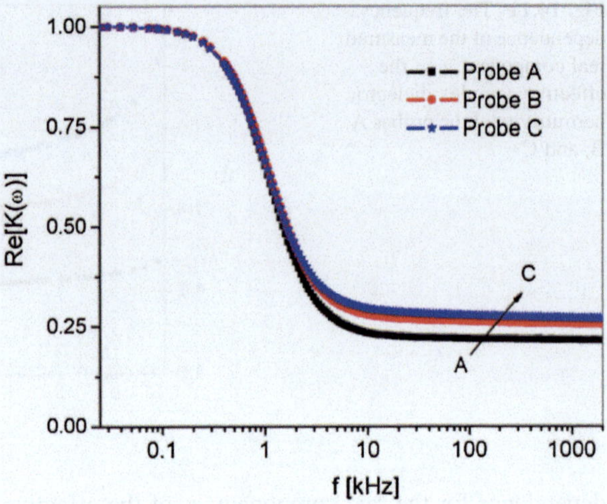

Fig. 14.12 The frequency dependence of the real component of the Clausius–Mossotti factor, for the probes A, B, and C

14.3.2 Numerical Simulations

14.3.2.1 Mathematical Model

The experimental results obtained by dimensional analysis and dielectric measurements are used as input data for the simulation of the transport phenomena inside a realistic DEP device for nanoparticle trapping from flue gas. First, we compute the pDEP force distribution inside a typical DEP device and then we move to the problem of determining the distribution profile of nanoparticles under the influence of dielectrophoresis. Finally, we analyze and discuss the obtained numerical results in terms of *Filtration rate*, a global quantity correlated with the concentration field, which offers a more suggestive characterization of the capabilities of the device regarding the separation process of nanoparticles from flue gas. All the numerical simulations were performed using the COMSOL Multiphysics program.

A typical DEP-based separation device with parallel interdigitated bar electrodes placed on the bottom surface is illustrated in Fig. 14.2. In most of the proposed models, due to the symmetry of the geometry and considering the electrodes much longer than their width, the problem is treated in two dimensions and the electrodes' height is neglected. In our study, in order to have a more realistic description of the experimental device, a 3D geometry and a detailed description of the electrodes shape are used. Anyway, taking into account the periodic distribution of the electrodes, the numerical calculations of the DEP force and the concentration field can be performed considering as computational domain only a so-called basic unit cell, which fully describes the entire system, except the vicinity of the walls. The geometry of the computational domain, together with the associated boundary conditions necessary to solve the Laplace equation for electric potentials, V_R and V_I, is presented in Fig. 14.13.

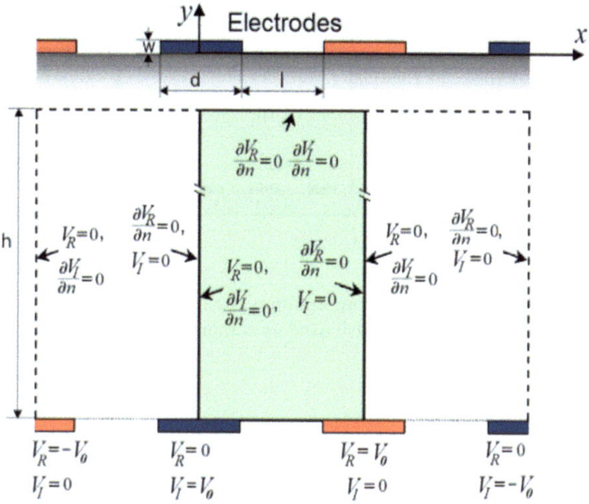

Fig 14.13 The geometry of the computational domain and the associated boundary conditions for the electric potentials. The *solid lines* indicate the basic unit cell

The macroscopic behavior of a suspension of spherical particles in a dense and viscous fluid can be modeled considering the mechanical equilibrium between an external spatially dependent force **F** and the Stokes drag. When the size of the particles, relative to the length L of the microchannel and the volume fraction C of particles is small, the dynamics of the two-phase system can be expressed by the following system of equations [35, 48]:

$$\mathbf{v}_{\mathrm{p}} = \mathbf{v}_{\mathrm{m}} + \frac{2a^2}{9\eta}\mathbf{F}, \tag{14.26}$$

$$\frac{\partial C}{\partial t} + \nabla \cdot \mathbf{j} = 0, \quad \text{where } \mathbf{j} = C\mathbf{v}_{\mathrm{p}} - D\nabla C. \tag{14.27}$$

Here $\mathbf{v}_{\mathrm{m}}$ and $\mathbf{v}_{\mathrm{p}}$ are the fluid and particle velocities, respectively, a the particle radius, η the viscosity of the fluid, t the time, $\mathbf{j}$ the particle flux, D the diffusion coefficient of the particles, and **F** denotes the dielectrophoretic external field.

The fluid flow field for the carrier medium is obtained by calculating the stationary solution of the Navier–Stokes equation in the compressible case, with no-slip condition imposed to the walls and on the electrodes' surface. A schematic representation of a typical computational domain corresponding to a microfluidic DEP separation device with an interdigitated electrode array, revealing its main geometric parameters, is given in Fig. 14.14a. The device has a rectangular cross section $(L_x = L, L_y = h)$ with parallel bar electrodes placed on the bottom surface. Very often, due to the symmetry of the device and considering the electrodes much longer than their width, the flow problem can be treated in two dimensions, and the Navier–Stokes equation is solved in this cross section of the device. If the height of electrodes can be neglected, a simplified computational domain can be considered,

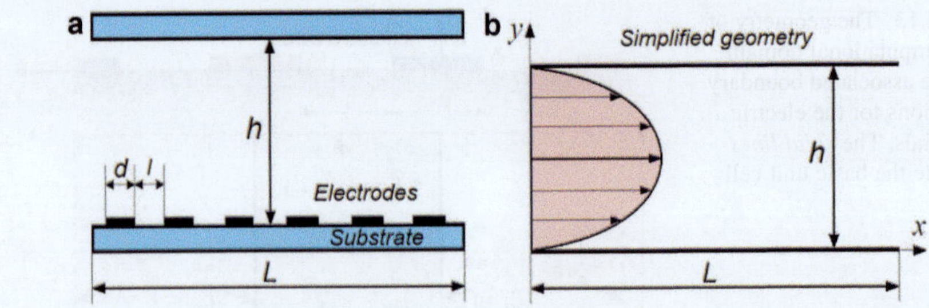

Fig. 14.14 Schematic representation of a dielectrophoretic chamber used for DEP separation (**a**), and the simplified geometry used as computational domain for the computation of the concentration field (**b**)

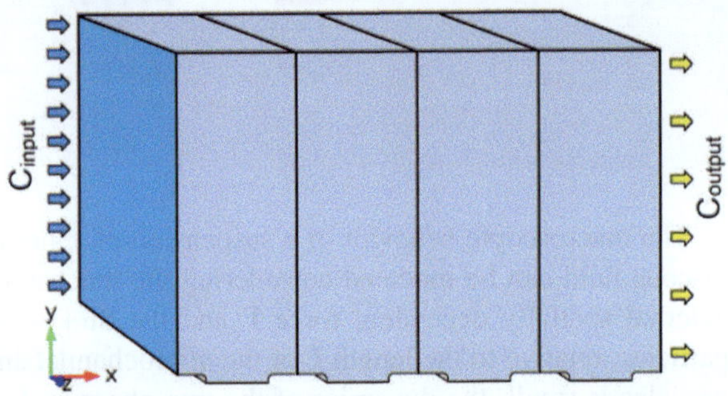

Fig. 14.15 Schematic representation of the separation device revealing the parameters used for defining the *Filtration rate*

as presented in Fig. 14.14b. In the latter case, one can assume that the flow field is described by a classical Poiseuille profile [8].

For a given DEP-force and fluid flow field, the particle concentration is evaluated by numerically solving the equation system (14.26)–(14.27). This calculated particle concentration field gives information at a local scale, showing how the particles are attracted on the margins of electrodes and the influence of the main parameters of the problem on this process. For the analysis of the filtration process we define the novel quantity named *Filtration rate* (F), which describes the process in terms of nanoparticles entrapment at the electrodes, related to the concentration distribution [49, 50]:

$$F = \frac{C_{\text{input}} - C_{\text{output}}}{C_{\text{input}}} = 1 - C_{\text{output}}/C_{\text{input}}\,[\%] \tag{14.28}$$

where C_{input} and C_{output} are the mean concentrations of suspended nanoparticles at the input and the output surfaces of the device, respectively, as schematically sketched in Fig. 14.15.

This proposed quantity gives the global information on the filtration process, and can be used in order to evaluate the efficiency of the filtration process.

14.3.2.2 Numerical Results

For the computation of the pDEP force, we first solved the Laplace equation for the real and imaginary components of the electric potential, together with the associated boundary conditions presented in Fig. 14.13. The computational domain consists of a unit cell described by the following set of geometric parameters: $l = d = 100$ μm, $H = 500$ μm, and $w = 100$ μm. The simulations were performed for particles with characteristic sizes $a = 50$ nm, $a = 100$ nm, and $a = 200$ nm, respectively, suspended in air. The dielectric response of the particles is characterized by the real part of the CM factor $K_R = 1$ and we considered the amplitude of the electric potential applied on the electrodes varying in the range $V_0 = 12 \div 24$ V.

In order to avoid numerical difficulties, due mainly to the extremely wide range of variation of the DEP force inside the computational domain, we chose to solve the model equations in the dimensionless form. If the electric potential is scaled with the applied electrode voltage V_0, the distances with the electrode width d, the time with d^2/D, the velocities with D/d, and the particle volume fraction with the initial average volume fraction C_0, the corresponding dimensionless form of the DEP force (14.11) is:

$$\langle \mathbf{F}_{\mathrm{DEP}} \rangle = F_{0\mathrm{DEP}} \nabla' \left(\left| \nabla' V_R' \right|^2 + \left| \nabla' V_I' \right|^2 \right). \tag{14.29}$$

We noted in the above equation $F_{0\mathrm{DEP}} = 2\pi a^3 \varepsilon_m K_R (V_0^2/d^3)$ a quantity that measures the intensity of the external field. The prime symbol above denotes the dimensionless quantities.

The magnitude of the vector $\nabla'(|\nabla' V_R'|^2 + |\nabla' V_I'|^2)$, proportional to the dimensionless DEP force given by Eq. (14.29), calculated in the vicinity of the electrodes, is presented in Fig. 14.16a in logarithmic scale. The results clearly show that the pDEP

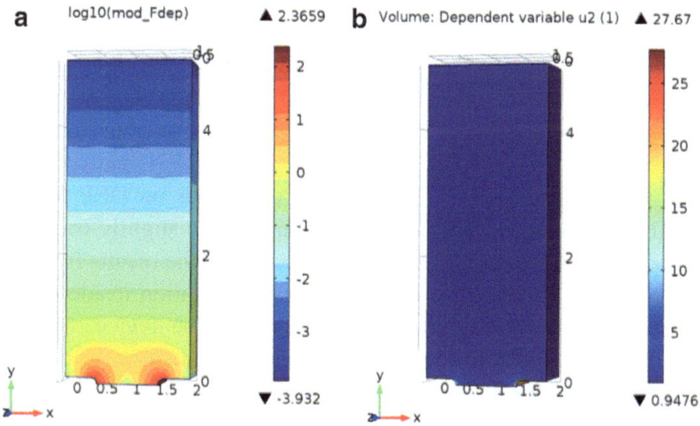

Fig. 14.16 (**a**) Spatial distribution of the calculated magnitude of the vector $\langle \mathbf{F}_{\mathrm{DEP}} \rangle / F_{0\mathrm{DEP}}$ in logarithmic scale, and (**b**) Calculated particle concentration distribution for a typical separation device

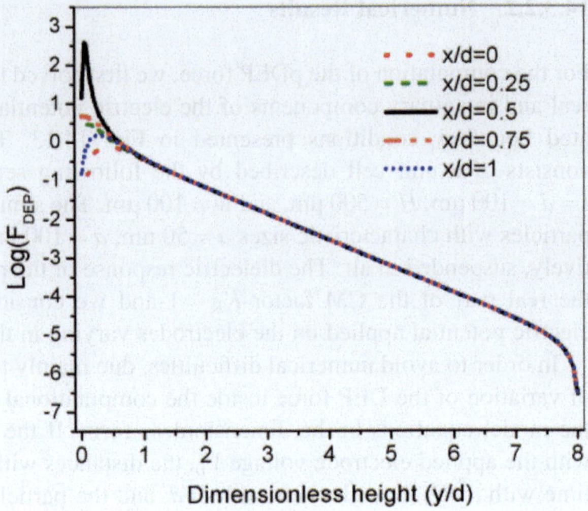

Fig. 14.17 The magnitude of the DEP dimensionless force $\langle \mathbf{F}_{DEP} \rangle / F_{0DEP}$ plotted along five vertical lines: $x/d = 0$, 0.25, 0.5, 0.75, 1 (Logarithmic scale)

force reaches its maximum near the electrodes' margins (a difference of at least two orders of magnitude), and diminishes rapidly with the height. Practically, at heights $y > d$ the dielectrophoretic effect is negligible. On the other hand, the results obtained by solving the problem (14.26)–(14.27) reveal that the nanoparticles subjected to the calculated pDEP force are strongly attracted to the electrodes edges, while their volume concentration in fluid diminishes, as showed in Fig. 14.16b.

The variation of magnitude of the DEP dimensionless force $\langle \mathbf{F}_{DEP} \rangle / F_{0DEP}$ plotted in Fig. 14.17 along five vertical lines, $x/d = 0, 0.25, 0.5, 0.75$, and 1, respectively, shows that above the value $y/d = 1$ this quantity decreases exponentially and remains constant with x across the array, independent of the shape of electrodes.

One observes that for values $y/d > 1$ the force practically not depends on the x coordinate, and decreases exponentially with y. For $y > d$ it holds very well the approximation:

$$\langle \mathbf{F}_{DEP} \rangle / F_{0DEP} = 0.195 \exp(-0.685 y/d) \tag{14.30}$$

which allows the evaluation of the levitation height in the case of negative dielectrophoresis.

The analysis of the effect of the height of electrodes on DEP force variation shows that the dependence on y is quite the same, only slightly translated with a value corresponding to electrode height, as revealed by the results from Fig. 14.18.

After this analysis of dielectrophoretic forces distribution inside the separation device, we move to the evaluation of the efficiency of the device.

The efficiency of the filtration process can be evaluated by calculating the *Filtration rate* (14.28) for different values of problem's parameters. The computation is performed using an iterative procedure: the output concentration in one unit cell is considered the input concentration for the next unit cell, in order to describe the cumulative effect of the filtration inside the dielectrophoretic device. This type of

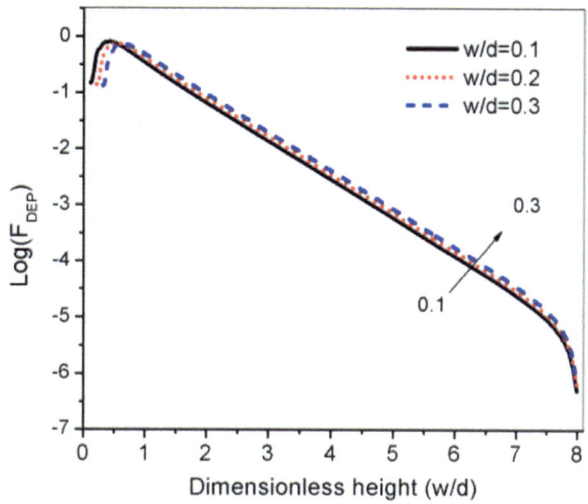

Fig. 14.18 Vertical variation of DEP force, for three different heights of electrodes

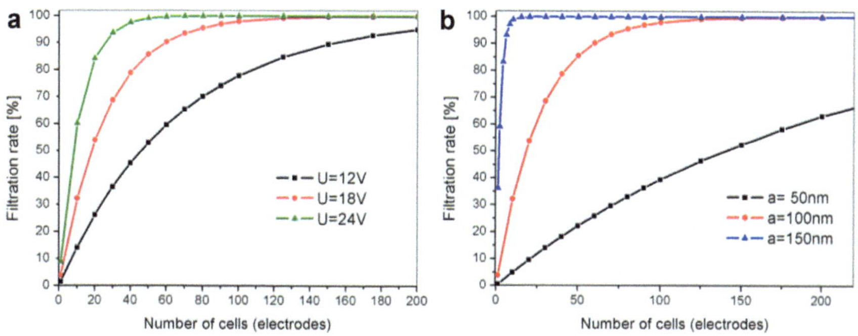

Fig. 14.19 Calculated filtration rate versus number of cells for (**a**) particles with $a = 100$ nm at three different applied voltages and (**b**) particles with three different radii at a fixed applied voltage of $V_0 = 18$ V ($d = l = 100$ μm, $w = 2$ μm)

analysis allows an estimation of the necessary number of cells (or electrodes) in order to obtain a certain desired filtration rate, when the other parameters of the problem are fixed. The results presented in Fig. 14.19a show that in the case of particle having size of 100 nm, a desired filtration rate of 90 % can be obtained by using about 30 electrodes when applying a voltage of 24 V, about 60 electrodes for 18 V, and nearly 200 electrodes for an applied voltage of 12 V.

Regarding the effect of particle radii on the filtration capacity, the results presented in Fig. 14.19b predict that, for example, when the applied voltage is 18 V, particles of 150 nm are completely captured after 10 electrodes. For particles of 100 nm we need about 150 electrodes for the complete capture, while for the particles of 50 nm are captured less than 60 % even if one use devices with 250 electrodes.

In conclusion, the simulations performed in the frame of the presented mathematical model allow an estimation of the performances of the filtration as a function of the geometric and physical parameters of the problem.

14.3.3 Preliminary Experimental Results

Finally, we focus on the validation of the mathematical model proposed in Sect. 14.3.2. Some preliminary but promising experimental results were already obtained. The tests performed with a DEP-based separation device having $l = d = 100$ μm and $H = 1{,}000$ μm reveal that in the absence of the applied voltage the particles are not at all attracted to the electrodes (Fig. 14.20a), while once applied an AC voltage of 12 V the dielectrophoretic effect appears (Fig. 14.20b). More than that, the concentration of captured particles clearly diminishes while we depart from the input region, which is in concordance with our simulations.

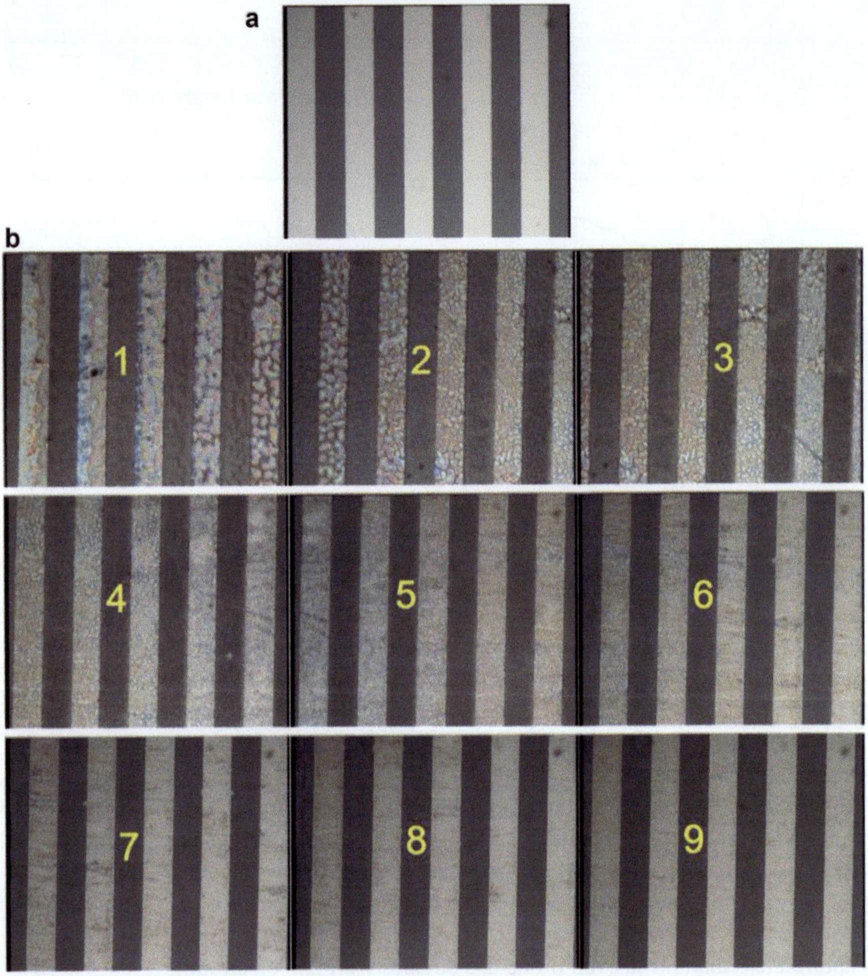

Fig. 14.20 DEP-based separation device with $l = d = 100$ μm (**a**) Before fumigation, and (**b**) After fumigation, successive snapshots, $U = 12$ V, AC, 50 Hz, time of fumigation $t = 30$ s

After this qualitative validation, a quantitative evaluation of the concentration of nanoparticles captured at the electrodes, at different distances from the input of the device, is necessary in order to give a solid validation of the model.

Conclusions

This study concerns the use of dielectrophoresis for the selective manipulation of nanoparticles. The DEP force depends on the gradient of the energy density, which changes on the length scale of the electrodes and is a short range effect. It can be modulated by changing the frequency and electrical properties of the suspending medium. After presenting the main theoretical aspects of the problem, the paper investigates the possibility to improve the filtering process of flue gas by separation of suspended nanoparticle using dielectrophoresis. The study focuses on the particles having an average radius of about 50–150 nm, that cannot be filtrated by classical techniques but have a harmful effect for environment and human health. We introduced a theoretical model and applied it to a physically realistic problem, using as input data the experimental results obtained by dimensional analysis and dielectric measurements performed on ashes sampled from the filters of Pro Air Clean Timisoara hazardous waste incinerator. The numerical solutions of the DEP force and particle concentration distribution for a typical interdigitated electrodes array are calculated using a proposed mathematical model. The calculations are performed using the COMSOL Multiphysics finite element solver. The functionality of a 3D DEP-based microsystem for the selective manipulation of nanometric particles is discussed in terms of *Filtration rate*, for different values of particle radius and applied voltage. The optimal parameters of the separation process can be determined through a detailed numerical study performed in the frame of the mathematical model and used for designing more efficient separation devices. Finally, we presented some preliminary experimental results regarding the trapping of nanoparticles from flue gas in a microfluidic device under DEP.

Acknowledgements This work was supported by a grant of the Romanian National Authority for Scientific Research, CNCS – UEFISCDI, project number PN-II-ID-PCE-2011-3-0762.

References

1. Neculae A, Giugiulan R, Lungu M and Strambeanu N (2013) Separation of nanoparticles from combustion gases wastes of incinerators. IMCET 2013, Kemer-Antalya, Turkey, 16–19 April, 2013, www.nanodep.com
2. Rickerby D, Morrison M (2007) Report from the workshop on nanotechnologies for environmental remediation, JRC Ispra 2007, www.nanowerk.com/nanotechnology/reports/reportpdf/report101.pdf

3. Minutolo P, Sgro L, Costagliola M, Prati M, Sirignano M, D'Anna A (2010) Ultrafine particle emission from combustion devices burning natural gas. Chem Eng Trans 22:239–244
4. Chang M, Huang C (2001) Characteristics of energy flow in municipal solid waste incinerator. J Environ Eng 127:78–81
5. Lin C, Wey M, Cheng H (2006) Effect of pressure fluctuations on the quality of fluidization and on the generation of particulate matters during incineration. J Environ Eng 132:960–966
6. Pethig R (2010) Review article—dielectrophoresis: status of the theory, technology and applications. Biomicrofluidics 4:022811-1–022811-34
7. Li M, Li WH, Zhang J, Alici G, Wen W (2014) A review of microfabrication techniques and dielectrophoretic microdevices for particle manipulation and separation. J Phys D Appl Phys 47:063001
8. Neculae A, Biris C, Bunoiu M, Lungu M (2012) Numerical analysis of nanoparticle behavior in a microfluidic channel under dielectrophoresis. J Nano Res 14:1–12
9. Zhang C, Khoshmanesh K, Mitchell A, Kalantar-Zadeh K (2010) Dielectrophoresis for manipulation of micro/nano particles in microfluidic systems. Anal Bioanal Chem 396:401–420
10. Pohl HA (1978) Dielectrophoresis. Cambridge University Press, Cambridge, UK
11. Ramos A (2011) Electrokinetics and electrohydrodynamics in microsystems, vol 530, CSIM courses and lectures. Springer, New York, NY
12. Hughes MP (2000) AC electrokinetics: applications for nanotechnology. Nanotechnology 11:124
13. Shklyaev S, Straube A (2008) Particle entrapment in a fluid suspension as a feedback effect. New J Phys 10:1–12
14. Barbaros C, Dongqing L (2011) Review – dielectrophoresis in microfluidics technology. Electrophoresis 32:2410–2427
15. Green NG, Morgan H (1998) Separation of submicrometre particles using a combination of dielectrophoretic and electrohydrodynamic forces. J Phys D Appl Phys 31:L25
16. Dickerson SJ (2007) Design of 3D integrated circuits for manipulating and sensing biological nanoparticles. PhD Thesis, University of Pittsburgh
17. Yuan Lin (2006) Numerical modeling of dielectrophoresis. Technical reports from Royal Institute of Technology KTH Mechanics SE-100 44 Stockholm, Sweden
18. Grebenkov DS, Filoche M, Sapoval B (2006) Mathematical basis for a general theory of Laplacian transport towards irregular interfaces. Phys Rev E Stat Nonlin Soft Matter Phys 73:021103
19. Lamoreaux SK (2005) The Casimir force: background, experiments and applications. Rep Prog Phys 68:201–236
20. Lapizco-Encinas BH, Rito-Palomares M (2007) Dielectrophoresis for the manipulation of nanobioparticles. Electrophoresis 28:4521–4538
21. Ying L, White SS, Bruckbauer A, Meadows L, Korchev YE, Klenerman D (2004) Frequency and voltage dependence of the dielectrophoretic trapping of short lengths of DNA and dCTP in a nanopipette. Biophys J 86:1018–1027
22. Tegenfeldt JO, Prinz C, Cao H, Huang RL, Austin RH, Chou SY, Cox EC, Sturm JC (2004) Micro- and nanofluidics for DNA analysis. Anal Bioanal Chem 378:1678–1692
23. Regtmeier J, Eichhorn R, Viefhues M, Bogunovic L, Anselmetti D (2011) Electrodeless dielectrophoresis for bioanalysis: theory, devices and applications. Electrophoresis 32:2253–2273
24. Nakano A, Ros A (2013) Protein dielectrophoresis: advances, challenges, and applications. Electrophoresis 34:1085–1096
25. Becker FF, Wang X-B, Huang Y, Pethig R, Vykoukal J, Gascoyne PRC (1995) Separation of human breast cancer cells from blood by differential dielectric affinity. Proc Natl Acad Sci U S A 92:860
26. Huang Y, Yang J, Wang X, Beckerand F, Gascoyne P (1999) The removal of human breast cancer cells from hematopoietic CD34+ stem cells by DEP field flow fractionation. J Hematother Stem Cell Res 8:481
27. Honegger T, Lecarme O, Berton K, Peyrade D (2010) 4-D dielectrophoretic handling of Janus particles in a microfluidic chip. Microelect Eng 87:756–759

28. Zhang L, Zhu Y (2010) Dielectrophoresis of Janus particles under high frequency ac-electric fields. App Phys Lett 96:141902
29. Perro A, Reculusa S, Ravaine S, Bourgeat-Lamic E, Duguet E (2005) Design and synthesis of Janus micro- and nanoparticles. J Mater Chem 15:3745–3760
30. Honegger T, Sarla S, Lecarme O, Berton K, Nicolas A, Peyrade D (2011) Selective grafting of proteins on Janus particles: adsorption and covalent coupling strategies. Microelect Eng 88:1852–1855
31. Gangwal S, Cayre OJ, Velev OD (2008) Dielectrophoretic assembly of metallodielectric janus particles in AC electric fields. Langmuir 24:13312–13320
32. Dimaki M, Bøggild P (2004) DEP of carbon nanotubes using microelectrodes: a numerical study. Nanotechnology 15:1095
33. Dimaki M, Bøggild P (2005) Frequency dependence of the structure and electrical behaviour of carbon nanotube networks assembled by dielectrophoresis. Nanotechnology 16:759
34. Wissner-Gross AD (2007) Dielectrophoretic reconfiguration of nanowire interconnects. Nanotechnology 17. http://alexwg.org/Nanotechnology2006.pdf
35. Lungu M, Neculae A, Bunoiu M (2010) Some considerations on the dielectrophoretic manipulation of nanoparticles in fluid media. J Optoelect Adv Mater 12:2423–2426
36. Morgan H, Green NG (2003) AC electrokinetics: colloids and nanoparticles, vol 50–62. Research Studies ltd, Baldock, Hertfordshire, pp 200–210
37. Lei U, Pei-Hou S, Pethig R (2011) Refinement of the theory for extracting cell dielectric properties from dielectrophoresis and electrorotation experiments. Biomicrofluidics 5:044109
38. Jones TB (2003) Basic theory of dielectrophoresis and electrorotation. IEEE Eng Med Biol Mag 22:33–42
39. Green NG, Ramos A, Morgan H (2002) Numerical solution of the dielectrophoretic and travelling wave forces for interdigitated electrode arrays using the finite element method. J Elstat 56:235–254
40. Falokun CD, Marks GH (2007) Electrorotation of beads of immobilized cells. J Elstat 65:475–482
41. Castellanos A, Ramos A, Gonzales N, Green N, Morgan H (2003) Electrohydrodynamics and dielectrophoresis in microsystems: scaling laws. J Phys D Appl Phys 36:2584
42. Dai H (2001) Nanotube growth and characterization. In: Avouris P, Dresselhaus MS, Dresselhaus G (eds) Carbon nanotubes synthesis, structures, properties, and applications, vol 80, Topics in applied physics. Springer Science & Business Media, Berlin
43. Green NG, Ramos A, González A, Morgan H, Castellanos A (2000) Fluid flow induced by nonuniform ac electric fields in electrolytes on Experimental measurements microelectrodes. Phys Rev E Stat Phys Plasmas Fluids Relat Interdiscip Topics E61:4011
44. Green NG, Ramos A, Gonzalez A, Castellanos A, Morgan H (2001) Electrothermally induced fluid flow on microelectrodes. J Electrostat 53(2):71–87
45. Einstein A (1905) The theory of the Brownian movement. Ann Phys 17:549
46. Huang Y, Wang XB, Becker FF, Gascoyne PRC (1997) Introducing dielectrophoresis as a new force field for field-flow fractionation. Biophys J 73:1118
47. Malaescu I, Giugiulan R, Lungu M and Strambeanu N, The Clausius-Mossotti factor in low frequency field of the powders resulted from waste combustion. The 13th International Balkan Workshop on Applied Physics, Constanta, July 4–6, 2013.
48. Holmes D, Green NG, Morgan H (2003) Microdevices for dielectrophoretic flow-through cell separation. IEEE Eng Med Biol Mag 22:85–90
49. Lungu M (2009) Separation of small nonferrous particles using a two successive steps eddy-current separator with permanent magnets. Int J Min Proc 93:172–178
50. Neculae A, Bunoiu M, Lungu A and Lungu M (2014) Flue gas filtration prediction in microfluidic devices using dielectrophoresis. The 14th International Balkan Workshop on Applied Physics IBWAP 2014, Constanta, July 2–4, 2014

Chapter 15
Multi-dimensional Electrophoresis: The March in Pharma Applications

Ranjita Shegokar and Sampada Sawant

Abstract Electrophoresis is widely used in pharmaceutical field for multiple applications. It is mainly employed for protein separation and identification. The information obtained on protein content and composition from multiple gel electrophoresis provides basic knowledge required for drug design, biomarker, vaccine, and antibiotic research. Furthermore, combination with advance analytical techniques like MALDI-TOF, mass spectroscopy provides added advantages in analysing protien samples. Electrophoresis by single and multiple dimensional methods is becoming very popular in nanoparticle research to identify protein corona around particle which can predict organ targeting when used in vivo. This technique can also be used to detect toxicity potential of nanoparticles and generated oxidative stress. Multi-dimensional electrophoresis has already been used in conventional pharmaceutical application and now is marching to nanoparticle and/drug delivery research area.

List of Abbreviations

1-DE	1-Dimensional electrophoresis
2-DE	2-Dimensional electrophoresis
2D-DIGE	2-Dimensional differential in-gel electrophoresis
CE	Capillary electrophoresis
CNTs	Carbon Nano Tubes
CB	Carbon black
CSF	Cerebrospinal Fluid
IEF	Immunofixation
IEF	Isoelectric focussing
NEPHGE	Nonequilibrium pH gel electrophoresis
PAGE	Polyacrylamide gel electrophoresis
pI	Isoelectric point

R. Shegokar (✉)
Department of Pharmaceutics, Biopharmaceutics and NutriCosmetics, Free University of Berlin, Kelchstr. 31, 12169 Berlin, Germany
e-mail: ranjita@arcsindia.com

S. Sawant
C.U. Shah College of Pharmacy, SNDT University, SantaCruz (W), 400049 Mumbai, India

M. Lungu et al. (eds.), *Nanoparticles' Promises and Risks*,
DOI 10.1007/978-3-319-11728-7_15

BAM	*N-tert*-butylacrylamide
PIC	Protease inhibitor cocktail
PMSF	Phenyl methyl sulfonyl fluoride
PTMs	Posttranslational modifications
SPE	Serum protein electrophoresis
SDS	Sodium dodecyl sulphate
TCA	Trichloroacetic acid
KOH	Potassium hydroxide
MPS	Mononuclear phagocytic system
MS	Mass spectroscopy
MWCNTs	Multi-walled carbon nanotubes
NIPAM	*N*-isopropylacrylamide
SILAC	Stable Isotope labelling by amino acids in cell culture
ALP	Alkaline phosphatase
LDH	Lactate dehydrogenase
CK	Creatine kinase
B. garinii	*Borrelia garinii*
C. trachomatis	*Chlamydia trachomatis*
E. coli	*Escherichia coli*
H. pylori	*Helicobacter pylori*

15.1 Introduction

15.1.1 Electrophoresis

Electrophoresis is widely used in separation and purification of macromolecules, e.g. proteins and nucleic acids of different size and charge. The charged molecules migrate towards either the positive or negative pole according to their charge in presence of electric field. The gel used in electrophoresis is composed of either agarose (0.5–2 %) or polyacrylamide (3.5–20 % in 14 × 14-cm) slabs or as minislabs (6 × 8-cm). Polyacrylamide gels have a rather small range of separation, but very high resolving power than agarose gels. The separation of biomolecules by electrophoresis mainly depends on the experimental system, properties of biomolecules, namely net charge, mass and shape. Gel placement could be as slab or as filled tube.

The protein separation can be native based on charge and mass denaturing (particle size dependant), buffer system based and other parameters inclined techniques (IEF based, 2DE). Native system can be divided into discontinuous and continuous mode, denaturing system can be classified as SDS based and urea formamide based to denature protein while others are based on separation based on IEF in pH gradient. The 2DE is a combination of IEF and SDS. Generally vertical apparatus is used in PAGE while horizontal apparatus in agarose, starch gel, and IEF. Evaluation of gel electrophoresis over the years (Fig. 15.1).

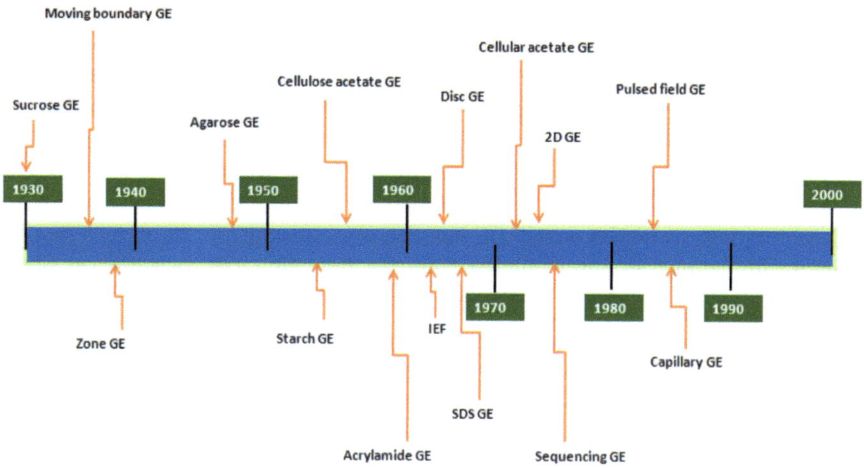

Fig. 15.1 Several decades of gel electrophoresis research developments and milestones as year scale. *GE* gel electrophoresis, *IEF* isoelectric focusing, *SDS* sodium dodecyl sulphate

15.1.2 One-Dimensional Gel Electrophoresis (1-DE)

1-DE application also known as sub-marine agarose gel electrophoresis can be used to perform separation based on one property, i.e. charge property. In general, 1D SDS PAGE is used to analyse less complex samples and to get idea of most abundant proteins. This technique provides feasibility to compare different samples on a single gel. Use of silver staining provides further refinement. This technique can be combined easily with other protein identification methods. The 1-DE analysis is commonly employed as a fast and easy tool to identify less complex samples. Feasibility of multiple sample analysis (up to 10) on a single gel of almost any size is possible. Analysis of sample before and after denaturation provides info on carbohydrate while radiolabelling can provide information on sulphate, phosphate and carbohydrate content. Identified protein can be recovered from gel by electroelution, electroblotting on polybrene coated derivative glass fibre sheet or polyvinylidene difluoride membrane filters method by applying further experimental steps [1]. Multiple casting of gels at one go can provide robust result by avoiding gel-to-gel variation. Mini gel slabs are becoming increasingly popular for isolation of peptide sequencing by taking unique advantage of speed and high resolution. The small gels are easily adapted to single concentration gradient and two-dimensional SDS-PAGE procedures [2].

15.1.3 Two-Dimensional Gel Electrophoresis 2-DE

It is abbreviated as 2-DE or 2-D electrophoresis and is a form of gel electrophoresis commonly used to analyse proteins. This technique separates soluble proteins with isoelectric points from pI 3–10 and Mw from 10 to 130 kDa. The separated protein bands can be further analysed by mass spectroscopy analysis and results can be predicted from database accession number. Typically 200–500 protein spots can be observed and identified in a silver stained 2D gel. 2D PAGE provides more detailed analysis of protein samples compared to 1D SDS PAGE and offers advantage of loading higher protein amount compared to the 1-DE gel. Main limitations are inability to identify hydrophobic membrane proteins. Mixtures of proteins are separated by two properties in two dimensions on 2D gels, i.e. charge and size. 2-D electrophoresis begins with 1-DE but then separates the molecules by a second property in a direction perpendicular from the first. In 1-DE, proteins (or other molecules) are separated in one dimension, so that all the proteins/molecules will align along a lane but that the molecules are spread out across a 2-D gel. Because it is unlikely that two molecules will be similar in two distinct properties, molecules are more effectively separated in 2-DE than in 1-DE. The two dimensions used to separate the proteins are isoelectric point and protein complex mass in the native state.

To separate the proteins by isoelectric point is called isoelectric focusing (IEF). Thereby, a gradient of pH is applied to a gel and an electric potential is applied across the gel, making one end more positive than the other. At all pH values other than their isoelectric point, proteins remain charged. If they are positively charged, they are pulled towards the more negative end of the gel and if they are negatively charged they are pulled to the more positive end of the gel. The proteins applied in the first dimension move along the gel and accumulates at their isoelectric point; that is, the point at which the overall charge on the protein is 0 (a neutral charge). For the analysis of the functioning of proteins in a cell, the knowledge of their cooperation is essential. Most often proteins act together in complexes to be fully functional. The analysis of this sub-organelle organisation of the cell requires techniques conserving the native state of the protein complexes. In native polyacrylamide gel electrophoresis (native PAGE), proteins remain in their native state and are separated in the electric field following their mass and the mass of their complexes, respectively. To obtain a separation by size and not by net charge, as in IEF, an additional charge is transferred to the proteins by the use of Coomassie Brilliant Blue or lithium dodecyl sulphate. After completion of the first dimension the complexes are destroyed by applying the denaturing SDS-PAGE in the second dimension, where the proteins of which the complexes are composed of are separated by their mass. As a separation, detection, and quantitation technique, 2-DE is an important tool, especially for clinical laboratories involved in the determination of protein expression levels and disease biomarker discovery (Fig. 15.2).

Before separating the proteins by mass, they are treated with sodium dodecyl sulphate (SDS) along with other reagents (SDS-PAGE in 1-DE). This denatures the proteins (that is it unfolds them into long, straight molecules) and binds a number of SDS molecules roughly proportional to the protein's length. Protein's length (when unfolded) is roughly proportional to its mass, and thus attaches a number of SDS molecules roughly proportional to the mass of protein. Since the SDS molecules are negatively charged, the result of this is that all of the proteins will have

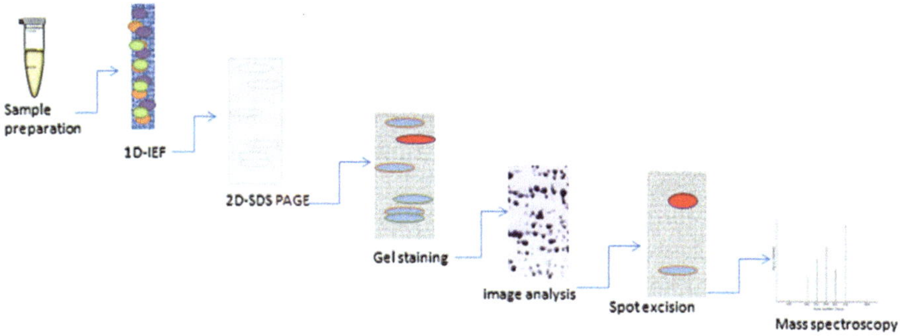

Fig. 15.2 Simplified 2D PAGE experimental setup. Image analysis is done by software and matched with fingerprint available in protein database. Last two steps are additional to separate and identify specific separated protein

approximately the same mass-to-charge ratio as each other. In addition, proteins will not migrate when they have no charge (a result of the isoelectric focusing step) therefore the coating of the protein in SDS (negatively charged) allows migration of the proteins in the second dimension (SDS-PAGE, is not compatible for use in the first dimension as it is charged and a nonionic or zwitter ionic detergent needs to be used). In the second dimension, an electric potential is again applied, but at a perpendicular angle from the first field. The proteins get attracted to the more positive side of the gel (because SDS is negatively charged) proportionally to their mass-to-charge ratio. This ratio remains nearly the same for all proteins. The proteins progress is be slowed by frictional forces. The gel therefore acts like a molecular sieve when the current is applied, separating proteins on the basis of their molecular weight with larger proteins being retained higher in the gel and smaller proteins being able to pass through the sieve and reach lower regions of the gel.

The result of this is a gel with proteins spread out on its surface. These proteins can then be detected by silver or Coomassie Brilliant Blue staining. Further identification of the proteins can be done using in-gel digestion of protein spots, and subsequent identification of the proteins by mass spectrometry.

15.1.4 2D PAGE History

The 2-Dimensional Electrophoresis (2-DE) based protein separation dates back to the late 1950s. The first 2-DE technique was developed by Smithies and Poulik in 1956 [3]. They used filter paper for separating proteins in the first dimension while in the second dimension proteins were separated on starch gel. Almost 20 human serum proteins were resolved. Around the same time, another technique was developed by Grabar and Williams [4]; they first performed agarose gel electrophoresis and in second step diffusion was performed against an anti-serum. Further method improvement led to the second step also being performed electrophoretically [5–7]. Introduction of polyacrylamide as matrix—the PAGE [8] improved the electrophoretic separation of proteins considerably. Raymond, in 1964, was the first who used PAGE for 2-DE separation of proteins [9]. The differences between the first and the second separation steps were the pH, the composition of the running buffer and the gel concentration.

Margolis and Kenrick [10] modified 2-D PAGE in order to separate the proteins first according to charge and then according to molecular weight (MW). A very low acrylamide concentration (2.7 %) was used in the first step, while an acrylamide gradient (4–24 %) in the second step. In 1970, Kaltschmidt and Wittmann [11] developed a 2-DE technique for separation of basic proteins that used Tris/EDTA-Na_2/borate-buffer, pH 9.6 and 8 % polyacrylamide in the first dimension and KOH/acetate buffer, pH 4.6 and 18 % polyacrylamide were used in the second. Extracts from 70S ribosomes of Escherichia coli were separated into more than 50 spots.

Disc electrophoresis was introduced in 1-DE technique by Ornstein [12] and Davis [13]. This technique composed of a gel system in which the proteins migrated throughout a stacking gel before entering the separation gel. The protein bands became surprisingly sharp, and consequently much more proteins could be detected.

Further two new methods that increased the resolution of proteins drastically were introduction of IEF in polyacrylamide gel and SDS-PAGE. IEF not only increased protein resolution but also the separation of proteins according to only one parameter, the isoelectric point (pI). In 1970, the disc electrophoresis system was used in the presence of SDS. SDS treatment gave proteins a negative charge and, therefore, separation in an SDS gel depended only on the molecular weight of the proteins. SDS-PAGE also improved the resolution of proteins considerably, similar to IEF. The two one dimensional techniques, IEF- and SDS-gel electrophoresis offered optimum conditions for the development of a 2-DE technique that led to high resolution in both directions, and separation according to different criteria, pI and MW. The development of the modern 2-DE began with the combination of IEF in the first and SDS PAGE in the second dimension, a technique independently developed by Dale and Latner, in 1969 [14] and Macko and Stegemann [15]. The analysis of the mixture of serum with this method, Dale and Latner detected about 60 protein spots. Domschke et al. [16] used a technique similar to that of Macko and Stegemann and separated serum proteins into about 90 spots. In the middle of the 1970s, several 2-DE techniques were published, all based on the concept to combine IEF, SDS-PAGE and disc electrophoresis. Anderson and Anderson [17] developed their own version and detected about 300 protein spots in 2-DE patterns of human serum (plasma) proteins.

After 1975, the application of 2-DE expanded considerably. The possibility for clinical applications made this method attractive. Other applications of 2-DE include protein separation from particular cellular fractions, such as erythrocyte membrane proteins [18], ribosomal proteins, chromosomal nonhistone proteins [19] and, moreover, proteins of economic interest such as potato proteins [20] or wheat proteins [21].

Studying cellular and tissue proteins from a global standpoint opened the door to many interesting questions: How do protein composition and expression change during embryonic development, cell differentiation and in the course of ageing? How do the proteins from different organs, tissues and cell types differ? Is it possible to study genetic variation between various individuals or different inbred mouse strains on a broad spectrum of proteins as revealed by 2-DE? And what is the effect on proteins if an organism is exposed to different environments or treatments? These questions are now being answered with the advances and technology developments in the 2-DE. The 2-DE technique was initially designed for the investigation of the chemically induced point mutation effects on a broad spectrum of tissue proteins. Mutations of single nucleotides frequently led to amino acid substitutions, which induced charge alterations in the corresponding protein. The mutated proteins can then be detected in 2-DE protein patterns by alterations in their electrophoretic position. To avoid any restrictions in the electrophoretic mobility of the native proteins carrying charge alterations due to mutations, SDS was not used for the second dimension. Under these electrophoretic conditions 275 protein spots were detected in 2-DE patterns from embryonic mouse liver [22]. Proteins were revealed by amido-black staining. In a modified version using SDS in the second dimension, 440 spots were detected in protein patterns from embryonic mouse liver [23]. Here Coomassie Blue staining was used for protein detection. Analysis of 2-DE protein patterns in the course of development of mouse embryos and studies on the effect of chemically induced mutations on proteins [24] were among the first investigations of the 2-DE technique. The 2-DE technique was established for separation of total proteins from *E. coli*. Cell culturing allowed labelling of proteins with radioactive amino acids and protein detection by autoradiography [25]. Using this method, O'Farrell obtained 2-DE patterns for 1,100 proteins which after staining with Coomassie Blue, however, decreased to only about 400 spots [25].

2-DE was used to resolve complex protein extracts from HeLa cells [26], human brain tissue [27] lymphocytes [28], trisomic cells [29], mouse embryos [30] and other cells and tissues. By the early 1980s it was confidently stated that it would be possible to detect 2,000–3,000 proteins with the present 2-DE technology [31]. For the evaluation of large numbers of protein spots, computer programs were developed for the first time [32]. In later years, a plan to catalogue every protein produced in the human body; "The Human Protein Index" was developed [33]. A systematic analysis of the mouse protein complement, identification of the relevant genes was performed and these probes were used to isolate the corresponding human genes [34].

The sensitivity of detecting separated proteins in 2-DE gels was drastically increased by the introduction of silver staining technique [35]. Later, fluorescence staining with CyDyes enhanced the sensitivity of protein detection further and allowed running test and control sample in the same gel. This improved the reproducibility and quantification of protein patterns considerably. Saturation labelling, a special, highly sensitive technique of the CyDye system, could be used to detect much less abundant proteins. Saturation labelling along with microdissection of single cells enabled analysis of single cell type proteins [36]. In 1982, the use of immobilised pH gradients (IPG) was introduced [37]. This method was then improved and adapted for 2-DE. It became commercially established with special convenient equipment and is the most commonly used 2-DE technique today. IPGs are available as strips and for different pH ranges.

15.1.5 Advancement in Two-Dimensional Gel Electrophoresis

In 1995, an improved version of original 2-DE technique was used [38]. The main feature of this method is the use of large (40–30 cm) but thin capillary (1-D: 0.9 mm, 2-D: 0.75 mm) gels. They used carrier ampholytes for pH gradients since IPGs in long distance gels formed protein bands rather than spots due to the fixed pH gradient. The problem was resolved by running IPGs in a series of strips with narrow pH gradients covering the whole gradient from pH 2–10 [39]. However, the resulting 2-DE patterns had overlapping border lines in neighbouring pH ranges making it difficult to recognise and eliminate the overlapping spots. In further development to the technique, protein sample was applied on the acidic side of the IEF gel and the run was stopped before the basic proteins moved out of the gel. In this way the very acid as well as the very basic proteins were retained in the 2-DE pattern, thereby saving the second 2-DE run for very basic proteins according to O'Farrell's NEPHGE method [40]. Using this large-gel 2-DE method, more than 10,000 protein spots could be detected in one gel depending on the kind of tissue or cells investigated, on the method of sample preparation and the protein detection method.

The development of software for detection and statistical analysis of 2-DE spots was another important expansion of the 2-DE technology. With the development of the 2-DE technology, the great gap between protein detection and further analysis of the proteins, i.e. large-scale protein identification, became obvious. This gap was overcome at the beginning of the 1990s by developments in mass spectroscopy (MS) allowing the identification of protein spots of 2-DE gels [41]. For 2-DE experts, the 2-DE protein pattern was the endpoint and MS was just a tool to identify all the protein spots resolved while for MS-experts, MS was the endpoint and 2-DE just a tool for prefractionation of complex protein samples. MS-experts pointed that 2-DE does not detect low-abundant proteins and hydrophobic membrane proteins, and that 2-DE cannot be standardised up to the level, where reproducibility is given from one laboratory to another.

Detection of low-abundant proteins by 2-DE can now be achieved by strategies like prefractionation of cell and tissue proteins, large-gel 2-DE and fluorescent labelling techniques for protein detection. For separating membrane proteins special 2-DE methods are available [42]. However, translating 2-DE techniques into an automated, high-throughput, MS adapted pipeline has not been feasible yet (Tecan project). Modern technologies of chromatographic prefractionation of proteins/peptides connected to MS offer the most promising way for total protein separation and identification till date.

15.1.6 Experimental Setup of 2DE

Electrophoresis equipment applies an electric charge to molecules, causing them to migrate towards their oppositely charged electrode. The technique is divided into gel and capillary techniques. Equipment includes horizontal gel electrophoresis units for DNA separation and vertical gel equipment for protein separation. 2-DE equipment includes equipment that separates proteins by charge and mass via separate components or complete automated workstations. Capillary electrophoresis systems exploit a separation polymer contained in a tiny capillary, used for applications such as DNA sequencing. Transilluminators are used to visualise molecules stained with signature dyes (Silver stain, Coomassie blue), and gel documentation systems are used to create permanent experimental records which can be analysed and compared with compatible data (Fig. 15.3).

Choice of electrophoresis equipment depends on the molecule of study, appropriate method of separation and downstream application.

A typical proteomics experiment (such as protein expression profiling) can be broken down into a series of steps. Firstly extraction, fractionation and solubilisation of proteins from a cell line, tissue or organism are carried out. This includes processing for partial or complete depletion of high-abundance proteins for increased sensitivity of downstream analyses. Labelling of proteins with differential dyes may also be carried out to delineate effects of various treatments. Then the gel-based separation of proteins in mixtures is carried out in one dimension alone (SDS-PAGE) or two successive dimensions for more complex mixtures (IEF followed by SDS-PAGE).

This is followed by gel staining or imaging analysis to allow for visualisation, isolation and relative quantitation of proteins. Protein spots/bands of interest are excised, digested with trypsin and identified by mass spectrometry. 2D-DIGE technology is 2D gel separation technique of proteins. In first step, proteins are separated based on isoelectric point and the second step involves separation based on apparent MW. In DIGE, proteins are visualised by fluorescence after specific chemical labelling with CyDye (Cy3 or Cy5).

Companies like Biorad, LG healthcare, and Life technologies now cater 2-DE equipments and reagents required. Details for the equipment setup can be found on the company websites and elsewhere in literature.

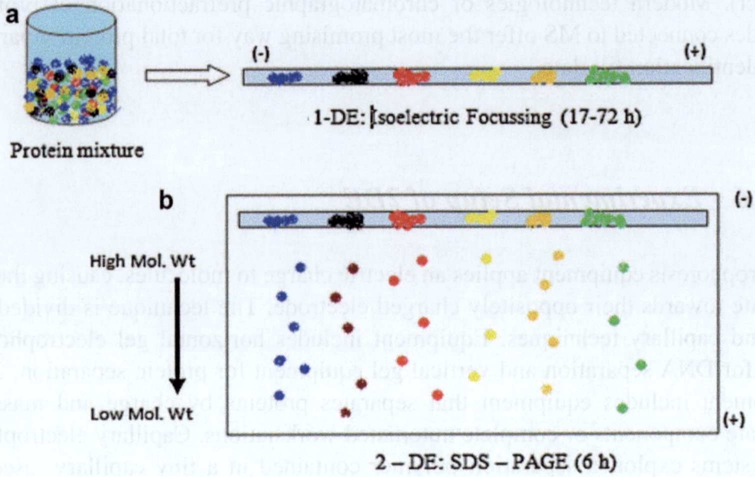

Fig. 15.3 Graphic illustration 2D PAGE. Panel (**a**) shows Iso electric focussing of the proteins based on the pI. Panel (**b**) illustrates SDS-PAGE wherein the proteins are separated based on their molecular weight

15.1.7 Advantages and Limitations of Method

The main advantage of using 2D-PAGE/DIGE is the large mass range and the large amount of proteins that can be analysed in a single run. 2D-PAGE/DIGE are particularly good for proteins within the mass range of 20–250 kDa and pI of 3–11. Proteins of particular pI and mass can be focused on by using narrow range pH isoelectric focusing strips and the percentage of acrylamide used in the second dimension PAGE gel, respectively. If sufficient sample is present on the gel (>300 μg of total protein) then proteins can be excised from the gel, subjected to in-gel proteolysis and analysed by mass spectrometry.

The 2-D SDS-PAGE and 2D-DIGE approaches to protein profiling are accessible and economical methods that possess high resolving power. The problem of reproducibility and quantisation has been greatly improved with the use of 2D-DIGE. IPGs have enhanced the resolution which enables the analyst to tailor the pH gradient for maximum resolution using ultra zoom gels with a narrow pH gradient range. With modern 2D-PAGE, it is also possible to resolve two proteins that differ in pI by 0.001 U. The use of imagers and computers allows fast data mining, acquisition, analysis, spot detection, normalisation, protein profiling, background correction and reporting and exporting of data.

Along with several advantages, 2-D PAGE also involves a few limitations. 2D-PAGE/DIGE is a lower throughput and time-consuming process (3–4 days per run) that involves several steps and requires high practical skills. 2D-PAGE/DIGE is not a good technique for the analysis of extremely acidic, basic or hydrophobic proteins such as membrane-bound proteins, and also smaller proteins and peptides (<15 kDa). The DIGE technique is also relatively expensive compared to silver or Coomassie staining of gels. Although 2D-PAGE has been limited by its inability to resolve proteins that are too basic or too acidic, too large or too small, this limitation is continuously being worked at. For example, the separation of basic proteins can be analysed using IPGs in the pH range of 4–12. The selection of type of stain used to visualise protein depends on the conditions and goals of experiment and the downstream applications.

15.2 Parameters affecting 2D PAGE efficiency

Several parameters of sample or nanoparticles affects the analysis of sample using 2D PAGE technique. In the following sections these parameters are discussed in detail one by one.

15.2.1 Particle Size and Porosity

The sizes and shapes of particle or nanoparticles are important factors that determine their physical and chemical properties. Capillary electrophoresis (CE) has proved to be one of the most powerful separation techniques [43] and has been applied to separate a variety of differently sized materials, including inorganic oxide [44], latex [45], polystyrene, silicate [46] and gold particles [47]. The separation mechanism relies upon surface charge of the material and hence their separation behaviour is similar to that of charged molecular species in CE. The variably sized particles display specific mobilities in a capillary column under the presence of an external applied voltage because they have different charge-to-size ratios. CE in an approach traditionally used to separate molecules in terms of their sizes and shapes [48]. For studying the effect of size and shape of the nanoparticles, silver nanoparticles were synthesised using microwave irradiation method [49]. An aqueous suspension comprising a mixture of silver nanorods (~6.2 and 55.5 nm) and nanoparticles (~80.2 nm) was obtained. Mafune and co-workers [50] and Chen and Yeh [51] have demonstrated that the surfactant SDS has a positive stabilising effect on silver nanoparticles.

During the CE separation process, samples prepared in a solution that differed from that of the running electrolyte were immediately enveloped by the running electrolyte. This blending offered the possibility of bringing particles into a defined environment; any subsequent changes in the effective charge-to-size ratios of the particles could be monitored by changing the composition of the running electrolyte. Concentrations of SDS surfactant in the range 0–40 mM were used to evaluate the concentration dependence of the separation of a mixture of 17.0—and 49.7 nm diameter silver nanoparticles in a single run. No separation occurred when a

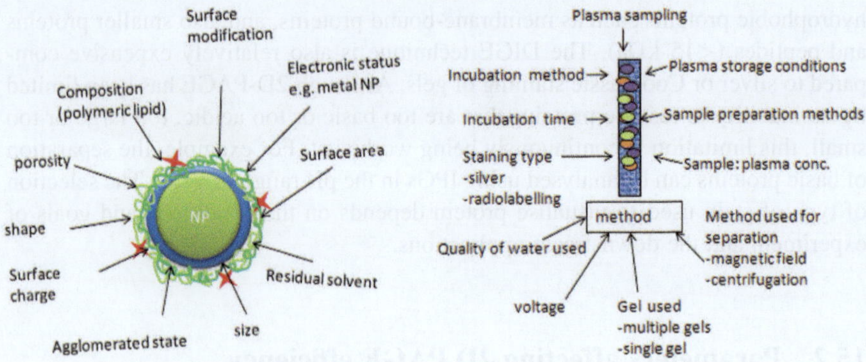

Fig. 15.4 Nanoparticle and method related parameters that affect the protein separation patterns when separated by gel electrophoresis

running electrolyte did not contain any SDS. The separation of the silver nanoparticles was improved upon the addition of SDS (20 mM) to the running electrolyte. Peak that migrated more quickly (10.5 min) were of the smaller diameter silver nanoparticles (17.0 nm); the more slowly eluting peak (12.2 min) corresponded to the larger diameter silver nanoparticles (49.7 nm).

Particle size of nanoparticles plays a major role in adsorption of the plasma protein corona around the nanoparticle. Molecules with molecular weight less than 5,000, or even higher for dense polymers like dendrimers, are removed from the body via the renal system. For particles with hydrodynamic radii of over 200 nm typically exhibit a more rapid rate of clearance than smaller particles. The biological responses to nanoparticles tend to scale with surface area rather than mass. As things become smaller, their surface areas shrink much more slowly than their volumes, causing nanoscale materials to have far greater surface-to-volume ratios than larger particles. A larger surface-to-volume ratio also implies more proteins will bind nanoparticle (relative to its mass) than a particle of larger size [52] (Fig. 15.4).

The effect of pore volume and size on the protein adsorption pattern using three ceramic materials viz. hydroxyapatite, zirconia and alumina was studied [53]. The pore volume study showed that particles of hydroxyapatite, zirconia and alumina were porous with pore volumes ranging from 0.29 cm^3/g up to 1.14 cm^3/g. The size of the pores (i.e. the pore diameters) varied somewhat among the three materials. Zirconia was found to have the smallest pores with maximum diameters of 0.2 mm. Hydroxyapatite and alumina had pores with up to 10 mm in diameter. The protein binding capacity study revealed that the maximum amount of plasma proteins adsorbed on the materials, after rinsing were 0.42 mg/m^2 on hydroxyapatite, 0.22 mg/m^2 on zirconia and 0.20 mg/m^2 on alumina. The low binding capacity of the materials was attributed to the fact that the material being porous in nature, the internal surface was not available for protein adsorption. The pore size distribution study revealed that the surfaces were available from a theoretical point of view, since no pores smaller than 30 nm were found in any material and all plasma proteins investigated had molecular diameters smaller than 15 nm. Thus, steric exclusion of proteins from the available surface could play a major role in low protein adsorption onto these surfaces.

15.2.2 Plasma Sampling

The task of collecting large numbers of clinical plasma samples that best represent, and most accurately reflect, the plasma composition in vivo is of great importance. Hence accurate plasma processing using standard protocol to yield reliable samples for proteomic analysis is essential.

Phenyl methyl sulfonyl fluoride (PMSF) and ethylene diamine tetraacetic acid (EDTA) affect the stability of serum samples [54], while Olivieri [55] has demonstrated that the protease inhibitors in red blood cell membrane lysates can have large effect on 2D-PAGE analysis. For standardised collection and processing of the plasma samples to obtain reliable and reproducible results via proteomic analysis, Hulmes et al. have described the role of protease inhibitor cocktail (PIC) in plasma stability [56]. They showed that the inclusion of PIC in the sample tubes provided stable and reliable human plasma samples that retained stability under less stringent processing, such as refrigeration for several hours before processing or one freeze–thaw cycle had little detrimental effect. Samples without PIC, from either heparin or EDTA plasma tubes, gave results that varied significantly from the control samples. Also, even with PIC present in blood tubes, it was important to quickly decant the separated plasma from the cellular components found in the blood tubes following centrifugation, as prolonged exposure again yielded different results from the standard procedure.

15.2.3 Plasma Storage Conditions

Storage temperature also has a profound effect on the protein profile of human plasma. In a study by Pasella et al. plasma samples were stored for 13 days at −80 °C, −20 °C, +4 °C and room temperature (20–25 °C) prior to proteomic analysis [57]. During storage for a short period (13 days) at four different temperatures plasma proteins were more affected by degradation processes at +4°C compared to the other temperatures studied. However, numerous protein spots (vitamin D binding protein, alpha-1-antitrypsin, serotransferrin, apoplipoprotein A-I, apolipoprotein E, haptoglobin and complement factor B) decreased in abundance with increasing temperature up to 4 °C, but at room temperature their intensity mean values were similar to those of time zero and −80 °C thus proving that these proteins are labile at 4 °C, but at the same time they are stable at room temperature (20–25 °C). Spots of serum albumin, fibrinogen gamma chain and haptoglobin were found to be more resistant to the higher temperatures tested, as they undergo changes in abundance only at room temperature; conversely, other spots of serum albumin, fibrinogen beta chain and serotransferrin are more labile as they undergo changes in abundance at all temperatures except at −80 °C.

15.2.4 Sample Preparation Method

Samples should have a high protein concentration (>10 mg/ml) and be free of salt and other disturbing factors, such as ionic detergents, nucleic acids, lipids, etc. that may interfere with the 2-DE. Precipitation followed by pellet uptake in isoelectric focusing (IEF) compatible sample solution is employed to concentrate and selectively separate proteins in the sample from the contaminating species. Protein concentration and desalting methods, however, are associated with artefacts, which can affect the results of the proteomic study, the most common being the incomplete protein recovery. Thus, the protein sample preparation method plays an important role in deducing the maximum information from 2-DE studies. Various methods like the ultrafiltration, precipitation using acetone, TCA, acetone/TCA, chloroform/methanol, ammonium sulphate have been studied to note their influence on protein resolution in 2-DE analysis [58]. In general, the precipitation methods and ultrafiltration delivered comparable results with no profound differences observed between untreated plasma and treated plasma samples. A relatively easy to perform concentration and desalting step using acetone precipitation resulted in a good recovery of all protein spots. Quantitative ammonium sulphate precipitation also resulted in an efficient precipitation of most proteins though spots representing certain proteins, like complement factor B, C3a and clusterin, were weak. The chloroform/methanol method yielded satisfactory results as well, although the spots representing C3a and clusterin are missing in the precipitate. Ultrafiltration resulted in recovery of practically all proteins. Thus, TCA precipitation, acetone precipitation and ultrafiltration delivered a higher protein recovery compared to ammonium sulphate and chloroform/methanol steps. Based on ease of performance, time required and cost effectiveness, precipitation using TCA/acetone was found to be the best method [59].

15.2.5 Surface Charge

Nanoparticles in a biological fluid (plasma, or otherwise) associate with a range of biopolymers, like proteins, organised into the "protein corona" that is associated with the nanoparticle and continuously exchanging with the proteins in the environment. Lundquist et al., have used six different polystyrene nanoparticles to study three different surface chemistries (plain polystyrene—neutral, carboxyl-modified—negatively charged, and amine-modified—positively charged) and two sizes of each (50 and 100 nm) to study the effect of surface properties and size on the detailed protein coronas [60]. Particles were incubated with plasma for 1 h followed by centrifugation to form pellets and extensive washing to remove all of the unbound proteins. The pellet size increased with increasing plasma concentration with subsequent decrease of the pellet size for all particles except the 100-nm amine-modified particles. The different behaviour for the amine-modified 100-nm particles compared with the other ones could be due to aggregation of the amine-modified 100-nm particles. Bound proteins were eluted from the particles and separated by 1D-PAGE. Selected bands were excised and the proteins were trypsin digested before detection by mass spectrometry.

For the 'neutral' particles, the major part of the "hard" corona consisted of fibrinogen, IgG, albumin and inter-alpha-trypsin inhibitor heavy chain [61]. The coronas around two different sized neutral polystyrene particles was very similar (80 % homology) suggesting that the molecular (e.g. hydrophobic) properties are more important than size for neutral nanoparticles. The homology between the coronas for the carboxyl-modified and amine-modified particles was only 50 %. Although many of the major highly abundant proteins in the corona are independent of size and surface charge, a whole range of proteins (many of them having quite distinct biological roles relevant to nanomedicine and nanosafety) form part of the corona. The nature of the proteins in the corona is determined by the chemical properties of the nanoparticle [62]. However, even for a fixed material type, the size of the particle and its surface modification are able to entirely change the nature of the biologically active proteins in the corona, and thereby possibly also the biological impacts.

15.2.6 Hydrophobicity and Shape

The adsorption of proteins on the particles surface after *i.v.* administration depends on their surface characteristics and is regarded as key factor for the in vivo organ distribution. Surface hydrophobicity highly influences the plasma protein adsorption patterns [63]. Latex particles with decreasing surface hydrophobicity were synthesised as model colloidal carriers to study the effect of hydrophobicity on plasma protein adsorption patterns. Surface hydrophobicity was determined by the adsorption of a hydrophobic dye Rose Bengal and the adsorption pattern was analysed employing 2-DE [64]. Acrolein particles with lowest hydrophobicity showed minimum protein adsorption while Styrene with highest hydrophobicity showed maximal adsorption. The major proteins adsorbed on the particles of each type were Fibrinogen (35 % of the overall pattern) and IgG (35.7 % of the overall pattern). Adsorption of Fibrinogen and the Apolipoproteins was affected by diminished surface hydrophobicity. The adsorption of the large proteins like Albumin, Fibrinogen and IgG decreased, but also the Apolipoproteins A-IV and J (MW 45 and 52 kDa) were strongly diminished with decreasing hydrophobicity of the nanoparticles. Thus the surface hydrophobicity and the functional groups present affect the protein adsorption patterns.

Nanoparticles of cross-linked copolymers of NIPAM and BAM (70–700 nm) and two comonomer ratios 50:50 and 85:15 were evaluated to study the effects of nanoparticle curvature and hydrophobicity on the nature and identity of adsorbed plasma proteins [65]. The hydrophobicity of the particle surface influenced both the amount and identity of the proteins bound to the particles [66]. The less hydrophobic 85:15 NIPAM–BAM copolymer particles showed virtually no protein adsorbed on them (except some HSA), in strong contrast to the more hydrophobic 50:50 particles which showed high levels of protein adsorption. The amount of bound protein varied with size, and scaled with the amount of available surface area. However, the protein pattern was the same for all sizes and apolipoprotein AI was always the most abundant protein recovered. This indicated that the surface curvature was not a major determining factor for the relative affinities of proteins for the particles.

15.2.7 Incubation Medium

The incubation medium for the particles can be blood, plasma or serum. Ideally, blood should be used, but it is very difficult to separate the nanoparticles from the blood cells. The monocytes present in the blood start to phagocytose the particles, thus the particles disappear. In vivo, this happens only to a limited extent, because particles are taken up, e.g. within 5 min by the liver macrophages or accumulate somewhere else. Therefore, in general, plasma or serum is used instead of blood.

However, serum and plasma yield different patterns, because, e.g. serum is depleted from proteins of the complement system due to the coagulation process (e.g. lack of C3 and fibrinogen). The use of plasma as an incubation medium for the in vitro evaluation of colloidal drug carriers is widely reported [67, 68]. However the plasma contains compounds to inhibit coagulation, e.g. citrate or heparin, which can also interfere with the protein adsorption process. To minimise such effects and to obtain comparable results, macromolecular heparins should be avoided. Hence, citrate plasma is often used as an incubation medium.

15.2.8 Ratio of Sample to the Incubation Medium

The ratio of particle suspension to incubation medium is a very crucial parameter. In case the ratio is too high (too many particles, too large of a surface area), some high affinity and lowly concentrated proteins in the plasma or serum might be depleted. They would have been adsorbed in a higher amount if more protein would have been available. The protocol “NCL Method ITA-4” [69] suggests to mix 1 mL plasma with 1 mL particle suspension (concentration 1 mg/mL = 0.1 %).

Reproducible patterns were obtained at a ratio of 3:1 plasma to suspension [70]. The particle concentration and related surface area for adsorption are important, and not the suspension volume. It is recommended to admix a particle suspension as highly as concentrated to plasma or serum to avoid too strong a dilution by the water of the particle suspension, to avoid reducing artificially the solubility of some proteins by the addition of water.

15.2.9 Method of Separation

The separation method is crucial to obtain reproducible results. Some of the currently used methods are centrifugation [71], chromatography [72] and separation by magnetic fields [73]. The separation method can affect the pattern. Chromatography might lead to desorption effects at the end of the separation process. The particle sediment from centrifugation needs to be washed to remove excess plasma in the pellet. Normally, several washing steps need to be applied.

15.2.10 Quality of Water Used

Organic contaminants affect the polymerisation of acrylamide gels, resulting to poor quality gels. The migration of proteins in the gel is dependent on ionic strength and pH of the running buffer. The ionic strength and pH of buffers used to prepare the gel and running buffer have to be maintained to assure reproducible results. High levels of ions could alter the ionic strength of the solutions and thus affect the migration rate of proteins. Water that is contaminated with bacteria and degradation by-products such as proteases that may degrade proteins affect the ionic strength of solutions leading to poor gel quality. Highly purified water is thus an essential component to obtain rich gel quality, with lower background and better spot resolution.

15.3 Applications of Gel Electrophoresis in Pharmaceuticals

15.3.1 Protein Identification

The initial goal of most proteomics projects is to identify and determine differential abundance of proteins or posttranslational modifications (PTMs) between samples followed by detailed analysis of individual proteins of interest. 2-DE followed by mass spectrometry provides direct visual confirmation of changes in protein/PTM

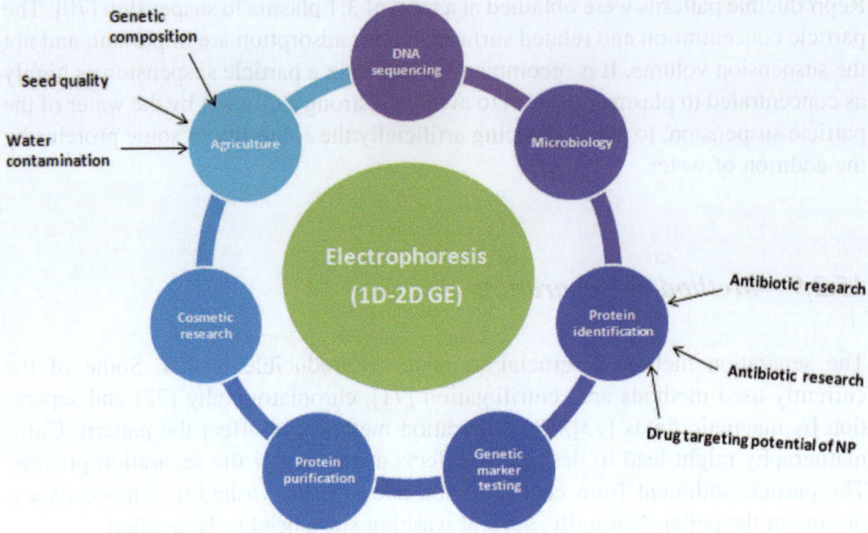

Fig. 15.5 Applications of (1D-2D) electrophoresis gel techniques

abundance, thus providing early justification for downstream analytical steps. A few applications of profiling include:

A critical modulator of biological activity in the proteome is posttranslational modifications, or PTMs. 2-D electrophoresis can be used effectively to study PTMs. 2-DE separates charge and size isomers of polypeptides. Hundreds to thousands of polypeptides can be resolved in a single 2-D PAGE gel and can be quantified, probed with antibodies (via blotting), tested for posttranslational modifications (using antibodies/chemical stains specific for each PTM) or extracted for mass spectrometric analysis. After enrichment of various PTMs of interest, they can be profiled via 2-DE. Some applications are as follows (Fig. 15.5):

- Phospho protein expression profiling is used in signal pathway studies. Phosphorylated protein pIs shift to a more acidic region on the gel.
- Acetylated protein expression profiling—Acetylation of proteins at their amino termini or on lysine side chains causes them to shift to a more acidic region on the gel. Acetylation of proteins can affect protein function, interactions and subsequent or additional posttranslational modifications.
- Methylated protein expression profiling—Methylation of lysine or arginine residues shifts their isoelectric points significantly towards the acidic spectrum. Analysis of protein methylation is particularly important in epigenetic studies.
- Glycosylated protein expression profiling—Glycosylation is both a cotranslational and posttranslational modification. Attachment of relatively simple glycans (O-glycosylation) or more complex glycans (N-glycosylation) can have varying effects on protein molecular weight, pI and protein functions.

Table 15.1 List of various nanoparticles (placebo and drug loaded) studied by 2-D and 1-D gel electrophoresis. Some interesting examples of capillary electrophoresis are also listed

Drug	Technique	Reference
Silicon oxide, dextranised silicon, polyurethane and dextranised polyurethane	2D PAGE	DOI: 10.1016/j.colsurfb.2011.01.006
Nevirapine	2D PAGE	DOI :10.1016/j.nano.2010.10.0122D PAGE
Polystyrene beads of mean diameter 100, 200 and 1,000 nm were used as model carriers	2D PAGE	DOI: 10.1002/1522-2683(20000701)21:12<2438::AID-ELPS2438>3.0.CO;2-F
Polystyrene beads, 60 nm Surface modified by adsorption of increasingly hydrophilic block copolymers, poloxamers 184, 188 and 407.	2D PAGE	Electrophoresis 1993, 14, 1382–1387
Oil in water emulsions for intravenous administration	2D PAGE	DOI: 10.1002/elps.1150190233
Poly(methoxypolyethyleneglycol cyanoacrylate-co-hexadecylcyanoacrylate) (PEG-PHDCA) nanoparticles	2D PAGE	DOI: 10.1002/elps.11501401214
Hydroxyapatite, alumina and zirconia	2D PAGE	Biomaterials. 2002 Feb; 23(4):1237–47
Egg lecithin - stabilised oil-in-water emulsions	2D PAGE	Zhang Y., Ph D Thesis, Free University Berlin, Germany
Haemoglobin particles modified with HAS	2D PAGE	
Rutin NC	2D PAGE	
PEG-PLA		DOI: 10.1016/S0927-7765(99)00156-3
PEG-PCL		
PEG-PLGA		
Polyanhydride microparticles		DOI: 10.1002/jbm.a.32815
Anion-substituted poly(vinyl alcohol) (PVA) membranes, arboxymethylated PVA (C-PVA), and sulfonated PVA (S-PVA) membranes	2D PAGE	DOI: 10.1007/BF03218975
Salbutamol sulphate (SS), sodium cromoglycate (SCG) and beclomethasone dipropionate (BDP)] in poly(D,L-lactic acid) (PLA) nanoparticles	Capillary electrophoresis	DOI: 10.1016/j.chroma.2007.11.041
Fluorescently (FL) labelled gelatin nanoparticles (gelatin NPs)	Chip electrophoresis	DOI: 10.1002/elps.201300074.
Metal nanoparticles	high-throughput two-dimensional-grid gel electrophoresis cell (2D-GEC)	DOI: 10.1021/nn405352v

(continued)

Table 15.1 (continued)

Drug	Technique	Reference
Poly(Lactic Acid) Nanoparticles, salbutamol sulphate	Capillary electrophoresis	Samuli Hirsjärvi, Faculty of Pharmacy of the University of Helsinki http://ethesis.helsinki.fi
Carboxylated or aminated latex particles(FluoSpheres®)	2D PAGE	DOI: 10.1016/j.ejps.2005.04.015
Logic gate nanoparticle composed of copolymer (poly-β-aminoester ketal-2) and PLGA loaded with DNA	Gel electrophoresis	DOI: 10.1155/2012/291219
Zinc oxide nanoparticles	SDS-PAGE	Int. J. of Pharma and Bio Sci, 2 (4), 2011
Carbon nanoparticles	Capillary electrophoresis	DOI: 10.1002/elps.201200627
Gold nanoparticles	native-PAGE	DOI: 10.4172/2324-8777.1000124
Silver Nanoparticles	2D-PAGE	DOI: 10.1007/BF03217491
Polymeric micelles	2D DIGE	DOI: 10.1002/pmic.201100602.
Uranium(U)-protein complexes	1-DE and 2-DE	DOI: 10.1016/j.talanta.2014.04.065

Table 15.1 lists the drug delivery systems studied for one or either reasons by electrophoresis technique especially 1-DE, 2D-PAGE and other advanced techniques.

15.3.2 Protein Purification

2-DE can be used to isolate contaminants during protein purification. Profiling samples at each step in the process can assess the purity of the end product. The Profiling of purified proteins, antibodies or samples during drug development as part of product characterisation was used to detect batch-to-batch differences. 2-D electrophoresis enables confirmation of uniform charge states between multiple batches.

15.3.3 Biomarker Discovery

Most of the research in medicine involves the search for biomarkers in which certain proteins change in expression with either a large increase or decrease as a result of disease [74].

The 2-DE remains the most widely used proteomic technique owing to the fact that it is currently the method that can separate and measure hundreds of proteins from a single biological sample simultaneously [75]. Analysis by 2D-PAGE has been used for proteomic studies of breast cancer [76], prostate cancer [77], hepatocellular carcinoma [78], renal cell carcinoma [79], rheumatoid arthritis [80] and hepatitis B infection [81]. Proteomic profiling allows for biomarker discovery based on the differences in protein expression levels between samples that have been analysed on a broad scale; that is, normal vs. disease model study, protein expression knockdown by siRNA or other conditions.

15.3.4 As a Tool for Exploring the Targeting Potential

The targeting potential of the nano/micro systems to the MPS cells that serve as microbial reservoirs can be studied using 2-DE. The adsorption patterns is qualitatively and quantitatively assessed. The relative adsorption by various organs can help to understand the organ distribution of the delivery systems.

The effect of surface (charge) modified submicron emulsions (cationic and anionic) on in vitro adsorption of plasma proteins via 2D PAGE has been carried out [82]. The presence of poloxamer 188 eliminated the adsorption of the larger proteins like immunoglobulins, fibrinogen, etc. The smaller proteins such as apolipoproteins and albumin were almost completely adsorbed onto the submicron emulsions. The adsorption of apolipoprotein, especially apoA-1, was approximately three times more on stearylamine- and oleylamine-based cationic emulsions and oleic acid-based anionic emulsions. In addition, the ratio between the apoA-1 and apoA-IV was found to be 1 for lipofundin MCT 10 % whereas it was about 0.26 for deoxycholic acid-based anionic emulsion and above 5 for oleic acid based anionic emulsions and cationic emulsions. This indicated that emulsions having similar surface/interfacial charge imparted by different anion-forming stabilisers (oleic or deoxycholic acids) exhibited markedly different protein adsorption patterns.

Shegokar et al., identified protein adsorbed on nanocrystals of nevirapine (antiretroviral drug) in vitro to predict in vivo fate of nanoparticle [83]. Protein pattern analysed showed macrophage targeting might be dominant for polysaccharide coated nanocrystals and brain targeting for albumin coated nanocrystals (Fig. 15.6). This predicted hypothesis was confirmed by in vivo results obtained. This type of prediction helps batch selection for in vivo experiments, thereby reducing the number of animal experiments required, time and cost as well. The data obtained on nevirapine confirm in vitro and in vivo correlation. Didanosine NLC showed targeting potential to brain when tested by 2D PAGE technique and can be used for HIV related dementia conditions and evaluation of the in vitro differential protein adsorption patterns of didanosine-loaded nanostructured lipid carriers (NLCs) for potential targeting to the brain [84].

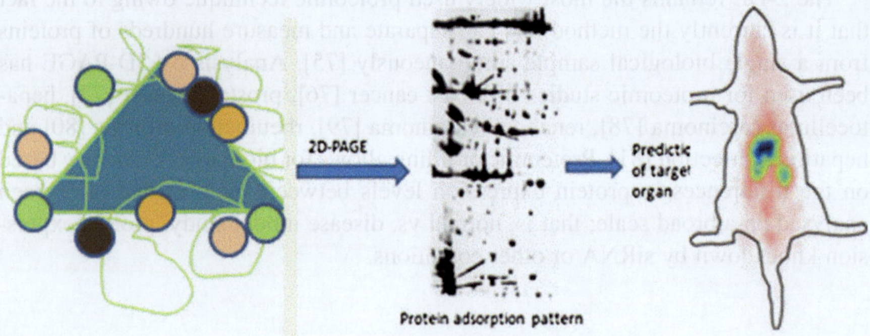

Fig. 15.6 Nevirapine nanocrystals functionalised with polysaccharide, serum albumin and PEG tested by 2D PAGE to determine the nature of protein corona to predict targeting potential. The adsorption pattern shown is for polysaccharide coated nevirapine nanocrystals. The in vitro protein prediction confirmed the targeting results obtained by gamma scintigraphic analysis when tested in rat (Shegokar, 2010 PhD Thesis)

Gref and co-workers determined the optimal PEG coat thickness and density, on biodegradable nanoparticles, to reduce plasma protein adsorption, surface charge, interaction with phagocytic cells, using 2D-PAGE technique [85]. They prepared PEG-PLA nanoparticles (160–270 nm) from diblock PEG-PLA, PEG-PLGA and PEG-PCL and from PEG-PLA: PLA blends. The plasma protein adsorption on the PEG-coated particles (~400 cpm) was drastically reduced as compared to the reference particles prepared from PLA (~1,600 cpm). Protein adsorption was decreased when the MW of PEG was increased from 2,000 to 5,000. Increase in the PEG MW above 5,000 showed no decrease in plasma protein adsorption. For PEG content 0.5–2 %, the overall protein amount (~1,600 cpm) was reduced by half as compared to the PLA reference particles (~800 cpm). The total amount of adsorbed proteins detected on the particles coated with 5 % PEG was three times lower than the nanoparticles with 2–5 %. PEG content. No considerable reduction of protein adsorption was achieved when further increasing the PEG content. The main proteins adsorbed on the nanoparticles were albumin, fibrinogen, IgG, Ig light chains and the apolipoproteins apoA-I and apoE. 2D PAGE studies thus helped elucidating the role of the type and amount of proteins adsorbed on the nanoparticles after i.v administration, as a function of their core. These results have been useful in the design of long circulating intravenously injectable biodegradable drug carriers endowed with protein resistant properties and low phagocytic uptake.

15.3.5 Size Separation Tool

Commercially available polystyrene particles with two different sizes (50 and 100 nm) and three different surface chemical modifications were studied for the effect of size and surface characteristics of the nanoparticles on plasma binding [86]. The size of nanoparticles defined the relative surface area and topological space for the binding of plasma proteins, while different chemical modifications on the nanoparticle surface presented significant surface property changes of charge and/or hydrophobicity. The nanoparticles with different sizes and surface modifications resulted in significant changes in the relative abundances of the nanoparticle bound proteins. Clustering analysis revealed that the surface chemical properties and particle size together played a significant role in the protein–nanoparticle interaction, and the surface property appeared to be more impactful on the protein absorption than did the particle size. The size separation is reported for synthetic (co)polymers, gold nanoparticles and nanoparticles [87].

15.3.6 Vaccine Developments

The 2D-PAGE technique in combination with Western blotting has been successfully applied in the discovery of antigens from *H. pylori*, *C. trachomatis* and *B. garinii*. Two-dimensional semi-preparative electrophoresis has provided complementary information regarding membrane protein expression in a strain of *H. pylori*. Through two-dimensional liquid chromatography tandem mass spectrometry, the most comprehensive information to date regarding protein expression in yeast was obtained. This technique is an important tool in vaccinology [88].

Proteomics complements genomics by offering abilities to rapidly identify the products of predicted genes, e.g. proteins in outer membrane preparations. The application of proteomics for identification of vaccine candidates for bacterium, *H. pylori* using two different approaches is described [89, 90]. The first involves rapid identification of a series of monoclonal antibody reactive proteins from N-terminal sequence tags. The other approach involves identification of proteins in outer membrane preparations by 2-D electrophoresis followed by trypsin digestion and peptide mass map analysis. Advanced proteomic technologies are currently employed to facilitate identification of changes in proteome accompanying tumorigenesis. Mapping of the protein and peptide spectra and comparison of proteomic profiles between tumorous and adjacent nontumorous tissue samples are currently implemented in the search for novel biomarkers with improved specificities for diagnosis, prognosis and treatment monitoring [91].

15.3.7 Toxicity Determination Tool

Use of stable isotope labelling by amino acids in cell culture (SILAC)-based quantitative proteomics to characterise the binding of human cellular proteins to two forms of carbon nanoparticles: namely multi-walled carbon nanotubes (MWCNTs) and carbon black (CB) [92]. The data showed MWCNTs and CB bind to vastly different sets of 750 proteins. The study proved that proteins with high affinity that bind with CNTs are potential targets after the latter enters the cell. The conformation and functions of these proteins may be affected, resulting in disruptions to cellular structure and functioning. Wang et al. in a cytotoxicity study showed that MWCNTs with larger diameters exhibit a higher level of toxicity to cells [93]. Interestingly, the toxicity of CNT was found to decrease after the protein binding [94] as these proteins saturated the protein binding sites of the CNT and prevented further interactions of cellular proteins. Thus, 2D-PAGE analysis of the serum protein binding to CNTs could account for its toxicity and thereby further use in medicine and therapy.

15.3.8 Other Applications

15.3.8.1 Applications in Food

Food technology has immensely benefited from the advances in current proteomic techniques. Proteomic analysis allowed to translate scientific knowledge into discovery and quantification of biomarkers that can allow to assess predisposition, efficacy and characterisation, and quantification of food bioactives [95], useful for the solutions of problems in food quality, safety and nutrition, and ultimately human health. Recent advancements in biochemistry and molecular biology have seen the use of proteins as potential candidate biomarkers for quality testing in various endeavours [96]. Nanoproteomics have been widely employed for biomedical applications for the identification of various biomarkers. Several parameters such as microbial contamination, allergens, toxins, adulteration and traceability of genetically modified organisms have been the focus of proteomic studies so far. Proteomic research is focused on the individuation of quality markers of various foods, from meat to vegetables, of biological or genetically engineered origin [97]. These protein quality biomarkers can be monitored throughout industrial production processes, in similar fashion to the actual application of proteomics in other fields of research [98]. A database of bioactive peptides has also been developed for determination of potential bioactivity of food proteins and their classification [99]. In the near future, the identification of potentially active food constituents and the demonstration of their bioavailability and efficacy will lead to develop "functional foods" for specific diseases or personalised nutrition. Biomarkers have been proven as being an important tool to understand and explain the responsiveness of humans to dietary components, whole food or individual compounds, and ultimately reveal the relationship between nutrition and health. Proteomic methods have been used to study the response of colon cancer cells to quercetin treatment [100] and allowed the discovery and development of several other nutraceuticals.

15.3.8.2 Applications in Microbiology

Protein expression profile of a single sample under different circumstances can be determined by the use of 2-DE. This profile can be compared to that of a control condition to examine the changes in protein expression of microbial/bacterial cells in response to a foreign stimulus resulting from chemical, physical or biological changes. A set of proteins that responds specifically to a particular status of the cell termed as a proteomic signature relates to a certain metabolic conditions such as the redox state, proton motive force, etc. This proteomic signature can be used to diagnose the cellular states of microbial organisms [101]. 2-DE has been extensively employed to differentiate and classify the microbial isolates required for the study of the molecular taxonomy and epidemiology of bacterial pathogens [102]. The proteome studies have been applied for the detection of antigens which may be used for diagnostics and vaccine candidate prediction [103]. 2-DE has been applied for the characterisation of proteins, characterising protein modifications, protein–protein interactions, metabolic engineering, characterisation of mutant proteins, microbial systematics and epidemiology, and evaluation of proteins involved in the toxic response [104].

V. cholera is a gram-negative bacterium with two different physiological states, in the aquatic environment and in the human small intestine. The whole cell proteome of the *V. cholera* strain N16961 under anaerobic conditions was separated by 2-DE and the protein spots compared with those in the aerobic environment. Under aerobic conditions, some proteins involved in substrate transport, amino acid metabolism and aerobic respiration were found to be abundant. The increased abundance of these proteins related to motility was observed when the bacterium was grown under anaerobic conditions, thus suggesting a correlation between *V. cholera* motility and pathogenesis [105]. This proteome analysis provided useful information for detection of the antigens by immunoproteomics for the purpose of vaccine development.

15.3.8.3 Clinical Chemistry

Gel electrophoresis is routinely used in clinical laboratories for screening protein abnormalities in various biological fluids (serum, urine, CSF). The technique is used for electrophoresis of serum, urine, CSF proteins, enzymes (ALP, LDH and CK), lipoproteins and haemoglobin [106]. Serum protein electrophoresis (SPE) is a very commonly used technique to evaluate changes in the relative concentration of fractions and thus allow easy recognition of pathological disorders associated with nephrotic syndrome, inflammatory reaction and hepatic diseases. Immunofixation (IFE) with use of specific antisera allows detection of the type of M component. For example, Bisalbuminemia as seen on the electrophoretogram as an albumin fraction split in two can be a sign of hereditary mutation or due to pancreatitis or a drug treatment such as high doses of beta-lactam in a patient with renal insufficiency.

Decrease in the α 1 globulin zone can be due to hepatocellular insufficiency, malnutrition or protein loss, congenital deficiency of α 1 anti trypsin whereas increase is mainly seen in inflammatory disorders.

Conclusions

Gel electrophoresis made significant advancement in protein separation and identification field. 2-D PAGE is a most commonly used technique in proteomic studies besides its long list of advances. Although, 2D-PAGE, like any fractionation scheme, has limitations, however the potential to explore though 2D PAGE is much more higher compared to other gel electrophoresis techniques. Literature confirms successful use of 2D PAGE in various pharmaceutical applications, e.g. identification of proteins, separation, microRNA research, vaccine and antibiotic development area. Various laboratories confirmed potential of 2D-PAGE for predicting targeting potential of nanoparticles. It is an essential technique for the characterisation of proteins 2-D PAGE holds promise in several pharma applications and, with no doubts it will be a method of choice in coming decades.

References

1. Hames BD, Rickwood D (eds) (1990) Gel electrophoresis of proteins: a practical approach, 2nd edn. Oxford University Press, New York, NY
2. Gallagher SR (ed) (2006) Current protocols in molecular biology 10.2.1. Wiley, New York, NY (analysis of proteins)
3. Smithies O, Poulik MD (1956) Two-dimensional electrophoresis of serum proteins. Nature 177:1033
4. Grabar P, Williams CA (1953) Méthode permettant l'étude conjuguée des proprietés électrophorétiques et immunochimiques d'un mélange de protéines; application au sérum sanguin. Biochim Biophys Acta 10:193–194
5. Bussard A, Huet JM (1959) Description of a technic simultaneously combining electrophoresis and immunological precipitation in gel: electrosyneresis. Biochim Biophys Acta 34:258–260
6. Laurell CB (1965) Antigen-antibody crossed electrophoresis. Anal Biochem 10:358–361
7. Ressler N (1960) Two-dimensional electrophoresis of protein antigens with an antibody containing buffer. Clin Chim Acta 5:795–800
8. Raymond S, Weintraub L (1959) Acrylamide gel as a supporting medium for zone electrophoresis. Science 130:711
9. Raymond S (1964) Acrylamide gel electrophoresis. Ann N Y Acad Sci 121:350–365
10. Margolis J, Kenrick KG (1969) Two-dimensional resolution of plasma proteins by combination of polyacrylamide disc and gradient gel electrophoresis. Nature 221:1056–1057
11. Kaltschmidt E, Wittmann HG (1970) Ribosomal proteins. VII: Two-dimensional polyacrylamide gel electrophoresis for fingerprinting of ribosomal proteins. Anal Biochem 36:401–412
12. Ornstein L (1964) Disc electrophoresis. I. Background and theory. Ann N Y Acad Sci 121:321–349
13. Davis BJ (1964) Disc electrophoresis. II. Method and application to human serum proteins. Ann N Y Acad Sci 121:404–427

14. Dale G, Latner AL (1969) Isoelectric focusing of serum proteins in acrylamide gels followed by electrophoresis. Clin Chim Acta 24:61–68
15. Macko V, Stegemann H (1969) Mapping of potato proteins by combined electrofocusing and electrophoresis identification of varieties. Hoppe Seylers Z Physiol Chem 350:917–919
16. Domschke W, Seyde W, Domagk GF (1970) Two dimensional separation of serum proteins by isoelectrical focussing and disk gel electrophoresis. Z Klin Chem Klin Biochem 8:319–320
17. Anderson L, Anderson NG (1977) High resolution two-dimensional electrophoresis of human plasma proteins. Proc Natl Acad Sci U S A 74:5421–5425
18. Bhakdi S, Knufermann H, Hoelzl Wallach DF (1974) Two dimensional separation of erythrocyte membrane proteins. Biochem Biophys Acta 345:448–457
19. MacGillivray AJ, Rickwood D (1974) The heterogeneity of mouse-chromatin nonhistone proteins as evidenced by two-dimensional polyacrylamide-gel electrophoresis and ion-exchange chromatography. Eur J Biochem 41:181–190
20. Stegemann H, Francksen H, Macko V (1973) Potato proteins: genetic and physiological changes, evaluated by one- and two-dimensional PAA-gel-techniques. Z Naturforsch C 28:722–732
21. Wrigley CW (1970) Protein mapping by combined gel electrofocusing and electrophoresis: application to the study of genotypic variations in wheat gliadins. Biochem Genet 4:509–516
22. Klose J (1975) Protein mapping by combined isoelectric focusing and electrophoresis of mouse tissues. Humangenetik 26:231–243
23. Klose J, von Wallenberg-Pachaly H (1976) Changes of soluble protein populations during organogenesis of mouse embryos as revealed by protein mapping. Dev Biol 51:324–331
24. Klose J (1979) Isoelectric focusing and electrophoresis combined as a method for defining new point mutations in the mouse. Genetics 92:s13–s24
25. O'Farrell PH (1975) High resolution two-dimensional electrophoresis of proteins. J Biol Chem 250:4007–4021
26. Bravo R, Celis JE (1982) Up-dated catalogue of HeLa cell proteins: percentages and characteristics of the major cell polypeptides labeled with a mixture of 16 14C-labeled amino acids. Clin Chem 28:766–781
27. Comings DE (1982) Two-dimensional gel electrophoresis of human brain proteins. I. Technique and nomenclature of proteins. Clin Chem 28:782–789
28. Goldman D, Merril CR, Polinsky RJ, Ebert MH (1982) Lymphocyte proteins in Huntington's disease: quantitative analysis by use of two-dimensional electrophoresis and computerized densitometry. Clin Chem 28:1021–1025
29. Klose J, Zeindl E, Sperling K (1982) Analysis of protein patterns in two-dimensional gels of cultured human cells with trisomy. Clin Chem 28:987–992
30. Klose J, Putz B (1983) Analysis of two-dimensional protein patterns from mouse embryos with different trisomies. Proc Natl Acad Sci U S A 80:3753–3757
31. Clark BF (1981) Towards a total human protein map. Nature 292:491–492
32. Schneider W, Klose J (1983) Analysis of two-dimensional electrophoretic protein patterns using a video camera and a computer. I. The resolution power of the video camera. Electrophoresis 4:284–291
33. Anderson NG, Anderson L (1982) The human protein index. Clin Chem 28:739–748
34. Klose J (1989) Systematic analysis of the total proteins of a mammalian organism: principles, problems and implications for sequencing the human genome. Electrophoresis 10:140–152
35. Merril CR, Goldman D, Sedman SA, Ebert MH (1981) Ultrasensitive stain for proteins in polyacrylamide gels shows regional variations in cerebrospinal fluid proteins. Science 211:1437–1438
36. Wilson KE, Marouga R, Prime JE, Pashby DP, Orange PR, Crosier S, Keith AB (2005) Comparative proteomic analysis using samples obtained with laser microdissection and saturation dye labelling. Proteomics 5:3851–3858

37. Bjellqvist B, Ek K, Righetti PG, Gianazza E, Gorg A, Westermeier R, Postel W (1982) Isoelectric focusing in immobilized pH gradients: principle, methodology and some applications. J Biochem Biophys Methods 6:317–339
38. Klose J, Kobalz U (1995) Two-dimensional electrophoresis of proteins: an updated protocol and implications for a functional analysis of the genome. Electrophoresis 16:1034–1059
39. Poland J, Cahill MA, Sinah P (2003) Isoelectric focusing in long immobilized pH gradient gels to improve protein separation in proteomic analysis. Electrophoresis 24:1271–1275
40. O'Farrell PZ, Goodman HM, O'Farrell PH (1977) High resolution two-dimensional electrophoresis of basic as well as acidic proteins. Cell 12:1133–1141
41. Aebersold R, Mann M (2003) Mass spectrometry-based proteomics. Nature 422:198–207
42. Nothwang HG, Schindler J (2009) Two-dimensional separation of membrane proteins by 16-BAC-SDS-PAGE. Methods Mol Biol 528:269–277
43. Jorgenson JW, Lukacs KD (1981) Zone electrophoresis in open-tubular glass capillaries. Anal Chem 53:1298–1302
44. Quang C, Petersen SL, Ducatte GR, Ballou NE (1996) Characterization and separation of inorganic fine particles by capillary electrophoresis with an indifferent electrolyte system. J Chromatogr A 732:377–384
45. Petersen SL, Ballou NE (1992) Effects of capillary temperature control and electrophoretic heterogeneity on parameters characterizing separations of particles by capillary zone electrophoresis. Anal Chem 64:1676–1681
46. Huff BV, McIntire GL (1994) Determination of the electrophoretic mobility of polystyrene particles by capillary electrophoresis. J Microcol 6:591–594
47. Templeton AC, Cliffel DE, Murray RW (1999) Redox and fluorophore functionalization of water-soluble, tiopronin-protected gold clusters. J Am Chem Soc 121:7081–7089
48. Liu FK, Koa FH, Huang PW, Wu CH, Chu TC (2005) Studying the size/shape separation and optical properties of silver nanoparticles by capillary electrophoresis. J Chromatogr A 1062:139–145
49. Liu FK, Huang PW, Chang YC, Ko FH, Chu TC (2004) Microwave-assisted synthesis of silver nanorods. J Mater Res 19:469–473
50. Mafune F, Kohno J, Takeda Y, Kondow T, Sawabe H (2000) Formation and size control of silver nanoparticles by laser ablation in aqueous solution. J Phys Chem B 104:9111–9117
51. Chen YH, Yeh CS (2002) Laser ablation method: use of surfactants to form dispersed Ag nanoparticles. Colloids Surf A 197:133–139
52. Aggarwal P, Hall JB, McLeland CB, Dobrovolskaia MA, McNeil SE (2009) Nanoparticle interaction with plasma proteins as it relates to particle biodistribution, biocompatibility and therapeutic efficacy. Adv Drug Del Rev 61:428–437
53. Rosengren A, Pavlovic E, Oscarsson S, Krajewski A, Ravaglioli A, Piancastelli A (2002) Plasma protein adsorption pattern on characterized ceramic biomaterials. Biomaterials 23:1237–1247
54. Marshall J, Kupchak P, Zhu W (2003) Processing of serum proteins underlies the mass spectral fingerprinting of myocardial infarction. J Proteome Res 2:361–372
55. Olivieri E, Herbert B, Righetti PG (2001) The effect of protease inhibitors on the two-dimensional electrophoresis pattern of red blood cell membranes. Electrophoresis 22:560–565
56. Hulmes JD, Deidra Bethea D, Ho K, Huang SP, Ricci DL, Opiteck GJ, Hefta SA (2004) An investigation of plasma collection, stabilization, and storage procedures for proteomic analysis of clinical samples. Clin Proteo Journal 1:17–31
57. Pasella S, Baralla A, Canu E, Pinna S, Vaupel J, Deiana M et al (2013) Pre-analytical stability of the plasma proteomes based on the storage temperature. Proc Natl Acad Sci U S A 11:10–20
58. Jiang L, Hea L, Fountoulakis M (2004) Comparison of protein precipitation methods for sample preparation prior to proteomic analysis. J Chromatogr A 1023:317–320

59. Ahmad Y, Sharma N (2009) An effective method for the analysis of human plasma proteome using two-dimensional gel electrophoresis. J Proteomics Bioinform 2:495–499
60. Lundqvist M, Stigler J, Elia G, Lynch I, Cedervall T, Dawson KA (2008) Nanoparticle size and surface properties determine the protein corona with possible implications for biological impacts. Proc Natl Acad Sci U S A 105(38):14265–14270
61. Gessner A, Lieske A, Paulke BR, Muller RH (2002) Influence of surface charge density on protein adsorption on polymeric nanoparticles: analysis by two-dimensional electrophoresis. Eur J Pharm Biopharm 54:165–170
62. Gessner A, Lieske A, Paulke BR, Muller RH (2003) Functional groups on polystyrene model nanoparticles: influence on protein adsorption. J Biomed Mater Res A 65A:319–326
63. Blunk T, Hochstrasser DF, Sanchez JC, Muller BW, Muller RH (1993) Colloidal carriers for intravenous targeting: plasma protein adsorption patterns on surface modified latex particles evaluated by two-dimensional polyacrylamide gel electrophoresis. Electrophoresis 14:1382–1387
64. Gessner A, Waicz R, Lieske A, Paulke BR, Mader K, Muller RH (2000) Nanoparticles with decreasing surface hydrophobicities: influence on plasma protein adsorption. Int J Pharm 196:245–249
65. Cedervall T, Lynch I, Foy M, Bergg T, Donnelly SC et al (2007) Detailed identification of plasma proteins adsorbed on copolymer nanoparticles. Angew Chem Int Ed 46:5754–5756
66. Cedervall T, Lynch I, Lindman S, Nilsson H, Thulin E, Linse S, Dawson KA (2007) Understanding the nanoparticle-protein corona using methods to quantify exchange rates and affinities of proteins for nanoparticles. Proc Natl Acad Sci U S A 104:2050–2055
67. Lück M, Schröder W, Paulke BR, Blunk T, Müller RH (1999) Complement activation by model drug carriers for intravenous application: determination by two-dimensional electrophoresis. Biomaterials 20:2063–2068
68. Leroux JC, Alleman E, de Jaeghere F, Doelker E, Gurny R (1996) Biodegradable nanoparticles: from sustained release formulations to improved site specific drug delivery. J Contr Release 39:339–350
69. Dobrovolskaia MA, Guszcysski T, Specht S, McLeland CB (2014) NCL method ITA-4—analysis of nanoparticle interaction with plasma proteins by 2D PAGE, Available from: http://ncl.cancer.gov/NCL_Method_ITA-4.pdf. Accessed on 09 July 2014
70. Göppert TM, Müller RH (2005) Adsorption kinetics of plasma proteins on solid lipid nanoparticles for drug targeting. Int J Pharm 302:172–186
71. Göppert TM, Müller RH (2003) Plasma protein adsorption of Tween 80- and poloxamer 188-stabilized solid lipid nanoparticles. J Drug Target 11:225–231
72. Göppert TM, Müller RH (2004) Alternative sample preparation prior to two-dimensional electrophoresis protein analysis on solid lipid nanoparticles. Electrophoresis 25:134–140
73. Thode K, Lück M, Semmler W, Müller RH, Kresse M (1997) Determination of plasma protein adsorption on magnetic iron oxides: sample preparation. Pharm Res 14:905–910
74. Hanash S (2003) Disease proteomics. Nature 422:226–232
75. Rabilloud T (2002) Two-dimensional gel electrophoresis in proteomics: old, old fashioned, but it still climbs up the mountains. Proteomics 2:3–10
76. Rui Z, Jian-Guo J, Yuan-Peng T, Hai P, BingGen R (2003) Use of serological proteomic methods to find biomarkers associated with breast cancer. Proteomics 3:433–439
77. Charrier JP, Tournel C, Michel S et al (2001) Differential diagnosis of prostate cancer and benign prostate hyperplasia using two dimensional electrophoresis. Electrophoresis 22:1861–1866
78. Steel LF, Shumpert D, Trotter M et al (2003) A strategy for the comparative analysis of serum proteomes for the discovery of biomarkers for hepatocellular carcinoma. Proteomics 3:601–609
79. Kellner R, Lichtenfels R, Atkins D et al (2002) Targeting of tumor associated antigens in renal cell carcinoma using proteome-based analysis and their clinical significance. Proteomics 2:1743–1751

80. Sinz A, Bantscheff M, Mikkat S et al (2002) Mass spectrometric proteome analyses of synovial fluids and plasmas from patients suffering from rheumatoid arthritis and comparison to reactive arthritis or osteoarthritis. Electrophoresis 23:3445–3456
81. He QY, Lau GK, Zhou Y et al (2003) Serum biomarkers of hepatitis B virus infected liver inflammation: a proteomic study. Proteomics 3:666–674
82. Tamilvanan S, Schmidt S, Muller RH, Simon B (2005) In vitro adsorption of plasma proteins onto the surface (charges) modified-submicron emulsions for intravenous administration. Eur J Pharm Biopharm 59:1–7
83. Shegokar R, Jansch M, Singh KK, Müller RH (2011) In vitro protein adsorption studies on nevirapine nanosuspensions for HIV/AIDS chemotherapy. Nanomed Nanotech Biol Med 7:333–340
84. Kasongo WK, Jansch M, Muller RH, Walker RB (2011) Evaluation of the in vitro differential protein adsorption patterns of didanosine-loaded nanostructured lipid carriers (NLCs) for potential targeting to the brain. J Liposome Res 21(3):245
85. Gref R, Luck M, Quellec P, Marchand M, Dellacherie E, Harnisch S, Blunk T, Muller RH (2000) 'Stealth' corona-core nanoparticles surface modified by polyethylene glycol (PEG): influences of the corona (PEG chain length and surface density) and of the core composition on phagocytic uptake and plasma protein adsorption. Colloids Surf B Biointerfaces 18:301–313
86. Zhang H, Burnum KE et al (2011) Quantitative proteomics analysis of adsorbed plasma proteins classifies nanoparticles with different surface properties and size. Proteomics 11:4569–4577
87. Kowalczyk B, Lagzi I, Grzybowski BA (2011) Nanoseparations: strategies for size and/or shape-selective purification of nanoparticles. Curr Opin Coll Inter Sci 16:135–148
88. Nilsson CL (2002) Bacterial proteomics and vaccine development. Am J Pharmacogenomics 2(1):59–65
89. Humphery-Smith I, Cordwell SJ, Blackstock WP (1997) Proteome research: complementarity and limitations with respect to the RNA and DNA worlds. Electrophoresis 18(8):1217–1242
90. Clayton RA, White O, Fraser CM (1998) Findings emerging from complete microbial genome sequences. Curr Opin Microbiol 1(5):562–566
91. Adamczyk-Poplawska M, Sergiusz Markowicz S, Jagusztyn-Krynick EK (2011) Proteomics for development of vaccine. J Proteom 74:2596–2616
92. Cai X, Ramalingam R et al (2013) Characterization of carbon nanotube protein corona by using quantitative proteomics. Nanomed Nanotech Biol Med 9:583–593
93. Wang X, Jia G, Wang H, Nie H, Yan L, Deng XY et al (2009) Diameter effects on cytotoxicity of multi-walled carbon nanotubes. J Nanosci Nanotechnol 9:3025–3033
94. Ge C, Du J, Zhao L, Wang L, Liu Y, Li D et al (2011) Binding of blood proteins to carbon nanotubes reduce cytotoxicity. Proc Natl Acad Sci U S A 108(41):16968–16973
95. Kussmann M, Panchaud A, Affolter M (2010) Proteomics in nutrition: status quo and outlook for biomarkers and bioactives. J Proteome Res 9:4876–4887
96. Cairns DA (2011) Statistical issues in quality control of proteomic analyses: good experimental design and planning. Proteomics 11:1037–1048
97. Gaso-Sokač D, Kovač S, Josić D (2010) Application of proteomics in food technology and food biotechnology: process development, quality control and product safety. Food Technol Biotechnol 48:284–295
98. Rinalducci S, D'Amici GM, Blasi B, Vaglio S, Grazzini G, Zolla L (2011) Peroxiredoxin-2 as a candidate biomarker to test oxidative stress levels of stored red blood cells under blood bank conditions. Transfusion 51:1439–1449
99. Minkiewicz P, Dziuba J, Iwaniak A, Dziuba M, Darewicz M (2008) BIOPEP database and other programs for processing bioactive peptide sequences. J AOAC Int 91:965–980
100. Mouat MF, Kolli K, Orlando R, Hargrove JL, Grider A (2005) The effects of quercetin on SW480 human colon carcinoma cells: a proteomic study. Nutr J 4:11

101. VanBogelen RA, Schiller EE, Thomas JD, Neidhardt FC (1999) Diagnosis of cellular states of microbial organisms using proteomics. Electrophoresis 20(11):2149–2159
102. Cash P (2009) Proteomics in the study of the molecular taxonomy and epidemiology of bacterial pathogens. Electrophoresis 30(S1):S113–S141
103. Jungblut PR, Hecker M (2007) Proteomics of microbial pathogens. Wiley-VCH Verlag GmbH & Co. KGaA, Weinheim, Germany
104. Tabandeh F, Shariati P, Khodabandeh M (2012) Application of two-dimensional gel electrophoresis to microbial systems. In: Magdeldin S (ed) Gel electrophoresis - principles and basics. Intech Publications, Croatia
105. Kan B, Habibi H, Schmid M, Liang W, Wang R, Wang D, Jungblut PR (2004) Proteome comparison of Vibrio cholerae cultured in aerobic and anaerobic conditions. Proteomics 4 (10):3061–3067
106. Giot JF (2010) Agarose gel electrophoresis – applications in clinical chemistry. J Mol Biol 29:9–14

Chapter 16
Tailored Bio-Polysaccharide Nanomicelles for Targeted Drug Delivery

Sabyasachi Maiti

Abstract For poorly water-soluble drugs, the dissolution time in the gastrointestinal contents may be longer than the transit time to the intended absorptive sites. Therefore, dissolution of drugs is quite often the rate-limiting step for drug absorption. This poses a major challenge for effective oral delivery of poorly soluble drugs. Recently, polymeric micelles composed of amphiphilic block or grafted copolymers have shown much advantage in drug delivery systems and attracted lots of interest due to its solubilization, low toxicity, long circulation, and passive targeting against tumor. Generally, amphiphilic copolymers can self-assemble to form nanosized spherical structures (10–200 nm) consisting of hydrophobic inner core and hydrophilic outer shell in aqueous medium. The hydrophobic cores can be used to entrap hydrophobic drugs, and release them in a controlled manner at a later stage, while the hydrophilic shell provides stabilization for the micelles with no need of additional stabilizers. Furthermore, the hydrophilic shell can be modified to have desirable properties, such as evading scavenging by the mononuclear phagocyte system (MPS) or obtaining active targeting. Natural polysaccharides are nontoxic, biodegradable, and easily amenable to chemical modifications to have better materials for drug delivery applications. In most of cases, a number of synthetic polymers have been investigated for their drug solubilizing capacity, loading efficiency, improved bioavailability, and targeting efficiency. However, the reports on natural polymer-based amphiphilic copolymer are limited in the literature. The purpose of this chapter is to illustrate recent advancements in the field of polymeric micelles emphasizing tailored bio-polysaccharide based micellar carrier systems.

S. Maiti (✉)
Department of Pharmaceutics, Gupta College of Technological Sciences, Ashram More, G.T Road, Asansol 713301, West Bengal, India
e-mail: sabya245@rediffmail.com

M. Lungu et al. (eds.), *Nanoparticles' Promises and Risks*,
DOI 10.1007/978-3-319-11728-7_16

16.1 Introduction

By some estimation, as much as 40 % of potentially valuable drug candidates identified by high throughput screening are rejected and never enter a formulation development stage due to their poor water solubility [1]. In order to attain expected therapeutic effect of such drug, it is usually required that a solubilized form of the drug be administered to a patient. For this purpose, a number of approaches have been adopted, which are based on the use of cosolvents, surfactants, soluble form of the drugs (salts), prodrugs, soluble polymer (β-cyclodextrin)-drug complexes, special drug carriers such as liposomes, micro- or nano-emulsions, and others. Each of the above methods is hampered by one or more particular problems. The capacity of the liposomal membrane and cyclodextrin inner cavity for carrying water insoluble molecules is limited and thus their solubilization capacity varies from drug to drug [2]. Unfortunately, most liposome formulations are readily disrupted by intestinal detergents, such as bile salts, and are subject to degradation by intestinal phospholipases. Both mechanisms trigger the premature release of the liposome-entrapped drug in the gastrointestinal (GI) fluid [3]. In addition to liposomes, other colloidal formulations, such as emulsions and microemulsions, are effective for oral administration of drugs with limited solubility in biological fluids [4]. Compared to emulsions, microemulsions offer several advantages as drug delivery systems, such as high solubilization capacity, thermodynamic stability, and ease of preparation. However, the administration of microemulsions, especially castor oil-based formulations has triggered adverse events, most frequently renal dysfunction, hypertension, and hepatotoxicity [5].

The methods based on the surfactant micelles to solubilize hydrophobic drugs have problems in that most of the surfactants are relatively toxic and that precipitation of hydrophobic drug occurs when subjected to dilution with aqueous solutions (such as physiological fluids upon a parenteral administration) [6]. Another option is to use certain micelle-forming surfactants in formulations of insoluble drugs. Surfactants cannot retain solubilized material at concentrations lower than their critical micelle concentration (CMC) value, which is typically rather high in the cases of conventional low molecular weight surfactants [7].

Currently in clinical practice, the excipients used in intravenous formulations for anticancer drugs are toxic, producing adverse reactions that prove lethal to patients. Cremophor EL (polyethoxylated castor oil) induces hypersensitivity reactions along with nephrotoxicity and neurotoxicity. This particular formulation has other problems, i.e., it tends to form minute precipitates which require the use of a filter in the injection line and the required period of administration is long about 24 h [8]. Dimethyl sulfoxide, ethanol, and Tween also produce toxic reactions on prolonged use. The formation of salts or pH adjustment facilitates the dissolution of poorly soluble drugs if they contain ionizable groups. However, in some cases, the extremes in pH that required for solubilization of drugs are not physiologically acceptable.

Low solubility in water appears to be an intrinsic property of many drugs. If the molecular target of a drug is located inside the cell, a drug molecule must have a certain degree of hydrophobicity in order to cross the cell membrane [9]. It has also been observed that a drug or drug candidate often needs a lipophilic group to show an affinity toward the target receptor [10]. Indeed, the GI tract presents a variety of hurdles for a drug, from morphological barriers (mucus layer, microvilli, etc.) to stringent physiological factors (a wide range of pH, enzymatic activities, specific transport mechanisms, etc.), which all conspire to limit intestinal absorption of drug. In the case of poorly water-soluble drugs, the dissolution time in the GI contents may be longer than the transit time to the intended absorptive sites [11]. Therefore, dissolution of drugs is quite often the rate-limiting step which, ultimately, controls the bioavailability of the drug. This poses a major challenge for effective delivery of poorly water-soluble therapeutics via the oral route. Intravenous administration of aggregates formed by insoluble drug can cause embolization of blood vessels resulting in side effects as severe as respiratory system failure [12]. The formation of drug aggregates can also result in high-localized concentrations at the sites of deposition associated with local toxicity and/or lowered systemic bioavailability. Some leading pharmaceutical companies make efforts to exclude poorly soluble compounds very early in their screening process regardless how active these compounds are toward their molecular targets [10].

To minimize drug degradation and loss upon administration, prevent harmful or undesirable side effects, and increase drug bioavailability and the fraction of the drug accumulated in the pathological zone, various drug delivery and drug targeting systems are currently being developed or under development. Among drug carriers one can find soluble polymers, microparticles made of natural and synthetic polymers, liposomes, and micelles. Each of these carrier types offers its own advantages and shortcomings, and all those carriers can be made slowly degradable, stimuli reactive (for example, pH or temperature sensitive), and even targeted (for example, by conjugating them with specific antibodies against certain characteristic components of the area of interest). In addition, drug carriers should be long circulating [13] since prolonged circulation allows for maintaining the required therapeutic level of pharmaceuticals in the blood for extended time intervals. Long-circulating, high molecular weight drugs or drug-containing microparticulates can also slowly accumulate in pathological sites with affected and leaky vasculature (such as tumors, inflammations, and infarcted areas) via the enhanced permeability and retention effect (EPR) and enhance drug delivery in these areas [14]. In addition, prolonged circulation can help to achieve a better targeting effect for specific ligand-modified drugs and drug carriers since it increases the total quantity of targeted drug/carrier passing through the target, and the number of interactions between targeted drugs and their targets [15]. Ideally, pharmaceutical drug carriers for parenteral administration are expected to be biodegradable, have small particle size, possess high loading capacity, demonstrate prolonged circulation, and accumulate in required pathological sites in the body [16]. The development of drug carriers meeting all these requirements for poorly soluble pharmaceuticals still represents a challenge.

Recently, the use of amphiphilic copolymer micelles has attracted much attention for solubilization of poorly soluble drugs. Suitable hydrophobic drugs which may be incorporated into the copolymer micelles can be any bioactive agents having limited solubility in aqueous environment. Without limiting the scope of this carrier systems, suitable hydrophobic agents include anticancer drugs, antipholgistic, anodynes, immunosupressants, hepatism remedies, homoine compositions, chemotherapeutics, metabolic pharmaceuticals, digestive disease remedies, peripheral disease remedies, and circulatory disease remedies.

The reports on synthetic amphiphilic polymer micelles are well addressed in the literature. A few excellent reviews have also been published [17–22] but they focused mainly on the commercially available Pluronics® amphiphilic polymer and other synthetic polymers decorated with hydrophilic poly (ethylene glycol) segment. Natural polysaccharides are known to be nontoxic, biodegradable, easily available, and amenable to various chemical modifications. Till date, the reports on polysaccharide-based nanomicellar systems are scarce in the literature. This chapter deals with the tailoring procedure for the natural polysaccharides, design of polysaccharide-based nanomicellar reservoirs for delivering poorly water soluble drugs to the target sites, its current state of the art, and future directions.

16.2 Surfactant and Polymer Micelles

Molecules or ions that are adsorbed at the interfaces are termed as surface active agents or surfactants. Alternatively, it is called amphiphiles meaning that the molecule or ion has a certain affinity for both polar and nonpolar solvents. Depending on the number and nature of the polar and nonpolar groups present, the amphiphiles may be predominantly hydrophilic, lipophilic, or reasonably well balanced between these two extremes [23]. Surface active agents associate to form colloidal aggregates termed micelles, which predominate in the system above a certain concentration, called critical micelle concentration (CMC). Micelles have the particle size normally within the 5–100-nm range. Micelles are formed when amphiphiles are placed in water. They consist of an inner core of assembled hydrophobic segments capable of solubilizing lipophilic substances and an outer hydrophilic corona serving as a stabilizing interface between the hydrophobic core and the external aqueous environment [24]. They physically entrap sparingly soluble pharmaceuticals and deliver them to the desired site of action at concentrations that can exceed their intrinsic water solubility and thus increase their bioavailability. In aqueous systems, nonpolar molecules are solubilized within the micelle core, polar molecules will be adsorbed on the micelle surface, and substances with intermediate polarity will be distributed along surfactant molecules in intermediate positions.

Kabanov et al. [25] suggested that they should be stable in vivo for a sufficiently long time and should not provoke any biological reactions. They should release a free drug upon contact with target tissues or cells; and, finally, the components of

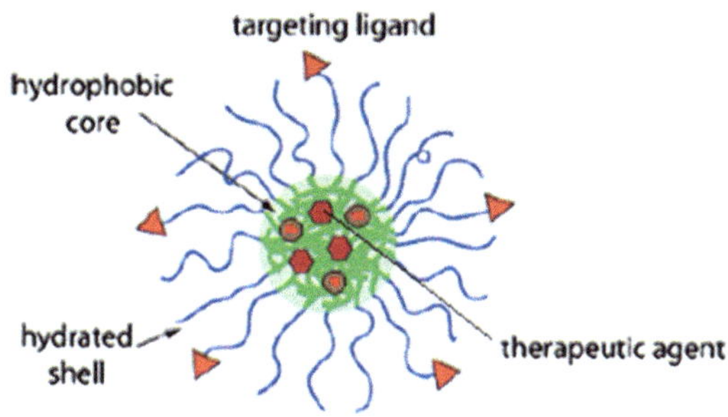

Fig. 16.1 Structure of polymeric micelles

the carrier should be easily removed from the body when the therapeutic function is completed. In broad terms, micelles as drug carriers provide a set of unbeatable advantages. The solubilization of drugs using micelle-forming surfactants (which results in formation of mixed micelles) leads to increased water solubility of a sparingly soluble drug and its improved bioavailability, reduction of toxicity and other adverse effects, enhanced permeability across the physiological barriers, and substantial changes in drug biodistribution [26]. Micelles may be made targeted by chemical attachment of targeting moiety to their surface. In the latter case, local release of the loaded drug from the micelles in the target organ should lead to increased efficacy of the drug. On the other hand, the drug is well protected in micellar form from possible inactivation under the effect of biological surroundings, and does not provoke undesirable side effects on nontarget organs and tissues. They can also be produced in a reproducible manner with a narrow particle size distribution (Fig. 16.1).

Owing to outer hydrophilic shell, the micelles remain stable in aqueous medium without the need of additional stabilizer. Further, the outer hydrophilic segment prevents easy recognition of the micelles by reticuloendothelial system (RES) and thus prolongs their circulation time in blood and allows drugs to be administered over prolonged periods of time. This is termed as stealth or long-circulating micelles. For nanoparticles, it is necessary to coat the particles with additional hydrophilic polymers (polyethylene glycol) or add nonionic surfactants (Tween 80, poloxamers) to impart it long circulating character. Hence, the design of micellar formulation is a best alternative. The size, charge, and surface characteristics can also be modified by variation in preparation technique.

However, one major disadvantage is that surfactant micelles rapidly break apart upon dilution, result in premature leakage of the drug, and ultimately cause its precipitation in situ. This limitation of surfactant micelles as drug delivery carriers triggered the search for micelles of significantly enhanced stability and solubilizing power. The design of biopolymer-based micelles has gained much attention because of their biocompatibility, biodegradability, and the multiplicity of functional groups they display for the conjugation of pilot molecules [27]. Like their low-molecular-weight counterparts, amphiphilic polymers associate in water to form polymeric micelles [28].

Polymeric micelles can be used as efficient carriers for compounds, which alone exhibit poor solubility, undesired pharmacokinetics, and low stability in a physiological environment. The hydrophilic shell contributes greatly to the pharmaceutical behavior of polymeric formulations by maintaining the micelles in a dispersed state, as well as by decreasing undesirable drug interactions with cells and proteins through steric stabilization effects. There are significant differences between the two types of assemblies from the physicochemical viewpoint. The polymer concentration at which the association first takes place, sometimes known as the critical association concentration (CAC), is lower by several orders of magnitudes than typical surfactant CMC values. They can increase drug bioavailability and retention, since the drug is well protected from possible inactivation under the effect of their biological surroundings [29]. Polymeric micelles have been studied extensively as delivery medium for injectable drug formulations of poorly soluble drugs such as paclitaxel, indomethacin, amphotericin B, adriamycin, and dihydrotestosterone. Overall, they proved to be highly effective drug delivery vehicles.

Driving force for micelle formation is mainly based on the hydrophobic interaction, hydrogen bonding, metal complexation, and electrostatic interaction. In order to modify these properties, functional groups are introduced to the core-forming segments. Previous experiments have prompted reports that the polymeric micelles from hydrophobic interactions remain stable in aqueous solutions for days. The preparation of the polymeric micelle also involves the induction of strong interactions between the core-forming blocks, such as electrostatic interaction, and metal complexation. Basically, the reduction of free energy is the key factor for preparing stable polymeric micelles. Therefore, it must be emphasized that entropy change is not always the major driving force for micelle formation. When strong cohesive forces such as hydrogen bonding and metal complexation are involved, enthalpy change plays a more important role in the stabilization of polymeric micelles. Indeed, the polymeric micelles prepared solely through the entropy-driven mechanism dissociate easily under the diluted conditions, and therefore, these micelles are not suitable for parenteral drug delivery.

The hydrophobic core of the polymeric micelles is expected to serve as the loading space for various lipophilic drugs. By design, given the nanometer size of the micelles, this space is limited. In order to exploit fully this loading space, one must manipulate the many factors that control loading capacity and efficiency. In devising a drug incorporation strategy, one must try to match as closely as possible the polarity of the hydrophobic micelle core to the solubility characteristics of the drug.

The hydrophilic corona of polymeric micelles can be composed of various types of polymers. PEG, the most commonly used shell-forming polymer, is one of the few synthetic polymers approved by the Food and Drug Administration (FDA) for internal use. Its biocompatibility and lack of toxicity have largely contributed to its acceptance. When hydrated, PEG forms a dense brush of polymer chains stretching out from the core of the micelle. Owing to its high aqueous solubility, high mobility, and large exclusion volume [30], PEG imparts steric stability by minimizing the interfacial free energy of the micellar core and by impeding hydrophobic intermicellar attractions [31].

16.3 Preparation of Drug-Loaded Polymer Micelles

In order to incorporate or solubilize one or more drugs physically into the block copolymer micelle, various methods have been proposed [32]:

(a) Stirring. A drug is added to an aqueous solution of a block copolymer, and stirred for 2–24 h to obtain micelles containing the physically entrapped drug.
(b) Heating. A drug and a block copolymer are dissolved in an organic solvent and the solvent is evaporated off at an elevated temperature (40–80 °C under a nitrogen atmosphere or by rotary evaporator under vacuum). The resulting mixture is kept at a temperature of 20–80 °C, preferably at 40–70 °C, for 2 h. Then, warm water (40–70 °C) is added thereto, and the mixture is stirred until a polymeric micelle containing drug is formed.
(c) Ultrasonic Treatment. A mixture of a drug and an aqueous solution of a block copolymer is subjected to ultrasonic treatment for a period ranging from about 1 s–1 h and then stirred at room temperature to obtain micelles containing the drug.
(d) Solvent Evaporation. A drug is dissolved in a water-immiscible organic solvent, for example, dichloromethane, chloroform, and the like, and then added to an aqueous solution of a block copolymer. Subsequently, the organic solvent is slowly evaporated off at 25–40 °C while stirring, and then filtered to remove undissolved drug. Drug along with the polymer is dissolved in a water-immiscible organic solvent like tetrahydrofuran, chloroform, acetone, or a mixture of solvents like chloroform and ethanol, and this solution is slowly added to the distilled water under vigorous stirring to form an emulsion with an internal organic phase and continuous aqueous phase, which rearranges the polymer to form micelles. Sometimes surfactants like polyvinyl alcohol are used in aqueous solution. This emulsion is then kept open to air with stirring so as to evaporate all the organic solvent.
(e) Dialysis. A drug and a block copolymer are dissolved in a water-miscible organic solvent. The solution is dialyzed against a buffer solution and then against water for the removal of organic solvent. In the dialysis method, suitable water-miscible organic solvents for dissolving drugs are selected from the group consisting of acetonitrile, dimethylformamide, dimethylsulfoxide, dioxane, dimethylacetamide, and the like.
(f) Solid dispersion method. In this method, drug along with the polymer is dissolved in the organic solvent, and a solid polymer matrix is obtained after the evaporation of solvent under reduced pressure. Drug-loaded PMs are obtained after the addition of water to the preheated polymer matrix [33].
(g) Microphase separation method. In this method the drug and polymer are dissolved in (organic solvent) tetrahydrofuran, and the solution is added dropwise in water under magnetic stirring. PMs are formed spontaneously, and drugs are entrapped in the inner part of the micelles. Organic solvent is removed under reduced pressure, and a blue-colored PM solution is formed [34].

One another innovative, one-step method of preparation of PMs consists of lyophilization of the solution mixture of drug and polymer in a water–*tert*-butanol system. Reconstitution of this freeze-dried cake of drug–polymer mixture with injectable vehicle catalyzes spontaneous formation of PMs [35].

16.4 Polymeric Micelles in Drug Targeting

Beyond solubilizing hydrophobic drugs, block copolymer micelles can also target their payload to specific tissues through either passive or active means. Drug-carrier delivery systems employ biologically inert macromolecules to direct a drug to its target site in the body. The major advantage of drug-carrier delivery systems is that the distribution of drugs in the body depends on the physicochemical properties of the carrier, not those of drugs. This implies that targeting can be manipulated by choosing an appropriate carrier, or by alterations in the physicochemical properties of the carrier. Targeting with drug-carrier systems can be divided into three types: passive, active, and physical and these are described as follows:

(a) Passive Targeting. Passive targeting relies on the natural distribution pattern of drug-carrier systems. For example, particles with a diameter of 5 μm or smaller are readily removed from the blood by macrophages of the RES when administered systemically. This natural defense mechanism of the RES thus provides an opportunity to target drugs to macrophages if they are encapsulated in or conjugated with an appropriate carrier system. Passive targeting also includes delivery of drug-carrier systems directly to a discrete region in the body (different regions of GI tract, eye, rectum, vagina, respiratory tract, etc.). This offers the opportunity for the treatment of diseases that require a persistent and sustained presentation of drugs at that site [36].

    Initially it was thought that receptor-mediated targeting was the only way to target the anticancer drugs to therapeutic sites; thus, many researchers have developed engineered polymer micelle conjugates in which various ligands have been attached to them. However, many recent studies have revealed that polymer-conjugated drugs and nanoparticles show prolonged circulation in the blood followed by passive accumulation in tumors even in the absence of targeting ligands [37], demonstrating the existence of a passive retention mechanism. Passive targeting of polymeric micelles to solid tumor can be achieved through the enhanced permeability and retention (EPR) effect [38]. In solid-tumor tissues, pathological, pharmacological, and biochemical studies show that solid tumor generally possesses such pathophysiological characteristics as hypervasculature, incomplete vasculature architecture, secretion of vascular permeability factors stimulating extravasation, and immature lymphatic capillaries [39]. Tumor vasculature has such characteristics as a high proportion of proliferative endothelial cells, increased tortuosity, pericyte deficiency, and aberrant basement membrane formation, resulting from the rapid vascularization necessary to provide oxygen and nutrients to proliferating tumors. These characteristics render tumor blood vessels permeable to macromolecules. Furthermore, the tumor lymphatic drainage system does not operate effectively as a result of immature lymphatic capillaries; thus, macromolecules are selectively retained for prolonged time periods in the tumor interstitium [40]. Thus, numerous studies have shown that the EPR effect causes passive accumulation of macromolecules and nanoparticulates in solid tumor, enhancing the therapeutic index while decreasing side effects. Also, it has been found that in most human tumors the effective pore size in the vasculature ranges from 200 to 600 nm in diameter, which allows for passive targeting to tumors [41]. It has been noted that secretion of various factors such as nitric oxide, prostaglandins, bradykinin, and basic fibroblast growth factor in tumor tissues and overexpression of genes such as vascular permeability factor or vascular endothelial growth factor (VEGF) [42, 43] cause hyperpermeability of tumor microvasculature. However, it should be noted that permeability of vessels in tumor varies during progression, tumor type, and anatomical location of tumor, whereas the physicochemical properties of polymer used affect the extravasations of polymeric nanoparticulates [44]. For passive targeting, micelles have to circulate in blood for longer periods, and their size determines their biological fate; thus, PMs smaller than 5 and 5–10 nm are easily eliminated through the renal glomeruli, whereas the larger micelles with size range of 50–100 nm are removed by the liver and spleen [45].

(b) Active Targeting. In cellular-specific targeting, pilot molecules are installed at the end of the hydrophilic segment so that they may extend outward from the micelle corona and readily encounter and interact with membrane receptors. The main purpose of functionalization of the hydrophilic corona is to modulate the biodistribution of polymeric micelles and induce specific cellular uptake by receptor-mediated endocytosis. Certain types of tissues are known to overexpress specific protein receptors on their surface. The localization of such site-specific receptors has contributed to several advancements in the field of targeted drug delivery. The tethering of a ligand to the outer shell of micelles is most often achieved through the postmodification of a copolymer with bifunctional spacer molecules [46] or via the direct synthesis of heterobifunctional block copolymers [47]. Targeted micelles generally exhibit greater cellular uptake and improved in vitro efficacy than their unmodified counterparts.

Active targeting employs a deliberately modified drug and drug-carrier molecule capable of recognizing and interacting with a specific cell, tissue, or organ in the body. Modifications of the carrier systems may include a change in the molecular size, alterations of the surface properties, incorporation of antigen-specific antibodies, or attachment of cell receptor specific ligands [48].

Here, the carrier can be engineered by means of ligand coupling with monoclonal antibodies, folate, transferrin, luteinizing hormone-releasing hormone, epidermal growth factor (EGF), or α_2-glycoprotein or addition of a pH-sensitive moiety according to the biological characteristics of the tumor tissue.

As with other delivery systems, the drug delivery potential of polymeric micelles may still be further enhanced by attaching targeting ligands to the micelle surface. The attachment of various specific ligands to the water-exposed termini of hydrophilic blocks could be used to improve the targeting of micelles and micelle-incorporated drugs. Among those ligands one can name various sugar moieties [49], transferrin [46] and folate residues [50] since many target cells, especially cancer cells, overexpress appropriate receptors (such as transferrin and folate receptors) on their surface. Thus, it was shown that galactose- and lactose-modified micelles made of PEGpolylactide copolymer specifically interact with lectins, thus modeling targeting delivery of the micelles to hepatic sites [51]. Transferrin-modified micelles based on PEG and poly (ethyleneimine) sized between 70 and 100 nm are expected to target tumors with overexpressed transferring receptors [46].

Active targeting makes use of characteristics shown by the tumor cells, such as overexpression of cell surface tumor-associated antigens that are at low levels in normal tissue cells, as well as of the tumor-specific antigens and relatively more acidic nature of tumor (pH 7.0) than normal tissue (pH 7.4). Specific interactions between the targeting components with antigens displayed on target tissues cause the selective accumulation of drug in the target tissue. Active targeting decreases adverse side effects, because the drug accumulates only in the tumor sites and it allows the drugs cellular uptake through endocytosis [52]. Micelles with surface-attached specific antibodies, also called immunomicelles, provide an opportunity for targeting in term of diversity of targets and specificity of interactions. Micelle attached antibodies retain their ability to specifically interact with their antigens.

Micelles equipped with folate ligands are used mainly for the intracellular transport of the drug. After intracellular transport of folate-equipped micelles, the drug carried by micelles should also be released intracellularly; this problem is solved by the pH sensitivity of PMs that enables selective drug release in the intracellular acidic compartment of endosomes with pH 5–6. Such micelles are crucial to the carriers for delivery of materials that should become pharmaceutically effective after entering the cell. Folate behaves as a ligand having high affinity for its receptors, folate-binding proteins that are selectively overexpressed on the surface of the cancer cells [53, 54]. Elevated levels of folate receptors are expressed on epithelial tumors of various organs such as colon, lung, prostate, ovaries, mammary glands, and brain [55]. Thus, the folate conjugates with an appropriate design can be directed to the tumor cells in the body and internalized in the target cells via receptor-mediated endocytosis.

Transferrin also acts as a suitable ligand for targeting tumor tissue, because the receptors for transferrin are overexpressed on cancer cells in correlation with the degree of malignancy [56]. Transferrin is an 80-kDa glycoprotein synthesized by liver that binds to endogenous iron in plasma. Transferrin, after interaction with its surface receptors, undergoes endocytosis into the acidic compartment. In an acidic environment iron dissociates from transferrin, and dissociated transferrin recycles after being released from cells. Enhanced cellular uptake and reduced exocytosis were shown by the transferrin-conjugated nanoparticles loaded with PTX. Nanoparticles consisted of a shell of biodegradable polymer (PLGA). Nanoparticles showed greater anti-cancer activity and more sustained effect compared to free drug or conjugated nanoparticles [57].

(c) Physical targeting. Other targeting approaches rely on the fact that many pathological processes present either a slight increase in temperature or decrease in pH. Physical targeting refers to delivery systems that release a drug only when exposed to a specific microenvironment, such as a change in temperature or pH or the use of an external magnetic field.

The tumor extracellular pH, which is slightly lower than normal tissue pH, is the distinguishing characteristic of most solid tumors. The tumor extracellular pH ranges from pH 5.7–7.8 measured with invasive microelectrodes, with a mean value of 7.0, but most of the pH values are below pH 7.2, with normal blood pH remaining constant at pH 7.4. This is mainly attributed to the higher rate of aerobic and anaerobic glycolysis compared to normal cells [58]. Cellular compartments, such as endosomes and lysosomes, exhibit even lower pH levels of approximately 5–6. pH-sensitive block copolymer micelles capable of dissociating in response to decreased pH levels have been designed to free their incorporated drug molecules upon accumulation at the tumor site and/or entry into the cytoplasm. Such systems are developed by using pH-sensitive polybase poly(L-histidine), using pH-sensitive linkages (hydrazone linkage), or introducing titrable groups (methacrylic acid) into the copolymer at the hydrophilic end while modifying PMs for pH-sensitive targeting. Micelles with a phase transition of approximately pH 7.0 are useful for tumor targeting, in that solid tumors have a pH of approximately 7.0, but for subcellular targeting, micelles that are responsive to pH values ranging from 5.0 to 6.0 (endosomal–lysosomal pH) are beneficial.

Lee and co-workers have developed such a system, in which the CMC of the PEG-*b*-poly (L-histidine) copolymer was found to increase, thus reducing thermodynamic stability as the pH of the medium was reduced [59]. An abrupt increase in the CMC occurs at pH 7.2, resulting in unstable micelles that release their content. Doxorubicin-loaded PEG-*b*-poly (L-histidine) micelles showed a significant increase in the in vitro intracellular accumulation of drug, in vivo tumor localization, and tumor growth inhibition in the A2780 ovarian tumor model [60] via extracellular pH-dependent targeting of the tumor. pH sensitivity was also obtained by linking a pH-sensitive moiety between the drug and copolymer. A hydrazone linkage between doxorubicin and PEG-*b*-(aspartate) copolymer had been used. Such micelles showed selective release of doxorubicin in the endosomes where pH ranges from 5.0 to 6.0. Released drug was found to localize in cytoplasm and eventually in the nucleus.

Anticancer drugs targeting to tumors can also be triggered by hyperthermia. This method has significant advantages in clinical applications because of its ease of implementation and reduction of side effects of chemotherapy. For such types of targeting, drugs are loaded into the thermosensitive polymer matrix with a low critical solution temperature (LCST) between 37 and 42 °C to release drug site specifically upon local heating [61]. However, because the LCST of most thermoresponsive materials is below 37 °C, PMs with such materials would precipitate once injected into the human body with a normal body temperature of 37 °C and thus lose their thermal targeting function. Locally lowering or increasing the temperature sometimes becomes difficult, because local application can only activate the micelles that are situated closely underneath the skin but not those that are deeply distributed as in the bloodstream. This obstacle can be overcome by the incorporation of metal or metal oxide nanoparticles such as gold and γ-Fe_2O_3, which can be activated by irradiation by infrared light or by exposure to an alternating magnetic field, respectively [62, 63].

16.5 Engineered Amphiphilic Bio-Polysaccharides

Polysaccharides are polymers of monosaccharides (sugars). They are found in abundance and are inexpensive and available in a variety of structures exhibiting a variety of properties [64]. They can be easily modified chemically and biochemically. Further, they are nontoxic and in addition biodegradable, which suggests their use in targeted drug delivery systems.

Usually, natural polysaccharides are used in the food industry and in developing pharmaceutical dosage forms. They are commonly used as binder, thickener, suspending agents, and matrix forming materials and regarded as safe for human consumption. Therefore, in recent years, much emphasis is being given to the polysaccharide biomaterials and their structural modifications via chemical pathway with the aim to have better materials for designing novel drug delivery systems. Some of the potential bio-polysaccharides and their design as micellar carriers have been discussed below:

(a) Gellan Gum. Gellan gum is an anionic, bacterial (*Sphingomonas clodea*) exopolysaccharide, and consists of repeating tetrasaccharide units of β-D-glucose, β-D-glucuronic acid, and α-L-rhamnose residues [65]. There are two chemical forms of gellan gum (1) native or natural form, which has high acyl contents, and (2) low or deacetylated form (Gelrite®). The native form contains two acyl substituents, namely, acetate and glycerate, both of which are located on the same glucose residue.

The synthesis of C_{18}-*g*-gellan copolysaccharide was described as follows. Initially, 20 ml of thionyl chloride was added to 5 % (w/v) solution of sorbitan monooleate in chloroform and refluxed for 2 h without heat for chlorination. A semisolid, blackish brown mass of chlorinated sorbitan monooleate was obtained. Later, a homogenous dispersion of gellan polysaccharide (3 %, w/v) was prepared in dimethyl formamide (DMF) and the temperature of the dispersion was maintained at 10 °C. To this, same amount of sodium hydride was added. After this, a dispersion of chlorinated sorbitan monooleate in DMF (0.02 %, w/v) was also added. Thereafter, the resultant mixture was stirred for 1 h at room temperature and the total reaction mixture was transferred into 50 ml of distilled water. Subsequently, the copolysaccharide layer was isolated and adjusted to pH 7.0. The copolysaccharide was purified in ethanol, filtered off, and air-dried.

The dispersion of the copolymer led to the formation of nanomicellar carrier (371–750 nm), which was capable of holding >95 % budesonide and exhibited a longer dissolution profile in simulated nasal fluid (pH 5.5) [66].

(b) Dextran. Dextrans are a class of polysaccharides with a linear polymer backbone with mainly 1,6-α-D-glucopyranosidic linkages. They are obtained from bacterial cultures of *Leuconostoc mesenteroides* NRRL B-512 [67]. Dextrans are colloidal, hydrophilic, and water-soluble substances, inert in biological system, and do not affect cell viability.

Francis and his coworkers reported this method for the synthesis of amphiphilic polysaccharides [68]. In order to impart dextran (DEX) an amphiphilic character, a small number of hydrophobic cetyl groups were linked to its backbone. The hydrophobic groups were attached to the polysaccharide hydrophilic backbone via a short PEO linker, yielding DEX grafted with PEO_{10}-C_{16} (DEX-*g*-PEO_{10}-C_{16}); the digit next to PEO refers to the number of ethylene oxide groups in PEO, and the digit next to the carbon refers to the number of carbons in the alkyl chains. In aqueous solution, DEX-C_{16} form micelles that can entrap up to 8.5 % w/w of Cyclosporine A, incorporated within the hydrophobic core of the micelle by a dialysis process. Micelles with average hydrodynamic diameters of ~11–14 nm are formed in aqueous solutions of DEX-C_{16}.

(c) Chitosan. Chitosan is a high molecular weight, polycationic polysaccharide derived from naturally occurring chitin by alkaline N-deacetylation [69].

Chitosan, a natural aminopolysaccharide, is a pharmaceutical excipient for drug delivery because it has favorable biological properties, such as biocompatibility, biodegradability, positive charge, nontoxicity, and bioadhesivity. However, in spite of many reported successes, the major drawback of chitosan is its poor solubility at physiological pH, whereas it is soluble only in acidic environments with pH values lower than 6.0; this limits its application in the pharmaceutical field.

N-Trimethyl chitosan chloride (TMC), a partially quaternized chitosan derivative, is soluble in the entire pH range. However, TMC is a hydrophilic polymer which lacks any hydrophobic segments, so that it does not form micelles for solubilization in aqueous media. Therefore, hydrophobically modified TMC were constructed by chemically conjugating the hydrophobic segments to the hydrophilic TMC backbone.

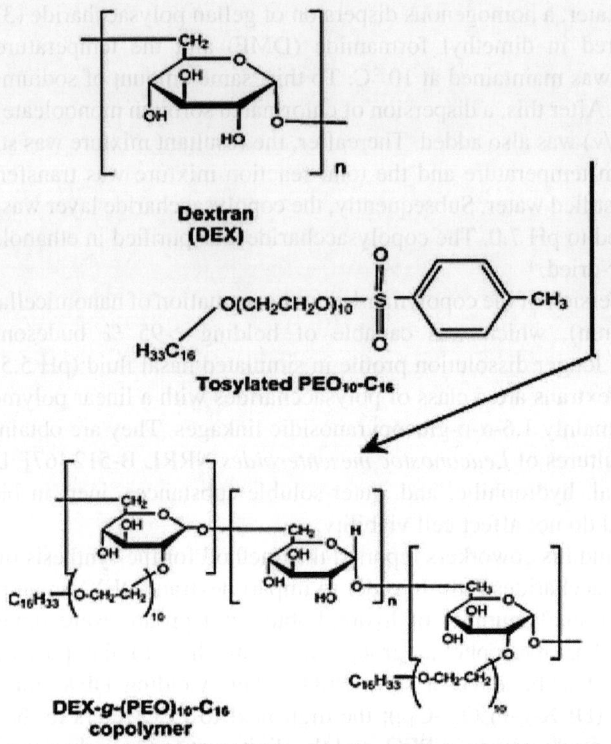

Fig. 16.2 Synthesis of DEX-*g*-PEO_{10}-C_{16} copolymer [68]

Trimethyl chitosan polymers can be synthesized by reductive methylation of chitosan by the chemical reaction between chitosan, *N*-methyl-2 pyrrolidinone (NMP), and iodomethane in the presence of sodium hydroxide (Figs. 16.2 and 16.3) [70].

A 100 ml three-neck flask, equipped with a reflux condenser and a dropping funnel, is loaded with decanoic acid (3.7 g) and heated at 84 °C with an oil bath to melt the acid. A mixture of thionyl chloride ($SOCl_2$, 7.85 g) and dimethyl formamide (0.5 mmol) is added slowly at 84 °C within 30 min. After 5 h, the product is recovered by removing the excess $SOCl_2$ under vacuum conditions.

The synthetic route leading to amphiphilic *N*, *O*-acyl group-*N*-trimethyl chitosan chloride from trimethyl chitosan (TMC) is as follows: TMC (1.0 g) is dissolved in 30 ml of the aqueous solution with trifluoroacetic acid (0.1 ml) [71, 72] and the acetone (80 ml) is added under intense stirring. Pyridine (3.9 g) dissolved in acetone (40 ml) is first added dropwise into the mixture and then a solution of decanoyl chloride (40 mmol) obtained by the above method in acetone (40 ml) is loaded later on. The reaction is lasted for 12 h and then the mixture is poured into ice water, subsequently the suspension is filtered and the filter liquor is extracted by diethyl ether for five times, dialyzed by dialysis membranes (MWCO 10,000) against

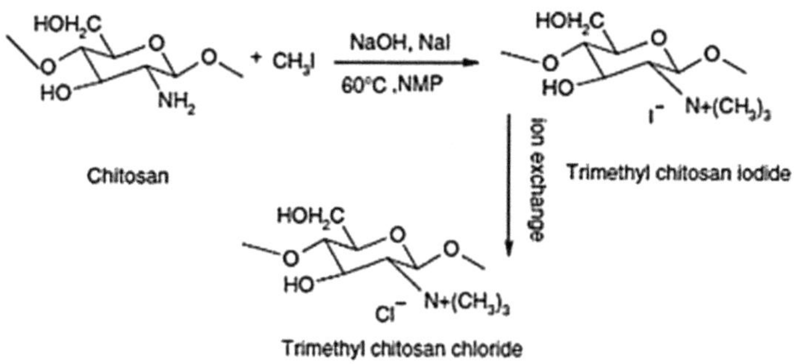

Fig. 16.3 Synthesis of trimethyl chitosan [70]

distilled water for 3 days, and lyophilized further. The same process can be repeated with lauric acid, myristic acid, palmitic acid, and stearic acid, respectively, to obtain desired product.

The solubilization of poor water-soluble drugs using *N*-acyl, *O*-acyl-*N*-trimethyl chitosan chloride (ATMC) micelles as a carrier system was investigated. Long-chain saturated fatty acids (C_{14}) were grafted. The ATMC micelles self-assemble and were used to encapsulate the poorly soluble drug, Cyclosporine A. These assemblies were prepared by a dialysis, wherein the drug loading capacity of the ATMC micelles ranged from 9.6 to 17.1 % and encapsulation capacity ranged from 35.8 to 69.8 %, with the mean micellar particle size of 288 nm. These ATMC micelles are being investigated as carriers to improve oral administration absorption of poorly permeable drugs [72].

ChitoPEG graft copolymer can be synthesized as reported previously [73]. Briefly, 100 mg of chitosan is dissolved in 0.2 ml of deionized water and diluted with 9.8 ml of DMSO. To this solution, calculated amount of methoxy poly (ethylene glycol) (MPEG)-*N*-hydroxysuccimide (NHS) dissolved in 2 ml of DMSO is added and reacted for overnight at nitrogen atmosphere. After that, the resulting solution is dialyzed using dialysis tube against a plenty of deionized water for 2 days followed by its lyophilization. Generally, chitosan is not soluble in dichloromethane (DCM) while MPEG is freely soluble in it. Therefore, unreacted MPEG-NHS is removed by suspending lyophilized solid into a plenty of DCM for three times. After that, ChitoPEG copolymer is fractionated into deionized water followed by its lyophilization. The methotrexate (MTX)-incorporated polymeric micelles of ChitoPEG copolymer has a particle size of around 50–100 nm. The drug contents of the polymeric micelle are around 4–12 % and the loading efficiency of MTX in the polymeric micelles is higher than 60 % (w/w) for all of the formulations (Fig. 16.4).

Stearic acid (SA) grafted chitosan oligosaccharide (CSOS-SA) can be synthesized by 1-ethyl-3-(3-dimethylaminopropyl) carbodiimide (EDC)-mediated coupling reaction. Low molecular weight chitosan (20 kDa molecular wt)-stearic acid (SA) is obtained via the reaction of carboxyl groups of SA with amino groups of CSOS in the presence of EDC, according to the previous reports [74, 75]. Briefly,

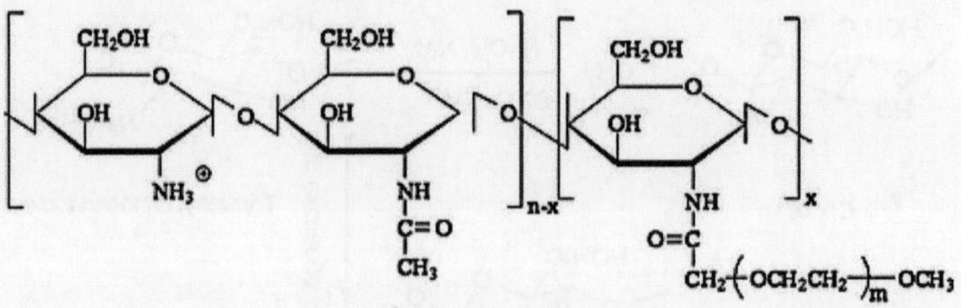

Fig. 16.4 Structure of methoxy poly (ethylene glycol) grafted chitosan copolymer [73]

CSOS (2.0 g) is dissolved in 50 ml distilled water. SA (1.75 g) and EDC (10 mol/mol SA) are dissolved in 40 ml ethanol. CSOS solution is heated up to 80 °C under vigorous stirring accompanied by dropwise addition of SA solution. The by-products were removed using the ultrafiltration by Millipore Labscale TFF system (MWCO 10,000) after 5 h. The size of CSOS-SA micelles is about 30 nm. The saturated solubility of 10-hydroxycamptothecin was 1.01 μg/ml in distilled water under 37 °C, and it is enhanced markedly to be 18.92 μg/ml due to encapsulation of the micelles.

The hydrophilic polysaccharides can be used as hydrophilic carrier in the design of nano-polysaccharide micelles. The polysaccharides which lacks sufficient hydration/water solubility and do not qualify for the use as hydrophilic part former of micellar structure, they can be modified chemically via etherification reaction to make them useful. Polymeric micelles are unstable at elevated temperature, at concentrations below their CMC, and to changes in solvent composition. Given the need for well-defined robust nanostructures for applications as vessels for targeted delivery vehicles, there has been significant interest in the stabilization of these micelles. The shell cross-linked nanostructures have been demonstrated to have great potential as encapsulation and delivery agents due to their higher core mobility and more rigid form compared with core cross-linked particles.

Covalent crosslinking is a common method for the generation of stable, hydrophilic shell-crosslinked micelles. In case, the water soluble is the hydrophilic shell, then by controlling the concentration of gelation metal ions, shell-crosslinked micelles can easily be obtained without the need of toxic, chemical reagents and solvents. This approach is ecofriendly and as the micelles are designed for targeted delivery, hence, there are no additional chances of remaining toxic reagents/solvents.

Further, galactose, mannose residues are also installed into PEG backbone of the micellar structure as ligand for drug targeting. Because, natural polysaccharides already contain sugar moieties there is no need to add such type of ligands in the micelles where polysaccharides have been used as hydrophilic corona.

Etherification involves the reaction of an alcohol (here a saccharide alcohol) with an alkylating agent in the presence of a base. Normally, a strong base will be used to deprotonate the alcohol to give the alkoxide. The hydroxyl groups of polysaccharides can be chemically modified via etherification reaction by using either dry method or wet method.

A simple nonaqueous method for the preparation of carboxymethyl (CM) derivatives of galactomannans and other biopolymers has been developed [76]. In a typical reaction, weighed samples (0.1–1 g) of galactomannan and finely powdered $NaHCO_3$ (1–5 g) are mixed well manually using a pestle and mortar and then, are added 0.1–0.2 ml of ethanol (100 % for surface wetting), and solid monochloroacetic acid (0.5–2 g). The reaction is carried out at ambient or elevated temperature (60, 80, 98 °C) for 2 h, with intermittent manual mixing with a glass rod, followed by the addition of dilute acetic acid to phenolphthalein end point. The salts formed are removed by repeated washings with 70 % aqueous ethanol followed by 100 % ethanol and solvent exchange drying. A number of CM derivatives of differing degree of substitution (DS) values (0.065–0.675) can be prepared by varying the ratio of catalyst and the reagent.

Polysaccharides that do not possess ionic gelling ability such as xanthan gum, guar gum, locust bean gum, pullulan, etc., can also be converted into their sodium carboxymethyl derivative by wet method [77]. The polysaccharides (2 g) are dispersed for 30 min in 6.72 ml of ice-cold deionized water containing 3.024 g of sodium hydroxide. Monochloroacetic acid (1.5 g) dissolved in 3.33 ml of deionized water was added slowly for a period of 1 h in the above mixture maintained at 15–18 °C. The temperature of the mixture is raised slowly to 60–65 °C and stirred for at least another 1 h. The wetted mass is washed with 80 % methanol for 15 min. The pH of the suspension is adjusted to neutrality with glacial acetic acid. Finally, it is washed with methanol and kept for drying at 50–60 °C until two consecutive weights were the same.

Once hydrophilic character has been imparted into the polysaccharides, these are further tailored chemically with hydrophobic counterparts to give amphiphilic character. This is achievable by any of the above synthetic pathways.

16.6 Future Directions

A lot of international publications and patents exists that deals with the polymeric micelles for drug delivery and targeting applications. In most of the cases, the copolymer contains poly(ethylene oxide) (also known as polyethylene glycol) as the hydrophilic component and poly(amino acids), poly(D-, L- and DL-lactic acids), poly(ε-caprolactone), poly(glycolic acid) as the hydrophobic component. In addition, most of the work focused mostly on micelles formed from commercially available Pluronic® triblock copolymers. Hitherto, the research efforts on polysaccharide-based micelles for drug delivery are limited in the scientific literature. Hence, there is a lot of opportunity to exploit the natural polysaccharides for their drug delivery potentials by decorating their chemical structures via simple chemistry.

Minimizing drug use would significantly reduce effective cost of drug which would give financial relief to the patients. Delivery systems increase commercial opportunity by distinguishing a drug from competitive threats posed by "me to" drugs and novel means of delivery particularly using nanocarriers can allow branded drugs to be rescued from abyss of generic competition (may be called resurrection of drugs). Priority should be given to drug delivery research due to its low investment and quick return. It is the best way to give technology leadership in pharmaceutical industry due to problems in new chemical entity (NCE) research and stricter patent laws. Because of biodegradability and nontoxicity of the bio-polysaccharides, they should be explored for their potential in designing nanomicellar carriers for targeted drug delivery. There is a lot of opportunity in this field for enhancing drug effects with the use of a minimum amount of drugs and thereby reducing the dose related toxicity of the drugs. Some drugs which possesses significant bioactivity as obvious from high throughput screening but are not considered for future development could find a new direction for their development.

References

1. Thompson D, Chaubal MV (2000) Cyclodextrins (CDS)-excipients by definition, drug delivery systems by function (Part I: Injectable applications). Drug Deliv Technol 2:34–38
2. Shabner BA, Collings JM (1990) Cancer chemotherapy. In: Principles and practice. Humana Press, Philadelphia, PA
3. Okada J, Cohen S et al (1995) In vitro evaluation of polymerized liposomes as an oral drug delivery system. Pharm Res 12:576–582
4. Toorisaka E, Ono H et al (2003) Hypoglycemic effect of surfactant-coated insulin solubilized in a novel solid-in-oil-in-water (S/O/W) emulsion. Int J Pharm 252:271–274
5. Uchida M, Kato Y et al (2000) Involvement of nitric oxide from nerves on diarrhea induced by castor oil in rats. Jpn J Pharmacol 82:168–170
6. Ray R, Kibbe AH et al (2003) Handbook of pharmaceutical excipients. APhA Publications, Washington, DC
7. Rosen MJ (2004) Surfactants and interfacial phenomena. Wiley Interscience, New York, NY
8. Kim S-C, Chang E-O et al. (2001) Biodegradable polymeric micelle-type drug composition and method for the preparation thereof. US Patent 6322805B1
9. Yokogawa K, Nakashima E et al (1990) Relationships in the structure-tissue distribution of basic drugs in the rabbit. Pharm Res 7:691–696
10. Lipinski CA (2000) Drug-like properties and the causes of poor solubility and poor permeability. J Pharmacol Toxicol Methods 44:235–249
11. Horter D, Dressman JB (2001) Influence of physicochemical properties on dissolution of drugs in the gastrointestinal tract. Adv Drug Deliv 46:75–87
12. Fernandez AM, Van Derpoorten K et al (2001) *N*-Succinyl-(beta-alanyl-Lleucyl-L-alanyl-L-leucyl) doxorubicin: an extracellularly tumor-activated prodrug devoid of intravenous acute toxicity. J Med Chem 44:3750–3753
13. Torchilin VP, Trubetskoy VS (1995) Which polymers can make nanoparticulate drug carriers long-circulating? Adv Drug Deliv Rev 16:141–155
14. Maeda H, Wu J et al (2000) Tumor vascular permeability and the EPR effect in macromolecular therapeutics: a review. J Cont Rel 65:271–284
15. Torchilin VP (1998) Polymer-coated long-circulating microparticular pharmaceuticals. J Microencapsul 15:1–19
16. Gref R, Minamitake J et al (1994) Biodegradable long-circulating polymeric nanospheres. Science 263:1600–1603

17. Lukyanov AN, Torchilin VP (2004) Micelles from lipid derivatives of water-soluble polymers as delivery systems for poorly soluble drugs. Adv Drug Deliv Rev 56:1273–1289
18. Francis MF, Cristea M et al (2004) Polymeric micelles for oral drug delivery: why and how. Pure Appl Chem 76:1321–1335
19. Torchilin VP (2004) Targeted polymeric micelles for delivery of poorly soluble drugs. Cell Mol Life Sci 61:2549–2559
20. Gaucher G, Dufresne MH et al (2005) Block copolymer micelles: preparation, characterization and application in drug delivery. J Cont Rel 109:169–188
21. Nishiyama N, Kataoka K (2006) Current state, achievements, and future prospects of polymeric micelles as nanocarriers for drug and gene delivery. Pharmacol Ther 112:630–648
22. Kedar U, Phutane P et al (2010) Advances in polymeric micelles for drug delivery and tumor targeting. Nanomed Nanotechnol Biol Med 6:714–729
23. Martin A, Bustamante P et al (2001) Interfacial phenomena. In: Physical pharmacy-physical chemical principles in the pharmaceutical sciences. Lippincott Williams and Wilkins, Philadelphia, PA
24. Kataoka K, Kwon GS et al (1993) Block-copolymer micelles as vehicles for drug delivery. J Cont Rel 24:119–132
25. Kabanov AV, Batrakova EV et al (1992) A new class of drug carriers; micelles poly (oxyethylene)-poly(oxypropylene) block copolymers as microcontainers for drug targeting from blood to brain. J Cont Rel 22:141–158
26. Jones M, Leroux J (1999) Polymeric micelles – a new generation of colloidal drug carriers. Eur J Pharm Biopharm 48:101–111
27. Andrianov AK, Payne LG (1998) Polymeric carriers for oral uptake of microparticulates. Adv Drug Deliv Rev 34:155–170
28. Kwon GS, Okano T (1996) Polymeric micelles as new drug carriers. Adv Drug Deliv Rev 21:107–116
29. Kwon GS (2002) Block copolymer micelles as drug delivery systems. Adv Drug Deliv Rev 54:167
30. Adams ML, Lavasanifar A et al (2003) Amphiphilic block copolymers for drug delivery. J Pharm Sci 92:1343–1355
31. Otsuka H, Nagasaki Y et al (2003) PEGylated nanoparticles for biological and pharmaceutical applications. Adv Drug Deliv Rev 55:403–419
32. Yokoyama M, Sakurai Y et al. (1996) Physical trapping type polymeric micelle drug preparation. US Patent 5510103
33. Taillefer J, Jones MC et al (2000) Preparation and characterization of pH-responsive polymeric micelles for the delivery of photosensitizing anticancer drugs. J Pharm Sci 89:52–62
34. Zhiang J, Wu M et al (2009) Anionic poly (lactic acid)-polyurethane micelles as potential biodegradable drug delivery carriers. Colloids Surf A Physicochem Eng Asp 337:200–204
35. Dufresne MH, Fournier E et al (2003) Block copolymer micelles-engineering versatile carriers for drugs and biomacromolecules. In: Gurny R (ed) Challenges in drug delivery for the new millennium. Bulletin Technique Gattefosse, Saint-Priest
36. Kumar V, Banker GS (2002) (2002) Target-oriented drug-delivery systems. In: Banger GS, Rhodes CT (eds) Modern pharmaceutics. Marcel Dekker, New York, NY
37. Duncan R (2003) The dawning era of polymer therapeutics. Nat Rev Drug Discov 2:347–360
38. Maeda H (2001) The enhanced permeability and retention (EPR) effect in tumor vasculature, the key role of tumor selective macromolecular drug targeting. Adv Enzyme Regul 41:189–207
39. Matsumura Y (2001) Drug delivery systems in cancer chemotherapy. In: Chiellini E, Sunamato J (eds) Biomedical polymers and polymer therapeutics. Springer, Berlin
40. Maeda H, Sawa T et al (2001) Mechanism of tumor targeted delivery of macromolecular drugs, including the EPR effect in solid tumor and clinical overview of the prototype polymeric drug SMANCS. J Cont Rel 74:47–61

41. Yuan F, Delian M et al (1995) Vascular permeability in a human tumor xenograft: molecular size dependence and cut off size. Cancer Res 55:3752–3756
42. Dvorak HF, Brown LF et al (1995) Vascular permeability factor/vascular endothelial growth factors, microvascular hyperpermeability and angiogenesis. Am J Pathol 146:1029–1039
43. Feng D, Nagy JA et al (1999) Pathways of macromolecular extravasation across microvascular endothelium in response to VPF/VEGF and other vasoactive mediators. Microcirculation 6:23–44
44. Kabanov AV, Batrakova EV et al (2002) Pluronic block copolymers as novel polymer therapeutics for drug and gene delivery. J Cont Rel 82:189–212
45. Iakaubov LZ, Torchillin VP (1997) A novel class of anti-tumor antibodies: nucleosome restricted antinuclear autoantibodies (ANA) from healthy aged non-auto-immune mice. Oncol Res 9:439–446
46. Vinogradov S, Batrakova E et al (1999) Polyion complex micelles with protein-modified corona for receptor mediated delivery of oligonucleotides into cells. Bioconjug Chem 10:851–860
47. Jule E, Nagasaki Y et al (2002) Surface plasmon resonance study on the interaction between lactose-installed poly (ethylene glycol)–poly (d, l-lactide) block copolymer micelles and lectins immobilized on a gold surface. Langmuir 18:10334–10339
48. Illum L, Davis SS (1985) Passive and active targeting using colloidal carrier systems. In: Buri P, Gumma A (eds) Drug targeting. Elsevier Science Publishers, Amsterdam
49. Nagasaki Y, Yasugi K et al (2001) Sugar-installed block copolymer micelles: their preparation and specific interaction with lectin molecules. Biomacromolecules 2:1067–1070
50. Leamon CP, Weigl D et al (1999) Folate copolymer-mediated transfection of cultured cells. Bioconj Chem 10:947–957
51. Jule E, Nagasaki Y et al (2003) Lactose-installed poly (ethylene glycol)-poly (d, l-lactide) block copolymer micelles exhibit fast-rate binding and high affinity toward a protein bed simulating a cell surface: a surface plasmon resonance study. Bioconj Chem 14:177–186
52. Bae Y, Jang WD et al (2005) Multifunctional polymeric micelles with folate mediated cancer cell targeting and pH-triggered drug releasing properties for active intracellular drug delivery. Mol Biosyst 1:241–250
53. Campbell IG, Jones TA et al (1991) Folate-binding protein is a marker for ovarian cancer. Cancer Res 51:5329–5338
54. Weitman SD, Lark RH et al (1992) Distribution of the folate receptor GP38 in normal and malignant cell lines and tissues. Cancer Res 52:3396–3401
55. Sudimack J, Lee RJ (2000) Targeted drug delivery via the folate receptors. Adv Drug Deliv Rev 41:147–162
56. Elliott RL, Elliott MC et al (1993) Breast carcinoma and the role of iron metabolism: a cytochemical, tissue culture and ultrastructural study. Ann N Y Acad Sci 698:159–166
57. Sahoo SK, Labhesetwar V (2005) Enhanced anti-proliferative activity of transferrin conjugated paclitaxel-loaded nanoparticle is mediated via sustained intracellular drug retention. Mol Pharm 2:373–383
58. Lee ES, Na K et al (2003) Polymeric micelle for tumor pH and folate-mediated targeting. J Cont Rel 91:103–113
59. Lee AL, Yong W et al (2009) The co-delivery of paclitaxel and Herceptin using cationic micellar nanoparticles. Biomaterials 30:919–927
60. Gao ZG, Lee DH et al (2005) Doxorubicin loaded pH-sensitive micelles targeting acidic extracellular pH of human ovarian A2780 tumor in mice. J Drug Target 13:391–397
61. Rijcken CJ, Snel CJ et al (2007) Hydrolysable core-crosslinked thermosensitive polymeric micelles: synthesis, characterisation and in vivo studies. Biomaterials 28:5581–5593
62. Sershen SR, Westcott SL et al (2000) Temperature-sensitive polymer nanoshell composites for photothermally modulated drug delivery. J Biomed Mater Res 51:293–296

63. Kato N, Oishi A et al (1998) Enzyme reaction controlled by magnetic heating due to the hysteresis loss of γ-FeO in thermosensitive polymer gels immobilized β-galactosidase. Mater Sci Eng C 6:291–296
64. Hovgaard L, Brondsted H (1996) Current applications of polysaccharides in colon targeting. Crit Rev Ther Drug Carrier Syst 13:185–223
65. Izumi Y, Kikuta N et al (1996) Phase diagrams and molecular structures of sodium-salt-type gellan gum. Carbohydr Polym 30:121–127
66. Maiti S, Chakravorty A et al (2014) Gellan co-polysaccharide micellar solution of budesonide for allergic anti-rhinitis: an in vitro appraisal. Int J Biol Macromol 68:241–246
67. Sinha VR, Kumria R (2001) Polysaccharides in colon-specific drug delivery. Int J Pharm 224:19–38
68. Francis MF, Cristea M et al (2005) Engineering polysaccharide-based polymeric micelles to enhance permeability of cyclosporin A across Caco-2 cells. Pharm Res 22:209–219
69. Felt O, Buri P et al (1998) Chitosan: a unique polysaccharide for drug delivery. Drug Dev Ind Pharm 24:979–993
70. Mourya VK, Inamdar NN (2009) Trimethyl chitosan and its applications in drug delivery. J Mater Sci Mater Med 20:1057–1079
71. Vasnev VA, Tarasov AI et al (2006) Synthesis and properties of acylated chitin and chitosan derivatives. Carbohydr Polym 64:184–189
72. Chen X, Ding S et al (2008) Synthesis of novel chitosan derivatives for micellar solubilization of cyclosporine A. J Bioactive Compatib Polym 23:563–578
73. Jeong Y-IL, Seo D-H et al (2009) Methotrexate-incorporated polymeric micelles composed of methoxy poly (ethylene glycol)-grafted chitosan. Macromol Res 17:538–543
74. You J, Hu FQ et al (2007) Polymeric micelles with glycolipid-like structure and multiple hydrophobic domains for mediating molecular-target delivery of PTX. Biomacromolecules 8:2450–2456
75. Li X, You J et al (2008) Preparation and characteristics of stearic acid grafted chitosan oligosaccharide polymeric micelle containing 10-hydroxycamptothecin. Asian J Pharm Sci 3:80–87
76. Parvathya KS, Susheelamma NS et al (2005) A simple non-aqueous method for carboxymethylation of galactomannans. Carbohydr Polym 62:137–141
77. Maiti S, Ray S et al (2007) Carboxymethyl xanthan microparticles as a carrier for protein delivery. J Microencapsul 24:743–756

MIX
Papier aus verantwortungsvollen Quellen
Paper from responsible sources
FSC® C105338

If you have any concerns about our products,
you can contact us on
ProductSafety@springernature.com

In case Publisher is established outside the EU,
the EU authorized representative is:
Springer Nature Customer Service Center GmbH
Europaplatz 3, 69115 Heidelberg, Germany

Printed by Libri Plureos GmbH
in Hamburg, Germany